BIOCHEMISTRY RESEARCH TRENDS

KAEMPFEROL

CHEMISTRY, NATURAL OCCURRENCES AND HEALTH BENEFITS

BIOCHEMISTRY RESEARCH TRENDS

Additional books in this series can be found on Nova's website
under the Series tab.

Additional e-books in this series can be found on Nova's website
under the e-book tab.

BIOCHEMISTRY RESEARCH TRENDS

KAEMPFEROL

CHEMISTRY, NATURAL OCCURRENCES AND HEALTH BENEFITS

GARION VILLERS
AND
YVES FOUGERE
EDITORS

New York

Library of Congress Cataloging-in-Publication Data

ISBN: 978-1-62618-515-9

Library of Congress Control Number: 2013935460

Published by Nova Science Publishers, Inc. † New York

Contents

Preface

Kaempferol is a flavonol that is relatively abundant in various plants and some edible berries. These compounds have antioxidative, anti-inflammatory, antihypertensive, lipolytic and anti-cancer effects. In this book, the authors discuss the chemistry, natural occurrences and health benefits of kaempferol. Topics include the effect of cyclodextrins complexation on the properties of kaempferol; kaempferitrin and kaempferol as nutraceutical compounds contributing to glucose homeostasis by acting at multiple biological sites; distribution of kaempferol glycosides and their function in plants; analytical approaches for kaempferol determination; chemoprotective effects of kaempferol in colorectal tumorigenesis; multipotent flavonoid kaempferol as molecular targets and mechanism of action in nanotechnology applications in cancer and human health; metal complexes of kaempferol and their speciation in human plasma; the action of kaempferol on oral cavity cancer and airway secretion; kaempferol as it regulates intestinal tight junction permeability; and identification of kaempferol in historical textiles.

Chapter 1 – The flavonoid kaempferol is a yellow compound with a low molecular weight (MW: 286.24 g/mol) that is commonly found in plant-derived foods and in plants used in traditional medicine. Because the enzymes involved in the biosynthesis of kaempferol are relatively common in the plant kingdom, it is not surprising that this flavonoid is widely distributed in plants. Kaempferol is relatively abundant in apples, grapes, gingko biloba, onions, leeks, citrus fruits, red wines, tea, cruciferous vegetables, and some edible berries. Many preclinical studies have shown that kaempferol and some glycosides of kaempferol have extensive pharmacological activities, including antioxidant, anti-inflammatory, anticancer, antimicrobial, cardioprotective, neuroprotective, antidiabetic, anti-osteoporotic, estrogenic/antiestrogenic, anxiolytic, analgesic and antiallergic activities. The above-mentioned antioxidant properties reside mainly in their radical-scavenging activity. Repeated studies have shown that flavonoids with many hydroxyl groups are extremely effective antioxidants. In fact, hydroxylated flavonols, such as myricetin, quercetin and kaempferol, have been demonstrated to be particularly effective antioxidants in many studies.

All these beneficial health effects attributed to kaempferol, make this natural compound interesting for the development of functional foods. The main problem is its low water solubility and its stability against various factors such as oxygen, light and temperature.

In recent years, complexation with cyclodextrins (CDs) has successfully been used to improve the solubility, chemical stability and bioavailability of a number of poorly soluble phenolics compounds, such as resveratrol, mirycetin and quercetin. CDs are a group of naturally occurring cyclic oligosaccharides derived from starch with six (α-), seven (β-) or eight (γ-cyclodextrins) glucose residues, linked by α(1 → 4) glycosidic bonds. In the pharmaceutical, cosmetics and food industries, CDs have been used as complexing agents to increase the water solubility of various compounds, such as drugs, vitamins and food colourants, increasing the solubility, stability and bioavailability of the guest molecule. Kaempferol can be included in CDs increasing the solubility, possibly its bioavailability and beneficial health effects.

Chapter 2 – Kaempferol is a flavonol that is relatively abundant in various plants and some edible berries. Kaempferol aglycone or its glycosilated forms contributes significantly to flavonoid intake in humans (about 33% from total flavonol intake), with intakes estimated around 10 mg per day in developed countries. These compounds have antioxidative, anti-inflammatory, antihypertensive, lipolytic and anti-cancer effects. Emerging evidence suggests that kaempferol exhibits nutraceutical properties since it produces physiological benefits and is able to combat disease processes, among them those concerned to diabetes mellitus. Several results have demonstrated that kaempferol can influence glucose metabolism acting as antihyperglycemic and especially as hypoglycemic agent through *in vitro* and *in vivo* approaches. Kaempferol aglycone and/or glycosides stimulate glucose uptake and glycogen synthesis in skeletal muscle, inhibit intestinal glucose absorption and act *per se* as insulin-mimetic, probably, by influencing the pleiotropic mechanisms of insulin signaling. Thus, these nutraceutical compounds have attracted the interest of both academia and industry as possible therapy options. This review cover the frontiers of the development of chemistry and biological properties of kaempferol (aglycone/glycosides), especially those associated with glucose homeostasis and carbohydrate metabolism able to ameliorate diabetes status. Moreover, updates the debate about the development of new approaches to study the interactions of nutraceutical compounds with target-sites of action combining *in vivo, ex vivo* and *in vitro* pharmacological studies. Also, the perspectives of controlled delivering of these substances in multiple and specific biological locations useful to diabetes mellitus therapy are highlighted.

Chapter 3 – Kaempferol is a flavonol that is relatively abundant in various plants and some edible berries. Kaempferol aglycone or its glycosilated forms contributes significantly to flavonoid intake in humans (about 33% from total flavonol intake), with intakes estimated around 10 mg per day in developed countries. These compounds have antioxidative, anti-inflammatory, antihypertensive, lipolytic and anti-cancer effects. Emerging evidence suggests that kaempferol exhibits nutraceutical properties since it produces physiological benefits and is able to combat disease processes, among them those concerned to diabetes mellitus. Several results have demonstrated that kaempferol can influence glucose metabolism acting as antihyperglycemic and especially as hypoglycemic agent through *in vitro* and *in vivo* approaches. Kaempferol aglycone and/or glycosides stimulate glucose uptake and glycogen synthesis in skeletal muscle, inhibit intestinal glucose absorption and act *per se* as insulin-mimetic, probably, by influencing the pleiotropic mechanisms of insulin signaling. Thus, these nutraceutical compounds have attracted the interest of both academia and industry as possible therapy options. This review cover the frontiers of the development of chemistry and biological properties of kaempferol (aglycone/glycosides), especially those associated with glucose homeostasis and carbohydrate metabolism able to ameliorate diabetes status. Moreover, updates the debate about the development of new approaches to study the interactions of nutraceutical compounds with target-sites of action combining *in vivo, ex vivo* and *in vitro* pharmacological studies. Also, the perspectives of controlled delivering of these substances in multiple and specific biological locations useful to diabetes mellitus therapy are highlighted.

Chapter 4 – The rapid development of technology and analytical instruments in the last 20 years allowed scientists to detect and quantify thousands of compounds which were difficult to analyse before. Kaempferol, a common biologically active phytochemical, occurs mostly in complex matrices together with other plant metabolites that complicate the analytical procedure. Various extraction and purification steps, including solid phase extraction, were therefore developed to facilitate the determination of kaempferol in complex biological samples. This chapter is an overview of currently used analytical instruments and their basic settings used to detect and quantify kaempferol in various plants, plant-derived sources and other biological samples.

Chapter 5 – Kaempferol is a bioactive polyphenol that has been shown to have chemoprotective properties in cell culture. The objective of this study was to evaluate the potential chemoprotective effects of kaempferol against colorectal tumorigenesis in humans and animal models. The association between kaempferol intake and colorectal adenoma recurrence was examined in the Polyp Prevention Trial (PPT), a 4-year randomized, clinical intervention trial which examined the effect of a high-fruit and vegetable, high-fiber, and low-fat diet on colorectal adenoma recurrence. Kaempferol consumption was estimated annually by self-administered food frequency questionnaires that were connected to the U.S. Department of Agriculture flavonoid database for foods.

Kaempferol intake was inversely associated with high-risk adenoma [odds ratio (OR) for individuals in the highest compared to those in the lowest quartile: 0.62; 95% confidence interval (CI): 0.40, 0.97; $P_{\text{Linear}} = 0.04$) and advanced adenoma recurrence (OR: 0.49; 95% CI: 0.26, 0.91; $P_{\text{Linear}} = 0.03$).

To examine whether the chemoprotective effect of kaempferol could be in part explained by its anti-inflammatory properties, the authors measured serum concentrations of cytokines at baseline, year 1, and year 3 in the intervention arm of the PPT. High intake of kaempferol was associated with lower serum interleukin 6 (IL-6) concentrations in year 1 and 3 (highest versus lowest kaempferol intake quartile: 1.86 versus 2.09 pg/mL; $P_{\text{Kaempferol}} = 0.05$). The greatest decrease in serum IL-6 concentrations was observed in participants with elevated serum IL-6 concentrations (> 2.37 pg/mL) at baseline (highest versus lowest intake quartile: 2.63 versus 3.23 pg/mL; $P_{\text{Kaempferol}} = 0.05$). Individuals with above median kaempferol consumption and equal or below median change in IL-6 concentrations had the lowest risk of advanced adenoma recurrence.

To examine whether the beneficial effects of kaempferol on colorectal tumorigenesis were associated with other compounds enriched in kaempferol-containing foods, the authors conducted a study in male *FVB/NCr* mice fed a diet with or without 500 ppm purified kaempferol for 85 days after colorectal tumor induction and promotion by the chemical carcinogen azoxymethane and the colonic irritant dextran sodium sulfate. In comparison to mice on the control diet, mice consuming kaempferol had fewer tumors (16 vs. 21 tumors; $P = 0.04$) and smaller tumors (11.6 vs. 7.4 mm^3/tumor; $P = 0.09$).

The authors results suggest kaempferol is a potential chemoprotective compound against colorectal tumorigenesis.

Chapter 6 – Kaempferol, a bioactive natural compound widely present in fruits and vegetables, is a polyphenol belonging to the class of flavonoids. Studies for the past several decades have identified the great potential of this flavonoid in the prevention and treatment of human chronic diseases, especially cancer. Kaempferol exerts this wide spectrum of effects by targeting several proteins/enzymes that are implicated in cancer and disease progression. Epidemiological studies have shown an inverse relationship between kaempferol intake and cancer. Kaempferol helps by augmenting the body's antioxidant defence against free radicals, which promote the development of cancer. At the molecular level, kaempferol has been reported to modulate a number of key elements in cellular signal transduction pathways linked to apoptosis, angiogenesis, inflammation, and metastasis. Significantly, kaempferol inhibits cancer cell growth and angiognesis and induces cancer cell apoptosis, while also appearing to preserve normal cell viability, in some cases exerting a protective effect. Because of its outstanding safety profile and its availability in highly purified forms, the clinical implications for kaempferol are growing rapidly. Even though kaempferol is one of the major flavonoids in the human diet, its bioavailability has been debated. Recent studies to improve its bioavailability by encapsulating it with PEO-PPO-PEO and PLGA nanoparticles are promising. This review focuses mainly on the molecular targets of kaempferol and the mechanisms by which kaempferol exerts its biological effects in the prevention and treatment of human chronic diseases and cancer. A discussion on the application of nanotechnology is also included.

Chapter 7 – Flavonoids are a large class of phenolic compounds which are sub classified as flavones, flavonols, isoflavones, flavanones and catechins, chalcones and anthocyanidins depending on phenyl substituent in the C_2 or C_3 position in *benzo-γ-pyrone* nucleus. Interest in the bioflavonoids is related to their diversity, biological significance as secondary plant metabolites and ecological role, use as chemotaxonomic markers, impact on fruit quality, physiological effects and industrial applications.

The flavonoids are potent antioxidants, free radical scavengers and metal chelators; they inhibit lipid peroxidation and exhibit various physiological activities, including anti-inflammatory, anti-allergic, anti-carcinogenic, antihypertensive and anti-arthritic activities.

It is well known that blood serum contains essential amino acids, 12 essential metal ions at least another 100 ligands as well as numerous low molecular weight complexes. In addition, there are α/β macro globulins such as albumin and transferin weighing up to 100000 Daltons. It has been well demonstrated that metal ions play a large number of important roles in biological systems. Metal ions which are considered essential to human life included calcium, magnesium, manganese, iron, cobalt, copper and zinc. Furthermore man is constantly being challenged by pollutants included the toxic metal ions mercury, cadmium and lead.

Chapter 8 – *Ginkgo biloba* extract has long been used as an anticancer drug and expectorant in Oriental medicine. Because *Ginkgo biloba* extract is a mixed compound comprising kaempferol and other components, it may be preferable to use only kaempferol for treatments due to possible adverse reactions with the other components. Furthermore, the action mechanism and potency of kaempferol compared to *Ginkgo biloba* extract have not been revealed. Therefore, this chapter gives a brief survey of the current knowledge on kaempferol and its potential use as an anti-oral cancer agent and also as an anti-hypersecretory agent.

In previous studies, *Ginkgo biloba* extract has been shown to induce apoptosis in oral cavity cancer cells. In the case of kaempferol, the results of cell proliferation assays on oral cavity cancer cell lines such as SCC-1483, SCC-25, and SCC-QLL1 showed that kaempferol significantly inhibited cellular proliferation. Flow cytometry showed that the anti-proliferative effect was due to increased apoptosis. Kaempferol showed cleavage of poly (ADP-ribose) polymerase (PARP). Therefore, kaempferol effectively induces caspase-3-dependent apoptosis of oral cavity cancer cells and can be considered a possible anti-oral cancer agent.

In addition to its anticancer effects, *Ginkgo biloba* extract suppresses IL-1β-induced *MUC5AC* gene expression in NCI-H292 cells, and therefore, kaempferol may be used as a suppressor of mucin production. The RT-PCR and real-time PCR results showed that kaempferol suppressed *MUC5AC* mRNA expression in a dose-dependent manner, with significant inhibition starting at 40 μM (equal concentration to about a 1/12or 1/13 dose of *Ginkgo biloba* extract).

Kaempferol suppressed phosphorylation of ERK and p38 kinases. Therefore, kaempferol may overcome the dose problem of *Ginkgo biloba* extract to play a clinically valuable role in controlling mucin hypersecretion in airway inflammation.

Chapter 9 – A primary function of gastrointestinal epithelial cells is to provide a physical barrier to the diffusion of pathogens, toxins, and allergens from the external environment into the tissues. Defects in the barrier function are involved in the pathogenesis of various intestinal diseases, such as inflammatory bowel diseases, irritable bowel disease, and food allergies. This intestinal barrier is organized through interactions among several barrier components, including the adhesive mucous gel layer, the mucosal immune system, and intercellular tight junctions (TJs). The TJ is a multi-protein complex, which locates around the apical end of the lateral membrane of polarized epithelial cells, and selectively regulates the paracellular passage of ions, molecules and water. The TJ complex is composed of transmembrane proteins that interact with the perijunctional actin cytoskeleton through intracellular plaque proteins. TJ permeability and assembly are dynamically regulated by various intracellular signaling molecules, such as protein kinase C, mitogen-activated protein kinases, and protein phosphatases. Kaempferol is one of the most abundant flavonoids in human diets and exhibits various biological effects, such as anti-inflammatory and anti-cancer activities. A recent study has demonstrated that kaempferol participates in the regulation of intestinal TJ barrier function. This chapter presents the recent advances in the authors understanding of the kaempferol-mediated suppressive effect on intestinal TJ permeability with a particular focus on intracellular molecular mechanisms.

Chapter 10 – The authors goal is to present, in brief, the historical use of kaempferol as a dyestuff material for textiles. For this reason the authors first review previously published reports that describe the identification of kaempferol in textiles of the cultural heritage. Then, a case study is presented where the identification of kaempferol in samples removed from three religious textiles, preserved in monasteries of Mount Athos, is described. Kaempferol is detected in mixture with several other colouring components using high performance liquid chromatography (HPLC) with diode-array detection (DAD). Finally, a brief overview of the various colourants found in historical textiles from Mount Athos is provided.

In: Kaempferol
Editors: Garion Villers and Yves Fougere
ISBN: 978-1-62618-515-9

Chapter 1

Biological Activities of Kaempferol: Effect of Cyclodextrins Complexation on the Properties of Kaempferol

M. T. Mercader-Ros, C. Lucas-Abellán, M. I. Fortea, A. Serrano-Martínez, J. A. Gabaldón and E. Núñez-Delicado
Department of Food Technology and Nutrition,
San Antonio Catholic University, Guadalupe. Murcia, Spain

Abstract

The flavonoid kaempferol is a yellow compound with a low molecular weight (MW: 286.24 g/mol) that is commonly found in plant-derived foods and in plants used in traditional medicine. Because the enzymes involved in the biosynthesis of kaempferol are relatively common in the plant kingdom, it is not surprising that this flavonoid is widely distributed in plants. Kaempferol is relatively abundant in apples, grapes, gingko biloba, onions, leeks, citrus fruits, red wines, tea, cruciferous vegetables, and some edible berries. Many preclinical studies have shown that kaempferol and some glycosides of kaempferol have extensive pharmacological activities, including antioxidant, anti-inflammatory, anticancer, antimicrobial, cardioprotective, neuroprotective, antidiabetic, anti-osteoporotic, estrogenic/antiestrogenic, anxiolytic, analgesic and antiallergic activities. The above-mentioned antioxidant properties reside mainly in their radical-scavenging activity. Repeated studies have shown that flavonoids with many hydroxyl groups are extremely effective antioxidants. In fact, hydroxylated flavonols, such as myricetin, quercetin and kaempferol, have been demonstrated to be particularly effective antioxidants in many studies.

All these beneficial health effects attributed to kaempferol, make this natural compound interesting for the development of functional foods. The main problem is its low water solubility and its stability against various factors such as oxygen, light and temperature.

In recent years, complexation with cyclodextrins (CDs) has successfully been used to improve the solubility, chemical stability and bioavailability of a number of poorly soluble phenolics compounds, such as resveratrol, mirycetin and quercetin. CDs are a group of naturally occurring cyclic oligosaccharides derived from starch with six (α-), seven (β-) or eight (γ-cyclodextrins) glucose residues, linked by α(1 → 4) glycosidic bonds. In the pharmaceutical, cosmetics and food industries, CDs have been used as complexing agents to increase the water solubility of various compounds, such as drugs, vitamins and food colourants, increasing the solubility, stability and bioavailability of the guest molecule. Kaempferol can be included in CDs increasing the solubility, possibly its bioavailability and beneficial health effects.

Introduction

Chemistry and Classification of Flavonoids

The structure of the flavonoids is based on the flavonoid nucleus, which consists of three phenolic rings named A, B, and C rings (Kühnau, 1976). The benzene ring A is condensed with a six-member ring (C), which in the 2-position carries a phenyl benzene ring (B) as a substituent. Ring C may be a heterocyclic pyran, which yields flavanols (catechins) and anthocyanidins, or pyrone, which yields flavonols, flavones and flavanones. The term 4-oxo-flavonoids is often used to describe flavonoids, such as flavanols (catechins), flavanones, flavonols and flavones, which carry a carbonyl group on C-4 of ring C (Aherne and O'Brien, 2002).

The chemical nature of the flavonoids depends on structural class, degree of hydroxylation, other substitutions or conjugations and degree of polymerization (Harborne, 1986). In plants, they are relatively resistant to heat, oxygen, dryness and moderate degre of acidity but can be modified by light (Kühnau, 1976). Photostability of the flavonoid molecule depends on the nature of the hydroxyl group attached to C-3 of ring C. The absence or glycosylation of this hydroxyl group, results in high photostability of the molecule (Smith et al., 2000; Aherne and O'Brien, 2002).

The biological activities of flavonoids and their metabolites depend on their chemical structure and spatial conformation. The basic structure of the flavonoid nucleus allows for a multitude of substitution patterns in the A, B and C rings, resulting in various subgroups. The flavonoids are classified -according to the oxidation level of C ring-, as: anthocyanidins, flavanols (catechins), flavones, flavonols, flavanones and isoflavonoids (Kühnau, 1976; Harborne, 1986), among others. Flavones and flavonols have been identified in almost all plants, the ones found most frequently being those with B ring hydroxylation in the C-3´ and C-4´ positions (Kühnau, 1976; Aherne and O'Brien, 2002).

Flavonoids are mainly present in plants as glycosides. Aglycones (the forms lacking sugar moieties) occur less frequently. At least 8 different monosaccharides or their combinations (di- or trisaccharides), can bind to the different hydroxyl groups of the flavonoid aglycone (Williams and Harborne, 1994).

The large number of flavonoids is a result of the different combinations of flavonoid aglycones and sugars, commonly D-glucose and L-rhamnose. The glycosides are usually O-glycosides, with the sugar moiety linked to the hydroxyl group at the C-3 or C-7 position (Erlund, 2004).

Occurrence of Flavonoids in Foods

Flavonoids are widely distributed in foods and beverages of plant origin, such as fruits, vegetables, tea, cocoa and wine.

In addition, a large body of literature exists regarding their content in differents foods. Within the subgroup of flavonols and flavones, the flavonol quercetin is the most frequently found in foods, followed by kaempferol, myricetin and the flavones apigenin and luteolin. Tea and onions are the main dietary sources of flavonols and flavones. Quercetin is the most abundant flavonol present in onions, whereas tea contain considerable amounts of both quercetin and kaempferol (Ross and Kasum, 2002).

Kaempferol has been found to have anti-oxidant and anti-inflammation effects, and recent studies evidenced that kaempferol has health promoting effects, since reduce arteriosclerosis, cardiovascular disease and cancer (Lau, 2008; Luo et al., 2010).

Kaempferol

Kaempferol (3,4´,5,7-Tetrahydroxyflavone) (Figure 1) is a natural compound firstly discovered in *Camellia sinensis* (common name tea tree) and is a flavonol naturally present in a variety of fruits and vegetables (Table 1).

Kaempferol can be isolated from tea, broccoli, witch-hazel, propolis, grapefruit and other plant sources (Yoshida et al., 2008). Generally, in plants kaempferol is present as glycoside, which determines their pharmacokinetic properties (He and Li, 2008) and it has been estimated that the human dietary intake of this polyphenol may be up to approximately 10 mg/day (Gates et al., 2007; Adebamowo et al., 2005; de Vries et al., 1997). Kaempferol is extracted in small amount from plant flowers, leaf or seeds, so it will be too expensive to produce for commercial trade (Park et al., 2006; Lau, 2008).

The history of tea began in ancient China over 5,000 years ago (Chen and Dou, 2008), and nowadays different varieties of tea are consume by 1/3 of the world people (Chen and Dou, 2008). Whereas *Camellia sinensis* is native from South and Southeast Asia, today is cultivated across the world. They grow in tropical and sub-tropical climates, with at least 50 inches of rainfall per year. Green tea (Figure 2) contain high amounts of threamines, polyphenols (Park et al., 2006) and kaempferol glycosides such as camelliaside A and B, having three sugar molecules linked to kaempferol (Park et al., 2006).

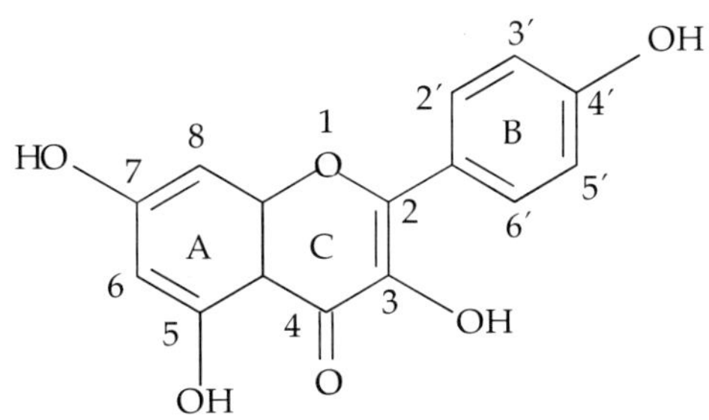

Figure 1. Chemical structure of Kaempferol.

Table 1. Concentration of Kaempferol in edible portion (mg/100g) of fruits and vegetables

Food matrix		Kaempferol	
	Name	min	max
Apples with skin (R)	*Malus domestica*	0.00	2.67
Apricots (R)	*Prunus armeniaca*	0.00	1.32
Asparagus (R)	*Asparagus officinalis*	0.00	5.20
Blackberries (R)	*Rubus spp.*	0.00	3.13
Broccoli (R)	*Brassica oleracea var. italica*	0.05	21.30
Capers (C)	*Capparis spinosa*	59.49	247.97
Capers (R)	*Capparis spinosa*	214.99	308.33
Carrots (R)	*Daucus carota*	0.00	1.53
Cherries (S) (R)	*Prunus avium*	0.00	0.67
Cranberries (R)	*Vaccinium macrocarpon*	0.00	0.61
Cucumber, with peel (R)	*Cucumis sativus*	0.00	0.76
Goji berry (D)	*Wolfberry*	6.20	6.20
Grape, Black	*Vitis vinifera*	0.00	0.18
Grape, red	*Vitis vinifera*	0.00	0.01
Grape, White or Green	*Vitis vinifera*	0.00	0.19
Kiwifruit, green (R)	*Actinidia deliciosa*	0.00	3.06
Lemons (R), without peel	*Citrus limon*	0.00	0.06
Lettuce, iceberg (R)	*Lactuca sativa var. capitata*	0.00	0.84
Melons, cantaloupe (R)	*Cucumis melo*	0.00	0.21
Onions (R)	*Allium cepa*	0.00	1.41
Onions, welsh (R)	*Allium fistulosum*	22.62	27.28
Oranges (R), all comercial varieties	*Citrus sinensis*	0.00	3.15
Peaches (R)	*Prunus persica*	0.00	0.65
Plum, yellow, whole (R)	*Prunus domestica*	0.06	0.17
Raisins, golden seedless	*Vitis vinifera*	0.57	2.47
Spices, saffron	*Crocus sativus*	146.75	318.35
Spinach (R)	*Spinacia oleracea*	0.00	55
Tea, green, brewed	*Camellia sinensis*	0.06	3.31
Tea, oolong, brewed	*Camellia sinensis*	0.90	0.90

Source: USDA. http://www.ars.usda.gov/SP2UserFiles/Place/12354500/Data/Flav/ Flav_R03.pdf.
(R) Raw; (C) canned; (S) Sweet; (D) Dried.

This discovery enables a mass production of kaempferol for commercial purposes, and could be isolated from kaempferol glycosides, using acids or bases or by enzymatic way (Park et al., 2006; Lau, 2008).

Biosynthesis of Kaempferol

Like other flavonoids, kaempferol has a diphenylpropane structure (C6-C3-C6) and is synthesized by condensation of 4-coumaroyl-CoA (C6-C3) with three molecules of malonyl-CoA (C6) (Winkel-Shirley, 2001; Winkel-Shirley, 2002; Calderón-Montaño et al., 2011). This reaction -catalyzed by the enzyme chalcone synthase (EC 2.3.1.74)-, results in the formation of the flavonoid naringenin chalcone (C6-C3-C6). This chalcone is transformed into the flavanone naringenin, by the enzyme chalcone isomerase (EC 5.5.1.6), which catalyzes the closure of C3 ring. In addition, the enzyme flavanone 3-dioxygenase (EC 1.14.11.9) introduces a hydroxyl group in naringenin at C3 position, to form dihydrokaempferol. Finally, the enzyme flavonol synthase (EC 1.14.11.23) insert a double bond in dihydrokaempferol at C2-C3, to produce kaempferol (Figure 3). Since the enzymes involved in the biosynthesis of kaempferol are relatively common in the plant kingdom, it is not surprising that this flavonoid is widely distributed in plants.

Different sugar molecules, such as glucose, rhamnose, galactose and rutinose are usually bound to kaempferol to form glycosides (Calderón-Montaño et al., 2011).

Figure 2. Camellia sinensis, green tea.

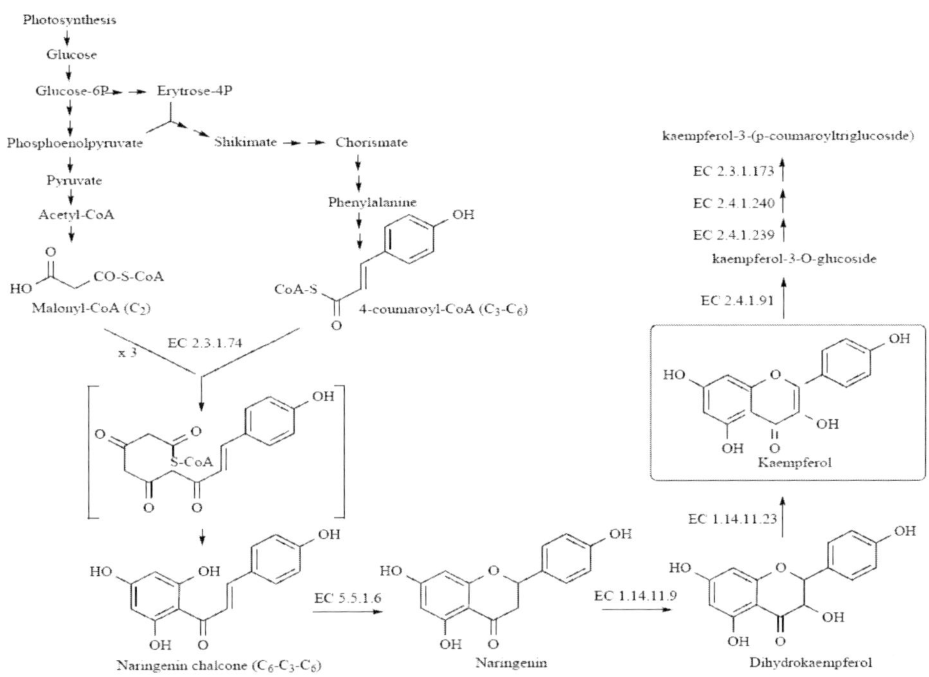

Figure 3. Biosinthesis of kaempferol and some kaempferol glycosides (see text for further details). EC 2.3.1.74: chalcone synthase; EC 5.5.1.6: chalcone isomerase; EC 1.14.11.9: flavanone 3-dioxygenase; EC 1.14.11.23: flavonol synthase; EC 2.4.1.91: flavonol 3-Oglucosyltransferase; EC 2.4.1.239: flavonol-3-O-glucoside glucosyltransferase; EC 2.4.1.240: flavonol-3-O-diglycoside glucosyltransferase; EC 2.3.1.173: flavonol-3-O-triglucoside p-coumaroyltransferase (Calderón-Montaño et al., 2011).

Bioavailability of Kaempferol

Generally, the biological activities of flavonoids such as kaempferol rely on the activity of their metabolites, due to rapid and extensive biotransformation (Barve et al., 2009). The most common metabolites or biotransformation products of flavonoids are methyl, sulfate or glucuronide conjugates (Scalbert and Williamson, 2000; Manach et al., 2005; Williamson and Manach, 2005). For these conjugates, variable rates of formation have been observed, depending on the species and organ/site of metabolism (Graefe et al., 2001; Steensma et al., 2006; Kuhnle et al., 2000; Jia et al., 2004; Hu et al., 2003; Zhang et al., 2003).

Thus, the relative contribution of methylation, sulfation or glucuronidation may vary according to the nature of the flavonoid, species and site of metabolism. Since the majority of flavonoids have high capacity to conjugation, they are poorly bioavailables (2–20%) (Scalbert and Williamson, 2000; Hu, 2007).

Barve et al. in 2009 evaluated the metabolism, oral bioavailability and pharmacokinetics of chemopreventive kaempferol in rats, scored the bioavailable fraction after the administration of 10mg/kg oral doses, found that the bioavailability was approximately 2% of the initial dose. Conjugated kaempferol and two most abundant metabolites, quercetin and isorhamnetin (30-O-methylated quercetin) and their glucuronides, were found in the systemic plasma. Similar metabolites were found in the portal plasma after oral administration of 100mg/kg kaempferol.

In this study, the analysis of urine following administration of kaempferol, revealed significant amounts of parent kaempferol and its glucuronide metabolites. However, insignificant amounts of quercetin and isorhamnetin were observed in urine from rats to which kaempferol was orally administered. Approximately, between 16–18% and 3–4% of the dose, was excreted as parent kaempferol after i.v. and oral administration, respectively (Barve et al., 2009).

Kaempferol and Health

The possible association between the ingestion of foods containing kaempferol and a reduced risk of developing several disorders, including cancer and cardiovascular diseases, has been evaluated in several epidemiological studies. Cui et al. (2008) conducted a clinical trial with 558 lung cancer volunteers and a group of 837 controls, and observed that the consumption of kaempferol-rich foods (i.e. tea, beans, broccoli, spinach, apples and strawberries; approximately 2 mg kaempferol/day) was inversely associated with lung cancer risk; this association was less clear among non-smokers. In another similar trial, 354 volunteers with gastric cancer, and 354 controls were recruited, associating consumption of kaempferol containing foods (i.e. onions, cruciferous, green beans, apples, grapes and strawberries) with the reduction of gastric cancer risk (García-Closas et al., 1999).

In addition, Nöthlings et al. (2007) conducted a cohort study to evaluate the possible association between the consumption of kaempferol-containing foods (i.e. onions, tea, apples, cruciferous and other vegetables) and pancreatic cancer risk in 183,518 people. After a follow-up period of 8 years, the authors found that kaempferol consumption was inversely associated with pancreatic cancer risk. The correlation between kaempferol-rich foods intake (i.e. broccoli, kale and tea) and the incidence of epithelial ovarian cancer was evaluated by a cohort study in 66940 women. The study reveals that kaempferol was associated with a decrease of 40% in epithelial ovarian cancer incidence (Gates et al., 2007).

However, other epidemiological studies not found a positive correlation between the consumption of kaempferol-containing foods and the risk of several other types of cancer, such as bladder cancer (Garcia et al., 1999), prostate cancer (McCann et al., 2005) or colorectal cancer (Lin et al., 2006).

Also, several case-control and cohort studies have evaluated the relationship between consumption of kaempferol-rich foods and the risk of cardiovascular diseases, related with a reduced mortality from coronary heart disease (Hertog et al., 1993; Lin et al., 2007) and a reduced incidence of myocardial infarction (Hertog et al., 1993; Geleijnse et al., 2002; Marniemi et al., 2005).

The association between flavonoid-containing foods intake (i.e. apples, grapefruit, onions, white cabbage, berries, juices and oranges), and the risk of several chronic diseases was assessed in a cohort study in 10,054 volunteers (Knekt et al., 2002). This study showed that the incidence of cerebrovascular disease was lower at higher kaempferol intakes (RR = 0.70; 95% CI: 0.56-0.86; p: 0.003; median intake was 0.5 mg kaempferol/day). These data suggest that consumption of foods containing kaempferol, may reduce the risk of developing some cardiovascular disorders.

In resume, epidemiological evidences (clinical trials and cohort studies) suggest that the consumption of kaempferol-rich foods, may reduce the risk of developing some types of cancer, including lung cancer (Cui et al., 2008; Garcia-Closas et al., 1998), gastric cancer (Garcia-Closas et al., 1999), pancreatic cancer (Nothlings et al., 2007) and ovarian cancer (Gates et al., 2007). Although few studies have evaluated the anti-carcinogenic activity of kaempferol in animal models (Yasukawa et al., 1990), numerous preclinical studies reveal that flavonoid act as cancer preventive and show therapeutic properties (Calderón-Montaño et al., 2011).

In addition to its chemopreventive properties, kaempferol show activities with relevance in cancer therapy. Numerous reports have shown that kaempferol and/or some kaempferol glycosides, induce cell death in a variety of cancer from different tissues, including lung (Nguyen et al., 2003; Kim et al., 2004; Conforti et al., 2009), breast (Brusselmans et al., 2005; Kang el al., 2009), colon (Kim et al., 2004; Li et al., 2009), prostate (Brusselmans et al., 2005; Knowles et al., 2000), liver (Mylonis et al., 2010), pancreas (Zhang et al., 2008), blood/lymph (Bestwick et al., 2005; Marfe et al., 2009), skin (Kim et al., 2004; Conforti et al., 2009; Li et al., 2007), esophagus (Zhang et al., 2008; Zhang et al., 2009), brain (Sharma et al., 2007; Jeong et al., 2009), uterus (Li et al., 2007; Xu et al., 2008), ovary (Sharma et al., 2007; Kim et al., 2004), thyroid (Yin et al., 1999) and bone (Huang et al., 2010). Many of these reports conclude that kaempferol induces cell death through apoptosis, and the possible mechanisms involved in this process have been elucidated.

Kaempferol and some kaempferol glycosides may also decrease triglyceride levels, cholesterol and/or reduce body weight (Belguith-Hadriche et al., 2010; Yu et al., 2006). Evidence suggests that some kaempferol glycosides and several kaempferol-containing plants have antidiabetic activity (Chen et al., 2010; Lee et al., 2009) and may prevent diabetic complications (Asgary et al., 2002; Ghaffari and Mojab, 2007).

Kaempferol -possibly through the metabolite kaempferol 3-neohesperidoside-, may increase glucose uptake into myocytes, by a process mediated by phosphatidylinositol 3-

kinase (PI3K) and protein kinase C (PKC) pathways (Zanatta et al., 2008; Cazarolli et al., 2009), being in this regard as potent as insulin (Jorge et al., 2004).

Numerous experimental data have shown that kaempferol possesses a wide range of biological activities involved in the prevention and treatment of these and other diseases.

Flavonoids have been labeled as "high level" natural antioxidants on the basis of their abilities to scavenge free radicals and active oxygen species (Klahorst, 2002; Fukumoto and Mazza, 2000; Unno et al., 2000). They contain conjugated ring structures and hydroxyl groups that have the potential to function as antioxidants *in vitro* or cell free systems by scavenging superoxide anion, singlet oxygen, lipid peroxyradicals, and stabilizing free radicals involved in oxidative processes through hydrogenation, or complexing with oxidizing species (Duthie and Dobson, 1999; 51 Birt et al., 2001).

In this sense, numerous reports reveal that kaempferol, some glycosides of kaempferol, and several kaempferol-containing plants exhibit *in vitro* and *in vivo* antioxidant activity (Kampkotter et al., 2007; Bonina et al., 2002; Sanz et al., 1994). This fact could be explained by the presence of a double bond at C2-C3, an oxo group at C4, and the presence of hydroxyl groups at C3, C5 and C4' (van Acker et al., 1996).

Kaempferol has been found to be a potent superoxide scavenger, with an IC_{50} of 0.5 μM (Wang et al., 2006). The ability of kaempferol to decrease superoxide levels, even at low concentrations, may play an important role in its antioxidant activity, since formation of superoxide anion is required for the normal production of most reactive oxygen and nitrogen species, involved in oxidative stress (Klaunig and Kamendulis, 2004). Wang et al. (2006) evaluated the antioxidant activity of several flavonoids, finding that kaempferol was one of the strongest scavengers for the Fenton-generated hydroxyl radical, with an IC_{50} of 0.5 μM. Likewise, Heijnen et al. (2001) observed that kaempferol was a potent peroxynitrite scavenger, with an IC_{50} of 0.35 μM.

Also, kaempferol can inhibit the activity of enzymes such as xanthine oxidase that generate ROS (Wang et al., 2006; Nagao et al., 1999). Like other flavonoids, kaempferol can also reduce the formation of hydroxyl radical through the Fenton's reaction by chelating ferrous or cuprous ions (van Acker et al., 1996).

In addition, kaempferol may induce antioxidant effects by increasing the expression or activity of antioxidant enzymes such as superoxide dismutase, catalase, and heme oxygenase-1 (Doronicheva et al., 2007). Other studies have revealed that kaempferol, and some glycosides of kaempferol prevent lipid peroxidation (van Acker et al., 1996; Maridonneau-Parini et al., 1986) the activity of kaempferol is greater than their respective glycosides, since show higher lipophilia and capacity to penetrate into lipid bilayers to exert this activity. Kaempferol can also prevent the oxidation of low-density lipoproteins (LDL), which may play a protective role in atherosclerosis (Tu et al., 2007; Yu et al., 2005).

The antioxidant activity of kaempferol has been evidenced in several *in vivo* studies (Kampkotter et al., 2007; Bonina et al., 2002; Sanz et al., 1994). For instance, kaempferol decrease intracellular ROS accumulation and increase the survival of *Caenorhabditis elegans* (Kampkotter et al., 2007). An extract of *Capparis spinosa* containing kaempferol derivatives, showed a significant antioxidant activity when applied topically in healthy human volunteers (Bonina et al., 2002). Sanz et al. (1994) treated mice with kaempferol-3-O-galactoside, after bromobenzene intoxication, observing a reduction of hepatic lipid peroxidation products and an increase in reduced glutathione levels. The intraperitoneal injection of an extract of Crassocephalum crepidioides containing kaempferol glycosides, showed a strong antioxidant effect and also the prevention of galactosamine- and lipopolysaccharide induced hepatotoxicity in rats (Aniya et al., 2005).

The anti-inflammatory activity of kaempferol may be mediated by several mechanisms of action. The activation of the nuclear factor kappa B (NF-κB) increases the expression of pro-inflammatory cytokines, chemokines and enzymes (e.g. TNF-α, IL-1, IL-6, IL8, COX-2, iNOS), and several works reveal that kaempferol inhibits NF-κB activity (Wang et al., 2006; Garcia-Mediavilla et al., 2007). Also, kaempferol can inhibit TNF-α activity (Kowalski et al., 2005; Chen et al., 2004) and the expression of IL-1β and IL-8 (Kowalski et al., 2005). Activator protein 1 (AP-1) is a transcriptional regulator composed by members of the Fos and Jun families, that participates in the inflammatory response, and evidenced that kaempferol inhibit the activation of AP-1 (Chen et al., 2004). Cyclooxygenases (COX), lipoxygenases (LOX) and inducible nitric oxide synthase (iNOS) enzymes play important roles in inflammation, by participating in the synthesis of eicosanoids (e.g. prostaglandins, leukotrienes) and in the production of reactive species. Some reports have shown that kaempferol can inhibit COX-2 (Garcia-Mediavilla et al., 2007; Mahat et al., 2010), LOX (Nakadate et al., 1984) and iNOS (Olszanecki et al., 2002; Garcia-Mediavilla et al., 2007). ROS are also known to participate in the inflammatory process and, as shown above, kaempferol can reduce the cellular levels of ROS. In summary, numerous *in vitro* and *in vivo* studies have evidenced that kaempferol show anti-inflammatory activity and several mechanisms to explain this fact have been proposed.

Kaempferol and its glycosides have been isolated from plants used in popular medicine, for their antimicrobial properties. Numerous papers have reported that kaempferol, its glycosides, or plants containing kaempferol have antibacterial, antiviral, antifungal and antiprotozoal activities. Kaempferol and its glycosides can also act synergistically with antibiotics (e.g. rifampicin, vancomycin, methicillin, erythromycin and clindamycin), against antibiotic-resistant bacteria (Lim et al., 2007; Otsuka et al., 2008). Also, kaempferol show anti-viral activity against several viruses, including herpes simplex virus (Amoros et al., 1992; Lyu et al., 2005), cytomegalovirus (Mitrocotsa et al., 2000), influenza virus (Jeong et al., 2009), and human immunodeficiency virus (HIV) (Min et al., 2001; 219 Mahmood et al., 1996).

Some studies reveal that kaempferol interferes with enzymes that are vital for growth or virulence of certain fungi (Hwang et al., 2001; Brunskole et al., 2009). Kaempferol has also been described as an antiprotozoal agent (Barbosa et al., 2007; Calzada et al., 1999). This flavonoid show activity against Leismania spp (Marin et al., 2009; Muzitano et al., 2006) and is also active against Entamoeba histolytica and Giardia lamblia (Calzada et al., 1999), which are known to cause diarrhea. The antigiardial activity of kaempferol has been observed *in vivo* in female CD-1 mice infected with Giardia lamblia (Barbosa et al., 2007).

In resume, epidemiological studies have found a positive association between the consumption of kaempferol containing foods and a reduced risk of developing cardiovascular diseases and some types of cancer. Numerous *in vitro* and some animal studies support a role of kaempferol in the prevention and/or treatment of these and other diseases, such as neurodegenerative diseases, infectious diseases, diabetes, osteoporosis, anxiety, allergies, inflammation and pain (Calderón-Montaño et al., 2011). However, many of these studies have been conducted at doses higher than those documented in humans and, therefore, it is difficult to predict from these results, the effects of kaempferol intake on the prevention of these diseases (Calderón-Montaño et al., 2011). Likewise, most *in vitro* studies have been carried out at concentrations far beyond those achieved in plasma and tissues, after the oral administration of kaempferol. This means that some biological effects induced by kaempferol *in vitro*, including some toxic effects, may not be relevant *in vivo* when this flavonoid is taken via oral. The low oral bioavailability and high metabolism of kaempferol in humans should be overcome (e.g. by using alternative routes and forms of administration), to maximize some therapeutic properties of kaempferol. Also, additional animal studies and clinical trials are needed to better understand the possible health effects of kaempferol, and to further evaluate its potential as a new drug (Calderón-Montaño et al., 2011).

Encapsulation in Food Industry

Encapsulation involves the incorporation of food ingredients, enzymes, cells or other materials in small capsules. Applications for this technique have increased in the food industry since the encapsulated materials can be protected from moisture, heat or other extreme conditions, thus enhancing their stability and maintaining viability.

In the food industry, the microencapsulation process could be applied for a variety of reasons, which have been summarized by Desai and Park (2005) as follows: (i) protection of the core material from degradation by reducing its reactivity to its outside environment; (ii) reduction of the evaporation or transfer rate of the core material to the outside environment; (iii) modification of the physical characteristics of the original material to allow easier handling; (iv) tailoring the release of the core material slowly over time, or at a particular time; (v) to mask an unwanted flavor or taste of the core material; (vi) dilution of the core material when only small amounts are required, while achieving uniform dispersion in the host material; (vii) to help separate the components of a mixture, thus avoiding that react together.

Usually, food ingredients such as acidulants, flavoring agents, sweeteners, colorants, lipids, vitamins and minerals, enzymes and microorganisms, are encapsulated by different technologies.

Limitations in many of the encapsulation techniques have occurred due to high costs of production and the lack of food-grade available materials. In this sense, research is necessary to eliminate these limitations.

Nowadays, the development of CDs has led to new products with longer shelf-life, reduced volatility and protection of heat-labile substances. There is a great deal of research that needs to be done concerning the use of CDs in the food industry. Unlike the pharmaceutical industry, which can tolerate high costs, manufacturing costs will have to be reduced for food applications (Gibbs et al., 1999).

Cyclodextrins

After their discovery, CDs were considered important substances and its capacity for complexes formation was only considered a scientific curiosity. Later on, research on CDs proved that they are not only non-toxic but they can be helpful for protecting flavours, vitamins and natural colours (Astray et al., 2009). The preparation process of CDs consists of four principal phases: (i) culturing of the microorganism that produces the CD glucosyl transferase enzyme (CGT-ase); (ii) separation, concentration and purification of the enzyme from the fermentation medium; (iii) enzymatical conversion of prehydrolyzed starch in mixture of cyclic and acyclic dextrins; and (iv) separation of CDs from the mixture, their purification and crystallization. CGT-ase enzymes degrade the starch and produce intramolecular reactions without the water participation. In the process, cyclic and acyclic dextrins are originated, which are oligosaccharides of intermediate size. The cyclic resulting products are CDs.

Structure and Properties of CDs

CDs are cyclic oligosaccharides that consist of (α-1, 4)-linked α-D-glucopyranose units and contain a somewhat lipophilic central cavity and a hydrophilic outer surface. Due to the chair conformation of the glucopyranose units, CDs are shaped like a truncated cone rather than perfect cylinders. The hydroxyl functions are orientated to the cone exterior with the primary hydroxyl groups of the sugar residues at the narrow edge of the cone and the secondary hydroxyl groups at the wider edge. The central cavity is lined by the skeletal carbons and ethereal oxygens of the glucose residues, which gives it a lipophilic character (Figure 4).

The polarity of the cavity has been estimated to be similar to that of an aqueous ethanolic solution (Loftsson et al., 2005).

The natural α-, β- and γ-CDs consist of six, seven, and eight glucopyranose units, respectively. The natural CDs, in particular β-CDs, are of limited aqueous solubility meaning that complexes resulting from interaction of apolar molecules with these CDs, can be of limited solubility resulting in precipitation of solid CDs complexes from water and other aqueous systems.

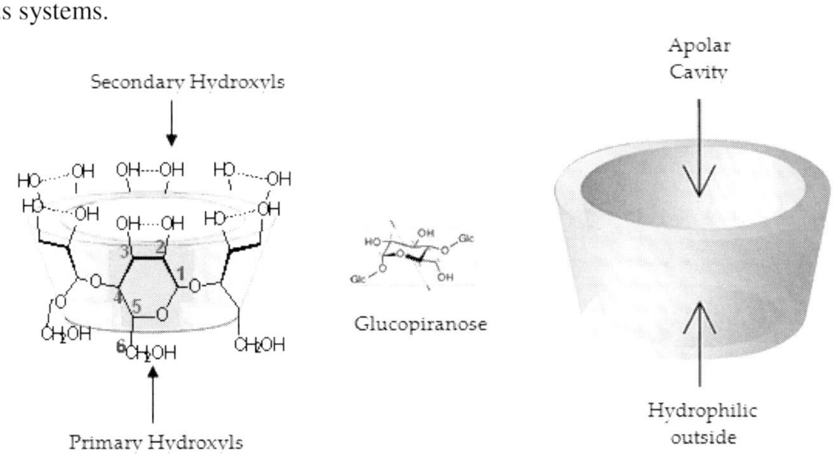

Figure 4. CD structure.

In fact, the aqueous solubility of the natural CDs is much lower than that of comparable acyclic saccharides. This is thought to be due to relatively strong intermolecular hydrogen bonding in the crystal state (Loftsson et al., 2005) (Table 2).

Substitution of any of the hydrogen bond forming hydroxyl groups, even by lipophilic methoxy functions, results in dramatic improvement in their aqueous solubility. CDs derivatives of pharmaceutical interest include hydroxypropyl derivatives of β- and γ-CDs, randomly methylated β-CDs, sulfobutylether β-CDs, and the so-called branched CDs such as glucosyl-β-CDs (Loftsson et al., 2005) (Figure 5).

Table 2. CDs properties

Properties	α-CDs	β-CDs	γ-CDs
Number of glucopyranose units	6	7	8
Molecular weight (g/mol)	972	1135	1297
Solubility in water at 25 ºC (%, w/v)	14.5	1.85	23.2
Outer diameter (Å)	14.6	15.4	17.5
Cavity diameter (Å)	4.7-5.3	6.0-6.5	7.5-8.3
Height of torus (Å)	7.9	7.9	7.9
Cavity volumen ($Å^{+3}$)	174	262	427

Martin del Valle, 2004.

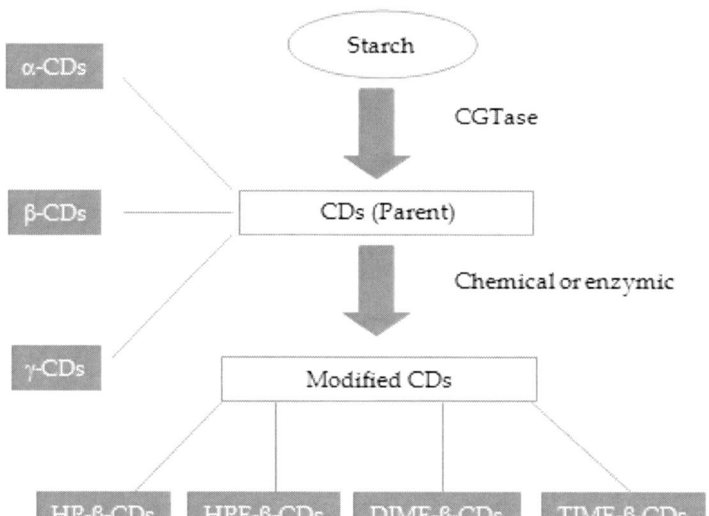

Figure 5. Natural CDs and derivatives CDs formations (Singh et al., 2002). HP-β-CDs: Hidroxypropyl-β-CDs; HPE-β-CDs: Hydroxyethyl-β-CDs; DIME-β-CDs: (2,6-dimethyl)-β-CDs; TIME-β-CDs: (2, 3, 6-trimethyl)-β-CDs.

The natural α- and β-CDs, unlike γ-CDs, cannot be hydrolyzed by human salivary and pancreatic amylases. However, both α- and β-CD can be fermented by the intestinal microflora. CDs are both large (MW ranging from almost 1,000 to over 2,000 Daltons) and hydrophilic, with a significant number of H-donors and acceptors and, thus, are not absorbed from the gastrointestinal tract in their intact form. Hydrophilic CDs are considered non-toxic at low to moderate oral dosages.

Lipophilic CD derivatives, such as the methylated CDs, are to some extent absorbed from the gastrointestinal tract into the systemic circulation, and have been shown to be toxic after parenteral administration (Loftsson et al., 2005).

Actually, oral administration of methylated β-CDs is limited by its potential toxicity (Loftsson et al., 2005).

About 30 different pharmaceutical products containing CDs are now on the market worldwide. In the pharmaceutical industry, CDs have mainly been used as complexing agents to increase the aqueous solubility of poorly water-soluble drugs, and to increase their bioavailability and stability.

In addition, CDs can be used to reduce or prevent gastrointestinal and ocular irritation, reduce or eliminate unpleasant smells or tastes, prevent drug-drug or drug-additive interactions, or to convert oils and liquid drugs into microcrystalline or amorphous powders (Loftsson et al., 2004).

Complex Formation and Drugs

CDs can be considered as empty capsules of a certain molecular size that can include a great variety of molecules in this cavity. In this case, a complex called ''inclusion complex'' is formed. Inclusion complexes are entities comprising two or more molecules. One of the molecule, the ''host'', includes totally or partly the ''guest'' molecules by physical forces.

Therefore, CDs are considered typical host molecules (Szente et al., 1998; García-Río et al., 2006; Brewster and Loftsson, 2007). The type of bond established between the guest and the host is no covalent. Figure 6 illustrates the approximate volumes of each capsule (α-, β-, and γ-CDs).

Complex formation in solution is a dynamic equilibrium process which can be illustrated by the equation (1), where CDs are the encapsulating agent, kaempferol is the guest molecule, and CDs-kaempferol is the inclusion complex.

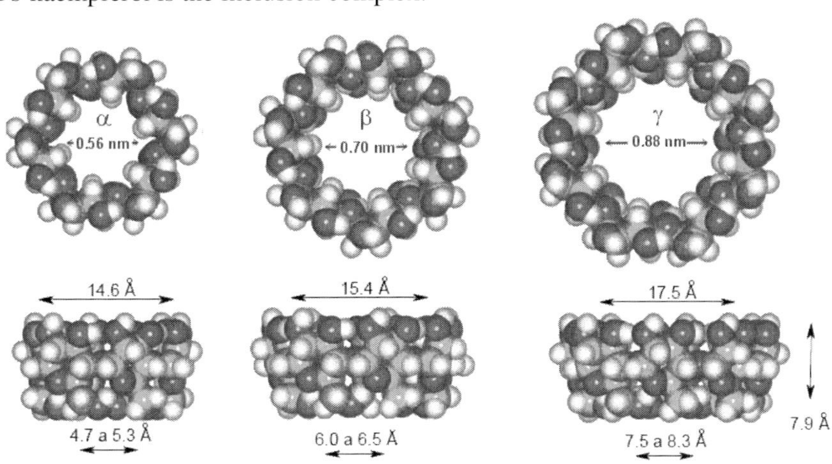

Figure 6. Outer and cavity diameters of differents CDs.

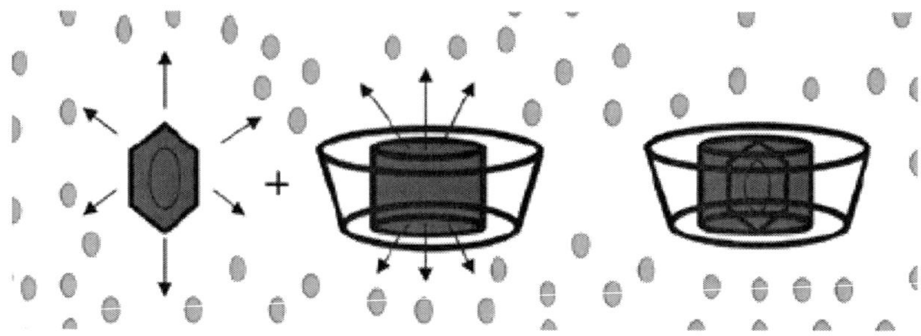

Figure 7. Inclusion complex of kaempferol-CDs.

The stability of the inclusion complex can be described by the complexation constant K_C (2).

$$Kaemp + CDs \xleftrightarrow{K_C} Kaemp - CDs \tag{1}$$

$$K_C = \frac{Kaemp - CDs}{[Kaemp]_L [CDs]_L} \tag{2}$$

Energy and Inclusion Mechanisms

The inclusion of a guest in a CDs cavity consists basically of a substitution of the included water molecules by the less polar guest (Figure 7). The process is energetically favoured by the interactions of the guest molecule with the solvated hydrophobic cavity of the host. In this process, entropy and enthalpy changes have an important role. In spite of the fact that the ''driving force'' of complexation is not yet completely understood, it seems that it is the result of various effects (Astray et al., 2009):

a) Substitution of the energetically unfavoured polar–apolar interactions (between the included water and the CDs cavity on the one hand, and between water and the guest on the other) by the more favoured apolar–apolar interaction (between the guest and the cavity), and the polar–polar interaction (between bulk water and the released cavity-water molecules).
b) CDs-ring strain release on complexation.
c) Van der Waals interactions and hydrogen bonds between host and guest.

Properties of CDs-Included Guests in Solution

When a more or less non-polar poorly water-soluble substance reacts with an aqueous CDs solution, the following consequences can occur (Liu et al., 1999):

1) The concentration of the guest in the dissolved phase increases significantly, and when solid complexes are simultaneously formed, the concentration of dissolved CDs decreases (except with ionized guests, or hydrogen bond establishing).
2) The spectral properties of the guest are modified.

3) The reactivity of the included molecule is modified. In most cases their reactivity decreases.
4) The formerly hydrophobic guest in complexed form becomes hydrophilic.

Use and Regulatory of CDs

The use of CDs has increased annually around 20–30%, of which 80–90% was in food products. The widespread utilization of CDs is reflected in pharmaceutical, food, chemical and other industrial areas (Szejtli, 1997). In the pharmaceutical industry, CDs and their derivatives have been used in drugs either for complexation or as auxiliary additives such as solubilizers, diluents, or tablet ingredients to improve the physical and chemical properties, or to enhance the bioavailability of poorly soluble drugs (Frömming and Szejtli, 1994). In the chemical industry, CDs and their derivatives are used as catalysts to improve the selectivity of reactions, as well as for the separation and purification of industrial-scale products (Hedges, 1998).

In the food, cosmetics, toiletry, and tobacco industries, CDs have been widely used either for stabilization of flavours and fragrances or for the elimination of undesired tastes, microbiological contaminations, and other undesired compounds (Martín del Valle, 2004; Singh et al., 2002).

However, the regulatory status of CDs in foods differs between countries. In the US α-, β-, and γ-CDs have obtained the GRAS list (FDA list of food additives that are 'generally recognized as safe') status, and can be commercialized as such. In Japan α-, β-, and γ-CDs are recognized as natural products and their commercialization in the food sector is restricted only by considerations of purity. In Australia and New Zealand α- and γ-CDs are classified as Novel Foods from 2004 and 2003, respectively (Cravotto et al., 2006). The recommendation of Joint FAO/WHO Expert Committee on Food Additives (JECFA) for a maximum level of β-CDs in foods is 5 mg/kg per day. For α- and γ-CDs no Acceptable Daily Intake (ADI) was defined because of their favourable toxicological profiles.

Moreover, in July 2005 the US Environmental Protection Agency (EPA) did away with the need to establish a maximum permissible level for residues of α-, β-, and γ-CDs in various food commodities (US Federal Register, 2005).

In Europe, α- and β-CDs has been approved as food products ingredients. The designation "α-CD" and "β-CD" must appear in the list of ingredients of food products containing it.

Applications of CDs

Since each guest molecule is individually surrounded by CDs, the molecule is micro-encapsulated from a microscopical point of view. This process can lead to advantageous changes in the chemical and physical properties of the guest molecules (Martín del Valle, 2004) such us:

- Stabilization of light -or oxygen-sensitive substances.
- Modification of the chemical reactivity of guest molecules.

- Fixation of very volatile substances.
- Improvement of solubility of substances.
- Modification of liquid substances to powders.
- Protection against degradation of substances by microorganisms.
- Masking of ill smell and taste.
- Masking pigments or the colour of substances.
- Catalytic activity of CDs with guest molecules.

These characteristics of CDs make them suitable for applications in analitycal chemistry, agriculture, the pharmaceutical field, in toilet articles and food industry (Singh et al., 2002).

The use of CDs in the food industry has increased significantly in the last years. These compounds are highly recommended for applications in food processing and as food additives. Thus, volatile compounds can be encapsulated in CDs in order to limit aroma degradation or loss during processing and storage (Lubbers et al., 1998).

Moreover, the use of CDs-flavour inclusion complexes allows using minor amounts of flavours.

On the other hand, CDs can also be used for deodorizing and removing undesirable components such as off flavours or bitter components present in the food in its natural form (Binello et al., 2004), or for improving the nutritional characteristics of many dairy foods such as milk (Kwak et al., 2004), mayonnaise (Jung et al., 2008), lard (Kim et al., 2007) or cream (Shim et al., 2003), through their complexation with cholesterol molecules.

In addition, CDs can play an important role as food colour modulators due to the ability of these compounds to form complexes with the polyphenol-oxidase substrates, which are responsible of enzymatic browning in foods (Martín del Valle, 2004; Núñez-Delicado et al., 2005).

Another interesting application of CDs is their incorporation in the food packaging materials as antiseptic or conserving agents (Wood, 2001).

Encapsulation of Kaempferol in Cyclodextrins

Research on the application of flavonoids, have recently attracted great interest in the functional foods, nutraceutical and pharmaceutical industries, due to their potential health benefits to humans.

However, the effectiveness of flavonoids depends on preserving their stability, boactivity and bioavailability. The utilization of encapsulated flavonoids, insteated of free compounds, can effectively solve these drawbacks. One outstanding advantage of the inclusion of flavonoids in CDs is related to improving the aqueous solubility, especially for less water soluble phytochemicals.

The inclusion of many polyphenols: hesperetin and hesperidin in HP-β-CDs (Tommasini et al., 2005), rutin in β-CDs (Haiyun et al., 2003), olive leaf extract (rich in oleuropein) in β-CDs (Mourtzinos et al., 2007), quercetin and myricetin in HP- β-CDs, G2-β-CDs, and β-CDs (Lucas-Abellán et al., 2008b), kaempherol in HP-β-CDs (Mercader-Ros et al., 2010b) and resveratrol in native and modifies CDs (Lucas-Abellán et al., 2007; Lucas-Abellán et al., 2008a) have been studied, and their solubility improved through encapsulation process.

The high hydrophobicity of kaempferol and its sensitivity to external agents such us air, light and oxidative enzymes, may constitute a serious problem for its bioavailability, formulation and manipulation in the elaboration of functional foods. In order to avoid this limitation, inclusion of kaempferol in CDs has been widely studied (Mercader-Ros et al., 2010b; Calabrò et al., 2004). The inclusion complex formation will modify physico-chemical properties such as solubility, stability and bioavailability of poorly water-soluble kaempferol.

In addition, its antioxidant activities increased in these encapsulated form. The improved antioxidant efficacy of the inclusion complex could be explained by the protection of kaempferol against rapid oxidation by free radicals (Mercader-Ros et al., 2010a), which may, in part, justify an increase in its solubility in the biological fluids (Ding et al., 2003).

The complexation of kaempferol with cyclodextrins (CDs) (β-, G2-β-, and HP-β-CDs) in basic medium was studied, and the complexation constants (K_C) were calculated by enzymatic, solubility and fluorometric methods. To establish the validity of the fluorometric method for determining the K_C between kaempferol and CDs, the same parameters were also determined by enzymatic and solubility methods.

The enzymatic method was carried out by using horseradish peroxidase as oxidative enzyme, and the K_C values obtained were similar to those obtained by using the solubility method. However, the fluorometric method underestimated the K_C value by about 1.2-fold with respect to the other methods used.

The oxidation of kaempferol by horseradish peroxidase in the presence of H_2O_2 at pH 9.0, led to the formation of an oxidation product with a maximum at 315 nm. In the presence of increasing concentrations of CDs, the oxidation rate of kaempferol by peroxidase decreased, indicating that kaempferol was complexed in the hydrophobic cavity of CDs, being free kaempferol the only form which peroxidase could use as substrate (Figure 8).

This result was observed whether β-, G_2-β-, or HP-β-CDs were used, although the inhibition was higher in the case of HP-β-CDs (Figure 8, solid circles). Fitting points in Figure 8, K_C values for β-, G_2-β-, and HP-β-CDs were calculated, giving the following values: 11,848 ± 258 M^{-1} for HP- β-CDs; 2,325 ± 312 M^{-1} for G_2-β-CDs; and 2,672 ± 245 M^{-1} for β-CDs (Table 3).

Table 3. Complexation constant (K_C) calculated by the enzymatic, solubility, and fluorometric methods

Methods	K_C β-CDs (M^{-1})	K_C G_2-β-CDs (M^{-1})	K_C HP-β-CDs (M^{-1})
Fluorimetric:			
Benisi-Hildebrand	1,622±259	1,056±120	9,872±332
Scatchard	1,555±268	998±109	9,367±325
Enzymatic	2,670±245	2,325±312	11,848±258
Solubility	2,747± 210	2,508±150	11,161±302

Phase solubility diagrams of kaempferol showed a linear relationship between the amount of kaempferol solubilized and the concentration of cyclodextrin in solution (AL type), indicating that the stoichiometry of complexes was 1:1 in all cases.

By comparison of the K_C values obtained for the complexation of kaempferol by β-, G_2-β-, and HP-β-CDs using the enzymatic, solubility, and fluorometric methods (Table 3), results showed that the enzymatic and solubility methods provided higher K_C values for all of the CDs studied than fluorometric one.

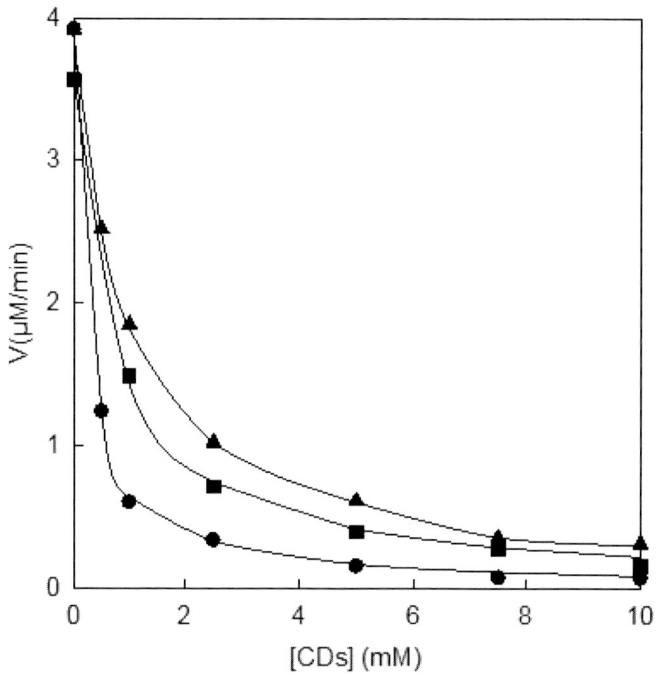

Figure 8. Effect of CDs on the oxidation of kaempferol by horseradish peroxidase in the presence of H2O2. The reaction medium at 25 ºC contained 100 mM sodium borate buffer, pH 9.0, 20.96 µM kaempferol, 400 µM H2O2, 0.37 units of horseradish peroxidase and increasing concentrations of CDs. (●) HP-β-CDs, (■) β-CDs, (▲) G2-β-CDs.

To determine which method is more accurate (enzymatic/solubility or fluorometric), the so-called CD assay (Lucas-Abellán et al., 2007; Lucas-Abellán et al., 2008a), in which the total CDs and kaempferol concentrations are adjusted so that the free kaempferol concentration remains constant, was carried out for all CDs studied. Using the K_C values determined enzymatically, by solubility studies or fluorometrically, the corresponding CD assay was performed. It was expected that when the concentration of effective free kaempferol remained constant, the rate of kaempferol oxidation by horseradish peroxidase (µM/min) could also be constant, regardless of the total kaempferol concentration. As shown in Figure 9, the results obtained using K_C values (solid symbols) determined enzymatically, or by solubility studies, were as expected (the enzymatic activity was independent of total CDs and kaempferol concentrations), but the result obtained using the fluorometrically determined K_C values (open symbols) deviated from expected results (the enzymatic activity did not remain constant), indicating that the free kaempferol concentration is not the same in all cases; therefore, the K_C values used are not correct.

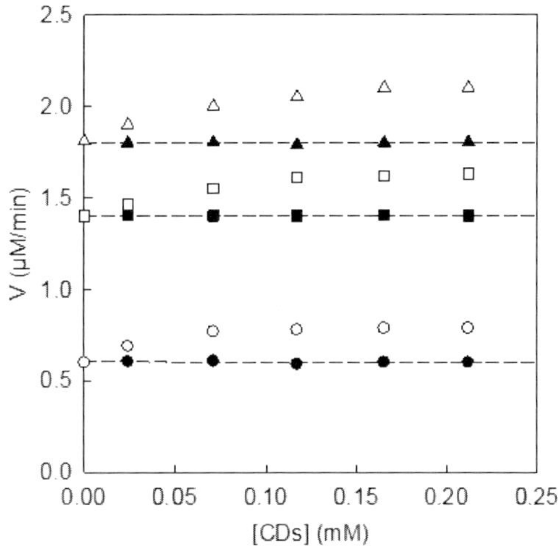

Figure 9. Cyclodextrin assay. The CDs and kaempferol total concentrations used in each case were calculated to obtain the free kaempferol concentrations, by using the KC values obtained with the enzymatic method ((●) 1.6 μM free kaempferol and increasing concentrations of HP-β-CDs; (■) 5.7 μM free kaempferol and increasing concentrations of β-CDs; (▲) 6.3 μM free kaempferol and increasing concentrations of G2-β-CDs) and the fluorimetric method: (○) 1.6 μM free kaempferol and increasing concentrations of HP-β-CDs; (□) 5.7 μM free kaempferol and increasing concentrations of β-CDs; (△) 6.3 μM free kaempferol and increasing concentrations of G2-β-CDs. The oxidation of free kaempferol was followed in 100 mM sodium borate buffer (pH 9.0), using 0.37 units of horseradish peroxidase.

Thus, the differences in the K_C values observed between the enzymatic/ solubility and fluorometricmethods (Table 3) are very important for knowing the real effective kaempferol concentration.

We conclude that if fluorescence decay of kaempferol is used, the K_C will be undervalued and the free kaempferol concentration will be higher than expected, which could explain the increase in horseradish peroxidase activity observed in Figure 9.

These results agree with those obtained for the complexation of resveratrol (Lucas-Abellán et al., 2008a).

In conclusion, kaempferol fluorescence decreases when it is complexed in the hydrophobic cavity of CDs. This fluorescence decrease cannot be used to determine the K_C between kaempferol and CDs because it is underestimated. Therefore, in the case of kaempferol and other compounds that can be oxidized by different enzymes, the best method for calculating its K_C is the enzymatic one, because of the high specificity of the enzymes, which act only against free substrate. When the enzymatic method could not be used, the solubility study is an optional method to calculate K_C values.

It is important to note that HP-β-CDs showed the highest K_C values, indicating that they are the most efficient in the formation of inclusion complexes with kaempferol, as occurs in the case of other flavonoids such as myricetin or quercetin (Lucas-Abellán et al., 2008b), and in the case of the stilbene resveratrol (Lucas-Abellán et al., 2008a).

Effect of CDs Complexation on the Antioxidant Activity of Kaempferol

The basic flavonoid structure is the flavan nucleus, which consists of 15 carbon atoms arranged in three rings (C6–C3–C6), labelled as A, B and C (See figure 1). Various classes of flavonoids (flavonols, flavones, flavanones, catechins or flavanols, anthocyanidins, isoflavones, dihydroflavonols and chalcones) differ in the level of oxidation and saturation of ring C, while individual compounds within a class differ in the substitution pattern of rings A and B. These differences in the structure and substitution will influence the phenoxyl radical stability and thereby, the antioxidant properties of the flavonoids (Huber et al., 2009).

The above-mentioned antioxidant properties reside mainly in their radical-scavenging activity. Repeated studies have shown that flavonoids with many hydroxyl groups are extremely effective antioxidants. In fact, hydroxylated flavonols, such as kaempferol, have been demonstrated to be particularly effective antioxidants in many studies (Cao et al., 1997; Soobrattee et al., 2005; Wojdylo et al., 2007).

As continuing research examines the effect of antioxidants on health, the testing for antioxidant protection has become focus of attention in the dietary and natural products industry. Researchers associated with the natural product industry have pushed for a standardised method for measuring antioxidant capacity in natural products (Honzel et al., 2008). A large number of methods have been developed to evaluate the antioxidant capacity in foods, one of the most popular and best standardised chemical antioxidant methods being the oxygen radical absorbance capacity (ORAC) test (Ou et al., 2001). The ORAC method is based on the inhibition of the peroxyl-radical-induced oxidation initiated by thermal decomposition of azo-compounds, like 2,20-azobis(2-amidino-propane) dihydrochloride (AAPH), and it is the only method that combines inhibition time and inhibition degree into a single quantity. This test is widely used for the evaluation and comparison of the antioxidant capacity of natural food products and plasma (Prior et al., 2007).

CDs can be used as a flavonol complexation agent, since they act as a substrate reservoir in a dose-controlled manner. The effect of complexing kaempferol with HP-β-CDs on its antioxidant capacity has been studied by means of the ORAC-fluorescein (ORAC-FL) assay (Mercader-Ros et al., 2010a). This complexation phenomenon increased the antioxidant activity of kaempferol, which reached a maximum level when kaempferol has been complexed in the hydrophobic cavity of CDs (Figure 10). The antioxidant activity increased because of the flavonol was protected against rapid oxidation by free radicals.

Conclusion and Future Trends

It is clear that the utilization of encapsulated kaempferol instead of free compound, can lead to an improvement in both, stability and bioavailability of the compound. Because there is still a lack of direct evidence for the use of polyphenols in preventing and treating of human diseases (Scalbert et al., 2005), most of polyphenols encapsulated are classified as ‘functional foods’ or ‘nutriceuticals’, which limits their potential markets. In food grade products, cost is an important factor for their industrialization.

Future research of polyphenol encapsulation is likely to focus on aspects of delivery and the potential use of co-encapsulation methodologies, where two or more bioactive ingredients can be combined to have a synergistic effect.

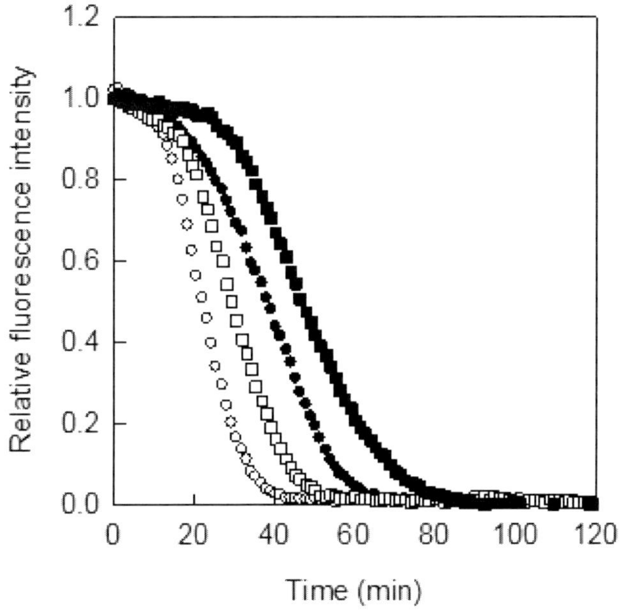

(a)

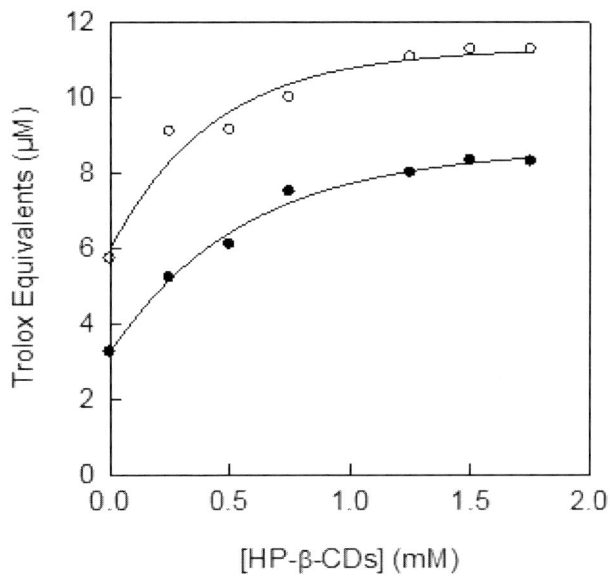

(b)

Figure 10. (A) FL fluorescence decay curves induced by AAPH in the presence of kaempferol alone (0.5 μM (○) and 0.75 μM (□)) and with HP-β-CDs 1.75 mM (kaempferol 0.5 μM (●) 0.75 μM (■)). **(B)** Effect of HP-β-CDs concentration on the μM Trolox equivalents of kaempferol 0.5 μM (●), 0.75 μM (○).

It is possible that the understanding of the health benefits of kaempferol, improvement in manufacturing technology, new strategy for stabilization of fragile nutraceuticals, and the development of novel approaches to site-specific carrier targeting, encapsulated kaempferol will play an important role in increasing the efficacy of functional foods or even pharmaceuticals, over the next decade.

References

Adebamowo, C. A.; Cho, E.; Sampson, L.; Katan, M. B.; Spiegelman, D.; Willett, W. C.; Holmes, M. D. Dietary flavonols and flavonol-rich foods intake and the risk of breast cancer. *Int. J. Cancer.*, 2005, 20, 628-33.

Aherne, S. A.; O'Brien, N. M. Dietary flavonols: chemistry, food content, and metabolism. *Nutrition.,* 2002, 18, 75-81.

Amoros, M.; Simoes, C. M.; Girre, L.; Sauvager, F.; Cormier, M. Synergistic effect of flavones and flavonols against herpes simplex virus type 1 in cell culture. Comparison with the antiviral activity of propolis. *J. Nat. Prod.*, 1992, 55, 1732-40.

Aniya, Y.; Koyama, T.; Miyagi, C.; Miyahira, M.; Inomata, C.; Kinoshita, S.; Ichiba, T. Free radical scavenging and hepatoprotective actions of the medicinal herb, Crassocephalum crepidioides from the Okinawa Islands. *Biol. Pharm. Bull.*, 2005, 28, 19-23.

Asgary, S.; Naderi, G. A.; Zadegan, N. S.; Vakili, R. The inhibitory effects of pure flavonoids on *in vitro* protein glycosylation. *J. Herb. Pharmacother.*, 2002, 2, 47-55.

Astray, G.; Gonzalez-Barreiro, C.; Mejuto, J. C.; Rial-Otero, R.; Simal-Gándara J. A review on the use of cyclodextrins in foods. *Food Hydrocolloids.*, 2009, 23, 1631–1640.

Barbosa, E.; Calzada, F.; Campos, R. *In vivo* antigiardial activity of three flavonoids isolated of some medicinal plants used in Mexican traditional medicine for the treatment of diarrhea. *J. Ethnopharmacol.,* 2007, *109*, 552-4.

Barve, A.; Chen, C.; Hebbar, V.; Desiderio, J.; Saw, C. L.; Kong, A. N. Metabolism, oral bioavailability and pharmacokinetics of chemopreventive kaempferol in rats. Biopharm. *Drug Dispos.*, 2009, 30, 356-65.

Belguith-Hadriche, O.; Bouaziz, M.; Jamoussi, K.; El Feki, A.; Sayadi, S.; Makni-Ayedi, F. Lipid-lowering and antioxidant effects of an ethyl acetate extract of fenugreek seeds in high-cholesterolfed rats. *J. Agric. Food Chem.*, 2010, 58, 2116-22.

Bestwick, C. S.; Milne, L.; Pirie, L.; Duthie, S. J. The effect of short-term kaempferol exposure on reactive oxygen levels and integrity of human (HL-60) leukaemic cells. *Biochim. Biophys. Acta*, 2005, 1740, 340-9.

Binello, A.; Cravotto, G.; Nano, G. M.; Spagliardi, P. Synthesis of chitosancyclodextrin adducts and evaluation of their bitter-masking properties. *Flavour Frag. J.*, 2004, 19, 394–400.

Birt, D. F.; Hendrich, S.; Wang, W. Dietary agents in cancer prevention: Flavonoids and isoflavonoids. *Pharmacol. Therapeut.,* 2001, 90, 157–77.

Bonina, F.; Puglia, C.; Ventura, D.; Aquino, R.; Tortora, S.; Sacchi, A.; Saija, A.; Tomaino, A.; Pellegrino, M. L.; de Caprariis, P. *In vitro* antioxidant and *in vivo* photoprotective effects of a lyophilized extract of Capparis spinosa L buds. *J. Cosmet. Sci.*, 2002, 53, 321-35.

Brewster, M. E.; Loftsson, T. Cyclodextrins as pharmaceutical solubilizers. *Adv. Drug Deliver. Rev.*, 2007, 59, 645–666.

Brunskole, M.; Zorko, K.; Kerbler, V.; Martens, S.; Stojan, J.; Gobec, S.; Lanisnik, R. T. Trihydroxynaphthalene reductase of Curvularia lunata-A target for flavonoid action? *Chem. Biol. Interact.*, 2009, 178, 259-67.

Brusselmans, K.; Vrolix, R.; Verhoeven, G.; Swinnen, J. V. Induction of cancer cell apoptosis by flavonoids is associated with their ability to inhibit fatty acid synthase activity. *J. Biol. Chem.*, 2005, 280, 5636-45.

Calabrò, M. L.; Tommasini, S.; Donato, P.; Raneri, D.; Stancanelli, R.; Ficarra, P.; Ficarra, R.; Costa, C.; Catania, S.; Rustichelli, C.; Gamberini, G. Effects of α- and β-cyclodextrin complexation on the physico-chemical properties and antioxidant activity of some 3-hydroxyflavones. *J. Pharma. Biomed. Ana.* 2004, 35, 365-77.

Calderón-Montaño, J. M.; Burgos-Morón, E.; Pérez-Guerrero, C.; López-Lázaro, M.A. Review on the Dietary Flavonoid Kaempferol. *Mini-Rev. Med. Chem.*, 2011, 11, 298-344.

Calzada, F.; Meckes, M.; Cedillo-Rivera, R. Antiamoebic and antigiardial activity of plant flavonoids. *Planta Med.*, 1999, 65, 78-80.

Cao, G.; Sofic, E.; Prior, R. Antioxidant and prooxidant behaviour of flavonoids: Structure–activity relationships. *Free Radical Biology and Medicine*, 1997, 22, 749–60.

Cazarolli, L. H.; Folador, P.; Pizzolatti, M. G.; Mena Barreto Silva, F. R. Signaling pathways of kaempferol-3-neohesperidoside in glycogen synthesis in rat soleus muscle. *Biochimie.*, 2009, 91, 843-9.

Chen, C. C.; Chow, M. P.; Huang, W. C.; Lin, Y. C.; Chang, Y. J. Flavonoids inhibit tumor necrosis factor-alpha-induced upregulation of intercellular adhesion molecule-1 (ICAM-1) in respiratory epithelial cells through activator protein-1 and nuclear factor-kappaB: structure-activity relationships. *Mol. Pharmacol.*, 2004, 66, 683-93.

Chen, D.; Dou, Q. P. Tea polyphenols and their roles in cancer prevention and chemotherapy. *Int. J. Mol. Sci.*, 2008, 9, 1196-206.

Chen, Q. C.; Zhang, W. Y.; Jin, W.; Lee, I. S.; Min, B. S.; Jung, H. J.; Na, M.; Lee, S.; Bae, K. Flavonoids and isoflavonoids from Sophorae Flos improve glucose uptake *in vitro*. *Planta Med.*, 2010, 76, 79-81.

Conforti, F.; Menichini, F.; Rigano, D.; Senatore, F. Antiproliferative activity on human cancer cell lines after treatment with polyphenolic compounds isolated from Iris pseudopumila flowers and rhizomes. Z. *Naturforsch. C.*, 2009, 64, 490-4.

Cravotto, G.; Binello, A.; Baranelli, E.; Carraro, P.; Trotta, F. Cyclodextrins as food additives and in food processing. *Curr. Nutr. Food Sci.*, 2, 2006, 343-350.

Cui, Y.; Morgenstern, H.; Greenland, S.; Tashkin, D. P.; Mao, J. T.; Cai, L.; Cozen, W.; Mack, T. M.; Lu, Q. Y.; Zhang, Z. F. Dietary flavonoid intake and lung cancer-A population-based case-control study. *Cancer.*, 2008, 112, 2241-8.

De Vries, J. H.; Janssen, P. L.; Hollman, P. C.; van Staveren, W. A.; Katan, M. B. Consumption of quercetin and kaempferol in freeliving subjects eating a variety of diets. *Cancer Lett.*, 1997, 19, 141-4.

Desai, K. G. H.; Park, H. J. Recent developments in microencapsulation of food ingredients. *Dry. Technol.*, 2005, 23, 1361-94.

Ding, H.; Chao, J.; Zhang, G.; Shuang, S.; Pan, J. Preparation and spectral investigation on inclusion complex of b-cyclodextrin with rutin. Spectrochimica Acta Part A: *Mol. Biomol. Spectros.*, 2003, 59, 3421-29.

Doronicheva, N.; Yasui, H.; Sakurai, H. Chemical structuredependent differential effects of flavonoids on the catalase activity as evaluated by a chemiluminescent method *Biol. Pharm. Bull.*, 2007, 30, 213-7.

Duthie, S. J.; Dobson, V. L. Dietary flavonoids protect human colonocyteDNAfrom oxidative attack *in vitro*. *Eur. J. Nutr.,* 1999, 38, 28–34.

Erlund, I. Review of the flavonoids quercetin, hesperetin, and naringenin. Dietary sources, bioactivities, bioavailability, and epidemiology. *Nutr. Res.,* 2004, 24, 851-874.

Frömming, K. H.; Szejtli, J. Cyclodextrins in pharmacy. Dordrecht: Kluwer Academic Publishers, 1994.

Fukumoto, L. R.; Mazza, G. Assessing antioxidant and prooxidant activities of phenolic compounds. *J. Agric. Food Chem.,* 2000, 48, 3597–604.

Garcia, R.; Gonzalez, C. A.; Agudo, A.; Riboli, E. High intake of specific carotenoids and flavonoids does not reduce the risk of bladder cancer. *Nutr. Cancer*, 1999, 35, 212-4.

Garcia-Closas, R.; Agudo, A.; Gonzalez, C. A.; Riboli, E. Intake of specific carotenoids and flavonoids and the risk of lung cancer in women in Barcelona, Spain. *Nutr. Cancer*, 1998, 32, 154-8.

Garcia-Closas, R.; Gonzalez, C. A.; Agudo, A.; Riboli, E. Intake of specific carotenoids and flavonoids and the risk of gastric cancer in Spain. *Cancer Causes Control*, 1999, 10, 71-5.

Garcia-Mediavilla, V.; Crespo, I.; Collado, P. S.; Esteller, A.; Sanchez-Campos, S.; Tunon, M. J.; Gonzalez-Gallego, J. The antiinflammatory flavones quercetin and kaempferol cause inhibition of inducible nitric oxide synthase, cyclooxygenase-2 and reactive C-protein, and down-regulation of the nuclear factor kappaB pathway in Chang Liver cells. *Eur. J. Pharmacol.*, 2007, 557, 221-9.

García-Río, L.; Hervés, P.; Leis, J. R.; Mejuto, J. C.; Pérez-Juste, J.; Rodríguez- Dafonte, P. Evidence for complexes of different stoichiometries between organic solvents and cyclodextrins. *Org. Biomol. Chem.*, 2006, 4, 1038–48.

Gates, M. A.; Tworoger, S. S.; Hecht, J. L.; De, V., I.; Rosner, B.; Hankinson, S.E. A prospective study of dietary flavonoid intake and incidence of epithelial ovarian cancer. *Int. J. Cancer*, 2007, 121, 2225-32.

Gates, M. A.; Tworoger, S. S.; Hecht, J. L.; De, V., I.; Rosner, B.; Hankinson, S.E. A prospective study of dietary flavonoid intake and incidence of epithelial ovarian cancer. *Int. J. Cancer*, 2007, 121, 2225-32.

Geleijnse, J. M.; Launer, L. J.; Van der Kuip, D. A.; Hofman, A.; Witteman, J. C. Inverse association of tea and flavonoid intakes with incident myocardial infarction: the Rotterdam Study. *Am. J. Clin. Nutr.*, 2002, 75, 880-6.

Ghaffari, M. A.; Mojab, S. Influence of flavonols as *in vitro* on low density lipoprotein glycation. *Iran. Biomed. J.,* 2007, 11, 185-91.

Gibbs, B. F., Kermasha, S., Alli, I. and Mulligan, C.N. Encapsulation in the food industry: A review. *International Journal of Food Sciences and Nutrition,* 1999, 50, 213-224.

Graefe, E. U.; Wittig, J.; Mueller, S.; Riethling, A. K.; Uehleke, B.; Drewelow, B.; Pforte, H.; Jacobasch, G.; Derendorf, H.; Veit, M. Pharmacokinetics and bioavailability of quercetin glycosides in humans. *J. Clin. Pharmacol.*, 2001; 41: 492–499.

Harborne, J. B. Nature, distribution, and function of plant flavonoids. In: Cody, V.; Middleton, E.; Harborne, J. B.; eds. Plant flavonoids in biology and medicine: biochemical, pharmacological, and structure–activity relationships. *New York: Alan R. Liss*, 1986, 213, 15.

He, X.-Z.; Li, W.-S.; Blount, J. W.; Dixon, R. A. Regioselective synthesis of plant (iso) flavone glycosides in Escherichia coli. *Appl. Microbiol. Biotechnol.*, 2008, 80, 253–60.

Hedges, R. A. Industrial applications of cyclodextrins. *Chem. Rev.*, 1998, 98, 2035–44.

Heijnen, C. G.; Haenen, G. R.; van Acker, F. A.; van der Vijgh, W. J.; Bast, A. Flavonoids as peroxynitrite scavengers: the role of the hydroxyl groups. *Toxicol. In vitro*, 2001, 15, 3-6.

Hertog, M. G.; Feskens, E. J.; Hollman, P. C.; Katan, M. B.; Kromhout, D. Dietary antioxidant flavonoids and risk of coronary heart disease: the Zutphen Elderly Study. *Lancet*, 1993, 342, 1007-11.

Honzel, D.; Carter, S. G.; Redman, K. A.; Schauss, A. G.; Endres, J. R.; Jensen, G. S. Comparison of chemical and cell-based antioxidant methods for evaluation of food and natural products: Generating multifaceted data by parallel testing using erythrocytes and polymorphonuclear cells. *J. Agric. Food Chem.*, 2008, 56, 8319–25.

Hou, L.; Zhou, B.; Yang, L.; Liu, Z. L. Inhibition of free radical initiated peroxidation of human erythrocyte ghosts by flavonols and their glycosides. *Org. Biomol. Chem.*, 2004, 2, 1419-23.

Hu, M.; Chen, J.; Lin, H. Disposition of flavonoids via recycling: mechanistic studies of disposition of apigenein in the Caco-2 cell culture model. *J. Pharmacol. Exp. Ther.*, 2003, 307: 314–21.

Hu, M. Commentary: bioavailability of flavonoids and polyphenols: call to arms. *Mol. Pharm.* 2007*; 4:* 803–6.

Huang, W. W.; Chiu, Y. J.; Fan, M. J.; Lu, H. F.; Yeh, H. F.; Li, K. H.; Chen, P. Y.; Chung, J. G.; Yang, J. S. Kaempferol induced apoptosis via endoplasmic reticulum stress and mitochondria-dependent pathway in human osteosarcoma U-2 OS cells. *Mol. Nutr. Food Res.*, 2010, 54, 1585-95.

Huber, G. M.; Rupasinghe, V.; Shahidi, F. Inhibition of oxidation of omega-3 polyunsaturated fatty acids and fish oil by quercetin glycosides. *Food Chem.*, 2009, 117, 290–5.

Hwang, E. I.; Ahn, B. T.; Lee, H. B.; Kim, Y. K.; Lee, K. S.; Bok, S. H.; Kim, Y. T.; Kim, S. U. Inhibitory activity for chitin synthase II from Saccharomyces cerevisiae by tannins and related compounds. *Planta Med.*, 2001, 67, 501-4.

Jeong, H. J.; Ryu, Y. B.; Park, S. J.; Kim, J. H.; Kwon, H. J.; Kim, J. H.; Park, K. H.; Rho, M. C.; Lee, W. S. Neuraminidase inhibitory activities of flavonols isolated from Rhodiola rosea roots and their *in vitro* anti-influenza viral activities. *Bioorg. Med. Chem.*, 2009, 17, 6816-23.

Jeong, J. C.; Kim, M. S.; Kim, T. H.; Kim, Y. K. Kaempferol induces cell death through ERK and Akt-dependent down-regulation of XIAP and survivin in human glioma cells. *Neurochem. Res.*, 2009, 34, 991-1001.

Jia, X., Chen, J., Lin, H., Hu, M. Disposition of flavonoids via enteric recycling: enzyme-transporter coupling affects metabolism of biochanin A and formononetin and excretion of their phase II conjugates. *J. Pharmacol. Exp. Ther.*, 2004, 310, 1103–13.

Jorge, A. P.; Horst, H.; de Sousa, E.; Pizzolatti, M. G.; Silva, F. R. Insulinomimetic effects of kaempferitrin on glycaemia and on 14C-glucose uptake in rat soleus muscle. *Chem. Biol. Interact.,* 2004, 149, 89-96.

Jung, T. H.; Ha, H. J.; Ahn, J.; Kwak, H. S. Development of cholesterol reduced mayonnaise with crosslinked β-cyclodextrin and added phytosterol. *Korean J. Food Sci. Anim. Resour.*, 2008, 28, 211–7.

Kampkotter, A.; Gombitang, N. C.; Zurawski, R. F.; Timpel, C.; Chovolou, Y.; Watjen, W.; Kahl, R. Effects of the flavonoids kaempferol and fisetin on thermotolerance, oxidative stress and FoxO transcription factor DAF-16 in the model organism Caenorhabditis elegans. *Arch. Toxicol.*, 2007, 81, 849-58.

Kang, G. Y.; Lee, E. R.; Kim, J. H.; Jung, J. W.; Lim, J.; Kim, S. K.; Cho, S. G.; Kim, K. P. Downregulation of PLK-1 expression in kaempferol-induced apoptosis of MCF-7 cells. *Eur. J. Pharmacol.*, 2009, 611, 17-21.

Kim, S. H.; Kim, H. Y.; Kwak, H. S. Cholesterol removal from lard with crosslinked b-cyclodextrin. *Asian Australas. J. Anim. Sci.*, 2007, 20, 1468–72.

Kim, Y. K.; Kim, Y. S.; Choi, S. U.; Ryu, S. Y. Isolation of flavonol rhamnosides from Loranthus tanakae and cytotoxic effect of them on human tumor cell lines. *Arch. Pharm. Res.*, 2004, 27, 44-7.

Klahorst, S. Exploring antioxidants. *Wd. Food Ingred.*, 2002, 54–59.

Klaunig, J. E. and Kamendulis, L. M. The role of oxidative stress in carcinogenesis. *Annu. Rev. Pharmacol. Toxicol.*, 2004, 44, 239-67.

Knekt, P.; Kumpulainen, J.; Jarvinen, R.; Rissanen, H.; Heliovaara, M.; Reunanen, A.; Hakulinen, T.; Aromaa, A. Flavonoid intake and risk of chronic diseases. *Am. J. Clin. Nutr.*, 2002, 76, 560-8.

Knowles, L. M.; Zigrossi, D. A.; Tauber, R. A.; Hightower, C.; Milner, J. A. Flavonoids suppress androgen-independent human prostate tumor proliferation. *Nutr. Cancer*, 2000, 38, 116-22.

Kowalski, J.; Samojedny, A.; Paul, M.; Pietsz, G.; Wilczok, T. Effect of apigenin, kaempferol and resveratrol on the expression of interleukin-1beta and tumor necrosis factor-alpha genes in J774.2 macrophages. *Pharmacol. Rep.*, 2005, 57, 390-4.

Kühnau, J. The flavonoids. A class of semi-essential food components: their role in human nutrition. *World. Rev. Nutr. Diet.*, 1976, 24, 117-191.

Kuhnle, G., Spencer, J. P., Schroeter, H., et al. Epicatechin and catechin are O-methylated and glucuronidated in the small intestine. *Biochem. Biophys. Res. Commun.,* 2000, 277, 507–12.

Kwak, H. S.; Kim, S. H.; Kim, J. H.; Choi, H. J; Kang, J. Immobilized β-cyclodextrin as a simple and recyclable method for cholesterol removal in milk. *Arch. Pharma. Research*, 2004, 27, 873–7.

Lau, T. A healthy way to live: The occurrence, bioactivity, biosynthesis and synthesis of kaempferol. *Chemistry*, 2008, 150, 1-3.

Lee, M. J.; Rao, Y. K.; Chen, K.; Lee, Y. C.; Tzeng, Y. M. Effect of flavonol glycosides from Cinnamomum osmophloeum leaves on adiponectin secretion and phosphorylation of insulin receptor-beta in 3T3-L1 adipocytes. *J. Ethnopharmacol.*, 2009, 126, 79-85.

Li, N.; Liu, J. H.; Zhang, J.; Yu, B. Y. Comparative Evaluation of Cytotoxicity and Antioxidative Activity of 20 Flavonoids. *J. Agric. Food Chem.*, 2008, 56, 3876-83

Li, W.; Du, B.; Wang, T.; Wang, S.; Zhang, J. Kaempferol induces apoptosis in human HCT116 colon cancer cells via the Ataxia-Telangiectasia Mutated-p53 pathway with the involvement of p53 Upregulated Modulator of Apoptosis. *Chem. Biol. Interact.*, 2009, 177, 121-7.

Li, Y. L.; Gan, G. P.; Zhang, H. Z.; Wu, H. Z.; Li, C. L.; Huang, Y. P.; Liu, Y. W.; Liu, J. W. A flavonoid glycoside isolated from Smilax china L. rhizome *in vitro* anticancer effects on human cancer cell lines. *J. Ethnopharmacol.*, 2007, 113, 115-24.

Lim, Y. H.; Kim, I. H.; Seo, J. J. *In vitro* activity of kaempferol isolated from the Impatiens balsamina alone and in combination with erythromycin or clindamycin against Propionibacterium acnes. *J. Microbiol.*, 2007, 45, 473-7.

Lin, J.; Rexrode, K. M.; Hu, F.; Albert, C. M.; Chae, C. U.; Rimm, E. B.; Stampfer, M. J.; Manson, J. E. Dietary intakes of flavonols and flavones and coronary heart disease in US women. *Am. J. Epidemiol.*, 2007, 165, 1305-13.

Lin, J.; Zhang, S. M.; Wu, K.; Willett, W. C.; Fuchs, C. S.; Giovannucci, E. Flavonoid intake and colorectal cancer risk in men and women. *Am. J. Epidemiol.*, 2006, 164, 644-51.

Liu, J.; Mendoza, S.; Román, R.; Lynn, M.; Xu, R.; Kaifer, A. Cyclodextrinmodified gold nanospheres. Host–guest interactions at work to control colloidal properties. *J. Am. Chem. Soc.*, 1999, 121, 4304–5.

Loftsson, T.; Jarho, P.; Másson, M.; Järvinen, T. Cyclodextrins in drug delivery. *Expert Opin. Drug Delivery*, 2005, 2, 335-51.

Loftsson, T.; Másson, M.; Brewster, M. E. Self-Association of cyclodextrins and cyclodextrin complexes. *J. Pharma. Sci.*, 2004, 93, 1091-9.

Lubbers, S.; Landy, P; Voilley, A. Retention and release of aroma compounds in food containing proteins. *J. Food Technol.*, 1998, 52, 68–74.

Lucas-Abellán, C.; Fortea, I.; Gabaldón, J. A.; Núñez-Delicado, E. Encapsulation of Quercetin and Myricetin in Cyclodextrins at Acidic pH. *J. Agric. Food Chem.* 2008b, 56, 255-9.

Lucas-Abellán, C.; Fortea, M. I.; López-Nicolas, J. M.; Núnez-Delicado, E. Cyclodextrins as resveratrol carrier system. *Food Chem.*, 2007, 104, 39–44.

Lucas-Abellán, C.; Fortea, M. I.; Gabaldón, J. A.; Núñez-Delicado, E. Complexation of resveratrol by native and modified cyclodextrins: Determination of complexation constant by enzymatic, solubility and fluorimetric assays. *Food Chem.*, 2008a, 111, 262-7.

Luo, H.; Daddysman, M. K.; Rankin, G. O.; Jiang, B. H.; Chen, Y. C. Kaempferol enhances cisplatin's effect on ovarian cancer cells through promoting apoptosis caused by down regulation of cMyc. *Cancer Cell Int.,* 2010, 10, 16.

Lyu, S. Y.; Rhim, J. Y.; Park, W. B. Antiherpetic activities of flavonoids against herpes simplex virus type 1 (HSV-1) and type 2 (HSV-2) *in vitro. Arch. Pharm. Res.,* 2005, 28, 1293-301.

Mahat, M. Y.; Kulkarni, N. M.; Vishwakarma, S. L.; Khan, F. R.; Thippeswamy, B. S.; Hebballi, V.; Adhyapak, A. A.; Benade, V. S.; Ashfaque, S. M.; Tubachi, S.; Patil, B. M. Modulation of the cyclooxygenase pathway via inhibition of nitric oxide production contributes to the anti-inflammatory activity of kaempferol. *Eur. J. Pharmacol.*, 2010, 642, 169-76.

Mahmood, N.; Piacente, S.; Pizza, C.; Burke, A.; Khan, A. I.; Hay, A. J. The anti-HIV activity and mechanisms of action of pure compounds isolated from Rosa damascena. *Biochem. Biophys. Res. Commun.*, 1996, 229, 73-9.

Manach, C.; Williamson, G.; Morand, C.; Scalbert, A.; Remesy, C. Bioavailability and bioefficacy of polyphenols in humans I. Review of 97 bioavailability studies. *Am. J. Clin. Nutr.*, 2005, 81(Suppl. 1), 230–42.

Marfe, G.; Tafani, M.; Indelicato, M.; Sinibaldi-Salimei, P.; Reali, V.; Pucci, B.; Fini, M.; Russo, M. A. Kaempferol induces apoptosis in two different cell lines via Akt inactivation, Bax and SIRT3 activation, and mitochondrial dysfunction. *J. Cell Biochem.*, 2009, 106, 643-50

Maridonneau-Parini, I.; Braquet, P.; Garay, R. P. Heterogeneous effect of flavonoids on K+ loss and lipid peroxidation induced by oxygen-free radicals in human red cells. *Pharmacol. Res. Commun.*, 1986, 18, 61-72.

Marin, C.; Boutaleb-Charki, S.; Diaz, J. G.; Huertas, O.; Rosales, M. J.; Perez-Cordon, G.; Guitierrez-Sanchez, R.; Sanchez-Moreno, M. Antileishmaniasis activity of flavonoids from Consolida oliveriana. *J. Nat. Prod.*, 2009, 72, 1069-74.

Marniemi, J.; Alanen, E.; Impivaara, O.; Seppanen, R.; Hakala, P.; Rajala, T.; Ronnemaa, T. Dietary and serum vitamins and minerals as predictors of myocardial infarction and stroke in elderly subjects. *Nutr. Metab. Cardiovasc. Dis.*, 2005, 15, 188-97.

Martín del Valle, E. M. Cyclodextrins and their uses: a review. *Process Biochem.*, 2004, 39, 1033–46.

McCann, S. E.; Ambrosone, C. B.; Moysich, K. B.; Brasure, J.; Marshall, J. R.; Freudenheim, J. L.; Wilkinson, G. S.; Graham, S. Intakes of selected nutrients, foods, and phytochemicals and prostate cancer risk in western New York. *Nutr. Cancer*, 2005, 53, 33-41.

Mercader-Ros, M. T.; Lucas-Abellán, C.; Gabaldón, J. A.; Fortea, M. I.; Martínez-Cachá, A.; Núñez-Delicado, E. "Kaempferol complexation in cyclodextrins at basic pH". *J. Agric. Food Chem.*, 2010b, 58, 4675-80.

Mercader-Ros, M. T.; Lucas-Abellán, C.; Fortea, M. I.; Gabaldón, J. A.; Núñez-Delicado, E. Effect of HP-β-cyclodextrins complexation on the antioxidant activity of flavonols. *Food Chem.*, 2010a, 118, 769–73.

Min, B. S.; Tomiyama, M.; Ma, C. M.; Nakamura, N.; Hattori, M. Kaempferol acetylrhamnosides from the rhizome of Dryopteris crassirhizoma and their inhibitory effects on three different activities of human immunodeficiency virus-1 reverse transcriptase. *Chem. Pharm. Bull. (Tokyo).*, 2001, 49, 546-50.

Mitrocotsa, D.; Mitaku, S.; Axarlis, S.; Harvala, C.; Malamas, M. Evaluation of the antiviral activity of kaempferol and its glycosides against human cytomegalovirus. *Planta Med.*, 2000, 66, 377-9.

Mourtzinos, I.; Salta, F.; Yannakopoulou, K.; Chiou, A.; Karathanos, V.T. Encapsulation of olive leaf extract in β-cyclodextrin. *J. Agric. Food Chem.* 2007, 55, 8088–94.

Muzitano, M. F.; Tinoco, L. W.; Guette, C.; Kaiser, C. R.; Rossi-Bergmann, B.; Costa, S.S. The antileishmanial activity assessment of unusual flavonoids from Kalanchoe pinnata. *Phytochemistry,* 2006, 67, 2071-7.

Mylonis, I.; Lakka, A.; Tsakalof, A.; Simos, G. The dietary flavonoid kaempferol effectively inhibits hif-1 activity and hepatoma cancer cell viability under hypoxic conditions. *Biochem. Biophys. Res. Commun.*, 2010, 398, 74-8.

Nagao, A.; Seki, M.; Kobayashi, H. Inhibition of xanthine oxidase by flavonoids. *Biosci. Biotechnol. Biochem.*, 1999, 63, 1787-90.

Nakadate, T.; Yamamoto, S.; Aizu, E.; Kato, R. Effects of flavonoids and antioxidants on 12-O-tetradecanoyl-phorbol-13-acetate-caused epidermal ornithine decarboxylase induction and tumor promotion in relation to lipoxygenase inhibition by these compounds. *Gann.*, 1984, 75, 214-22.

Nguyen, T. T.; Tran, E.; Ong, C. K.; Lee, S. K.; Do, P. T.; Huynh, T. T.; Nguyen, T. H.; Lee, J. J.; Tan, Y.; Ong, C. S.; Huynh, H. Kaempferol-induced growth inhibition and apoptosis in A549 lung cancer cells is mediated by activation of MEK-MAPK. *J. Cell Physiol.*, 2003, 197, 110-21.

Nothlings, U.; Murphy, S. P.; Wilkens, L. R.; Henderson, B. E.; Kolonel, L. N. Flavonols and pancreatic cancer risk: the multiethnic cohort study. *Am. J. Epidemiol.*, 2007, 166, 924-31.

Nothlings, U.; Murphy, S. P.; Wilkens, L. R.; Henderson, B. E.; Kolonel, L. N. Flavonols and pancreatic cancer risk: the multiethnic cohort study. *Am. J. Epidemiol.*, 2007, 166, 924-31.

Núñez-Delicado, E.; Serrano-Mejías, M.; Pérez-López, A. J.; López-Nicolás, J.M. Polyphenol oxidase from Dominga table grape. *J. Agric. Food Chem.,* 2005, 53, 6087-6093.

Olszanecki, R.; Gebska, A.; Kozlovski, V. I.; Gryglewski, R. J. Flavonoids and nitric oxide synthase. *J. Physiol Pharmacol.*, 2002, 53, 571-84.

Otsuka, N.; Liu, M. H.; Shiota, S.; Ogawa, W.; Kuroda, T.; Hatano, T.; Tsuchiya, T. Anti-methicillin resistant Staphylococcus aureus (MRSA) compounds isolated from Laurus nobilis. *Biol. Pharm. Bull.,* 2008, 31, 1794-7.

Ou, B.; Huang, D.; Hampsch-Woodill, M.; Prior, R. L. Development and validation of an improved oxygen radical absorbance capacity assay using fluorescein as the fluorescent probe. *J. Agric. Food Chem.*, 2001, 49, 4619–26.

Park, J. S.; Rho, H. S.; Kim, D. H.; Chang, I. S. Enzymatic Preparation of Kaempferol from Green Tea Seed and Its Antioxidant Activity. *J. Agric. Food Chem.*, 2006, 54, 2951-6.

Prior, R. L.; Gu, L.; Wu, X.; Jacob, R. A.; Sotoudeh, G.; Kader, A. A.; Cook, R. A. Plasma antioxidant capacity changes following a meal as a measure of the ability of a food to alter *in vivo* antioxidant status. *J. Am. Coll. Nutr.*, 2007, 26, 170–181.

Ross, J. A.; Kasum, C. M. Dietary flavonoids: bioavailability, metabolic effects, and safety. *Annu. Rev. Nutr.*, 2002, 22, 19-34.

Sanz, M. J.; Ferrandiz, M. L.; Cejudo, M.; Terencio, M. C.; Gil, B.; Bustos, G.; Ubeda, A.; Gunasegaran, R.; Alcaraz, M. J. Influence of a series of natural flavonoids on free radical generating systems and oxidative stress. *Xenobiotica*, 1994, 24, 689-99.

Scalbert, A.; Williamson, G. Dietary intake and bioavailability of polyphenols. *J. Nutr.* 2000, 130 (Suppl. 8S), 2073–85.

Scalbert, A.; Manach, C.; Morand, C.; Rémésy, C.; Jiménez, L. Dietary polyphenols and the prevention of diseases. *Crit. Rev. Food Sci.*, 2005, 45, 287-306.

Sharma, V.; Joseph, C.; Ghosh, S.; Agarwal, A.; Mishra, M. K.; Sen, E. Kaempferol induces apoptosis in glioblastoma cells through oxidative stress. *Mol. Cancer Ther.*, 2007, 6, 2544-53.

Shim, S. Y.; Ahn, J.; Kwak, H. S. Functional properties of cholesterol removed whipping cream treated by β-cyclodextrin. *J. Dairy Sci.,* 2003, 86, 2767–72.

Singh, M.; Sharma, R.; Banerjee, U. C. Biotechnological applications of cyclodextrins. *Biotech. Advances*, 2002, 20, 341–59.

Smith, G.; Thomsen, S. J.; Markham, K. R.; Andary, C.; Cardon, D. The photostabilities of naturally occurring 5-hydroxyflavones, flavonols, their glycosides and their aluminium complexes. *J. Photochem. Photobiol. A.*, 2000, 136, 87.

Soobrattee, M. A.; Neergheen, V. S.; Luximon-Ramma, A.; Aruoma, O. I.; Bahorum, T. Phenolics as potential antioxidant therapeutic agents: Mechanism and actions. *Mutation Research*, 2005, 579, 200–213.

Steensma, A.; Faassen-Peters, M. A.; Noteborn, H. P.; Reitjen, I. M. Bioavailability of genistein and its glycoside genistin as measured in the portal vein of freely moving unanaesthetized rats. *J. Agric. Food Chem.*, 2006, 54, 8006–12.

Szejtli, J. Utilization of cyclodextrins in industrial products and processes. *J. Mater. Chem.*, 1997, 7, 575–87.

Szente, L.; Mikuni, K.; Hasimoto, H.; Szejtli, J. Stabilization and solubilization of lipophilic natural colorants with cyclodextrins. *J. Inclus. Phenom. Mol.,* 1998, 32, 81–9.

Tommasini, S.; Calabrò, M. L.; Stancanelli, R.; Donato, P.; Costa, C.; Catania, S.; Villari, V.; Ficarra, P.; Ficarra, R. The inclusion complexes of hesperetin and its 7-rhamnoglucoside with (2-hydroxypropyl)-β-cyclodextrin. *J. Pharm. Biomed. Anal.*, 2005, 39, 572–80.

Tu, Y. C.; Lian, T. W.; Yen, J. H.; Chen, Z. T.; Wu, M. J. Antiatherogenic effects of kaempferol and rhamnocitrin. *J. Agric. Food Chem.*, 2007, 55, 9969-76.

Unno, T., Sugimoto, A., Kakuda, T. Scavenging effect of tea catechins and their epimers on superoxide anion radicals generated by a hypoxanthine and xanthine oxidase system. *J. Sci. Food Agric.,* 2000, 80, 601–6.

USDA:http://www.ars.usda.gov/SP2UserFiles/Place/12354500/Data/Flav/Flav_R03.pdf

Van Acker, S. A.; van den Berg, D. J.; Tromp, M. N.; Griffioen, D. H.; van Bennekom, W. P.; van der Vijgh, W. J.; Bast, A. Structural aspects of antioxidant activity of flavonoids. *Free Radic. Biol. Med.*, 1996, 20, 331-42.

Wang, L.; Tu, Y. C.; Lian, T. W.; Hung, J. T.; Yen, J. H.; Wu, M. J. Distinctive antioxidant and antiinflammatory effects of flavonols. *J. Agric. Food Chem.*, 2006, 54, 9798-804.

Williams, C. A.; Harborne, J. B. Flavone and flavonol glycosides. In: J. B. Harborne, Editor. The Flavonoids: Advances in Research Since 1986, *Chapman and Hall*, London. 1994, 337–85.

Williamson, G.; Manach, C. Bioavailability and bioefficacy of polyphenols in humans II. Review of 93 intervention studies. *Am. J. Clin. Nutr.,* 2005, 81 (Suppl. 1), 243–55.

Winkel-Shirley, B. Biosynthesis of flavonoids and effects of stress. *Curr. Opin. Plant Biol.*, 2002, 5, 218-23.

Winkel-Shirley, B. Flavonoid biosynthesis. A colorful model for genetics, biochemistry, cell biology, and biotechnology. *Plant Physiol.*, 2001, 126, 485-93.

Wojdylo, A.; Oszmianski, J.; Czemerys, R. Antioxidant activity and phenolic compounds in 32 selected herbs. *Food Chem.*, 2007, 105, 940–9.

Wood, W. E. Improved aroma barrier properties in food packaging with cyclodextrins. TAPPI – Polymers, *Laminations and Coatings Conference* 2001, 367–77.

Xu, W.; Liu, J.; Li, C.; Wu, H. Z.; Liu, Y. W. Kaempferol-7-O-betad-glucoside (KG) isolated from Smilax china L. rhizome induces G(2)/M phase arrest and apoptosis on HeLa cells in a p53-independent manner. *Cancer Lett.*, 2008, 264, 229-40.

Yasukawa, K.; Takido, M.; Takeuchi, M.; Sato, Y.; Nitta, K.; Nakagawa, S. Inhibitory effects of flavonol glycosides on 12-Otetradecanoylphorbol-13-acetate-induced tumor promotion. *Chem. Pharm. Bull.* (Tokyo)., 1990, 38, 774-6.

Yin, F.; Giuliano, A. E.; Van Herle, A. J. Growth inhibitory effects of flavonoids in human thyroid cancer cell lines. *Thyroid*, 1999, 9, 369-76.

Yoshida, T.; Konishi, M.; Horinaka, M.; Yasuda, T.; Goda, A. E.; Taniguchi, H.; Yano, K.; Wakada, M.; Sakai, T. Kaempferol sensitizes colon cancer cells to TRAIL-induced apoptosis. *Biochem. Biophys. Res. Commun.*, 2008, 375, 129-33.

Yu, J.; Wang, L.; Walzem, R. L.; Miller, E. G.; Pike, L. M.; Patil, B. S. Antioxidant activity of citrus limonoids, flavonoids, and coumarins. *J. Agric. Food Chem.*, 2005, 53, 2009-14.

Yu, S. F.; Shun, C. T.; Chen, T. M.; Chen, Y. H. 3-O-beta-Dglucosyl-(1-->6)-beta-D-glucosyl-kaempferol isolated from Sauropus androgenus reduces body weight gain in Wistar rats. *Biol. Pharm. Bull.,* 2006, 29, 2510-3.

Zanatta, L.; Rosso A.; Folador, P.; Figueiredo M. S. R. B.; Pizzolatti, M. G.; Leite, L. D.; Silva, F. R. M. B. Insulinomimetic effect of kaempferol 3-neohesperidoside on the rat soleus muscle. *J. Nat. Prod.*, 2008, 71, 532–5.

Zhang, Y.; Hendrich, S; Murphy, P. A. Glucuronides are the main isoflavone metabolites in women. *J. Nutr.*, 2003, 133, 399–404.

Zhang, Q.; Zhao, X. H.; Wang, Z. J. Cytotoxicity of flavones and flavonols to a human esophageal squamous cell carcinoma cell line (KYSE-510) by induction of G(2)/M arrest and apoptosis. *Toxicol. In vitro,* 2009, 23, 797-807.

Zhang, Q.; Zhao, X. H.; Wang, Z. J. Flavones and flavonols exert cytotoxic effects on a human oesophageal adenocarcinoma cell line (OE33) by causing G2/M arrest and inducing apoptosis. *Food Chem. Toxicol.*, 2008, 46, 2042-53.

Zhang, Y.; Chen, A. Y.; Li, M.; Chen, C.; Yao, Q. Ginkgo biloba extract kaempferol inhibits cell proliferation and induces apoptosis in pancreatic cancer cells. *J. Surg. Res.,* 2008, 148, 17-23.

In: Kaempferol
Editors: Garion Villers and Yves Fougere

ISBN: 978-1-62618-515-9

Chapter 2

Kaempferol and Kaempferitrin: Nutraceutical Compounds Contribute to Glucose Homeostasis by Acting at Multiple Biological Sites

Fátima Regina Mena Barreto Silva**[*1]**, Leila Zanatta**[†2]**,
Marisa Jádna Silva Frederico**[‡1]**,
Moacir Geraldo Pizzolatti**[§3]
and Angela Machado De Campos**[‖4]

[1]Departamento de Bioquímica, Centro de Ciências Biológicas, Universidade Federal de Santa Catarina, Campus Universitário, Bairro Trindade, Florianópolis, SC, Brazil
[2]Área de Ciências da Saúde, Universidade Comunitária da Região de Chapecó, Chapecó, SC, Brazil
[3]Departamento de Química, Centro de Ciências Físicas e Matemáticas, [4]Departamento de Química Farmacêutica, Centro de Ciências da Saúde, Universidade Federal de Santa Catarina, Campus Universitário, Florianópolis, SC, Brazil

Abstract

Kaempferol is a flavonol that is relatively abundant in various plants and some edible berries. Kaempferol aglycone or its glycosilated forms contributes significantly to flavonoid intake in humans (about 33% from total flavonol intake), with intakes estimated around 10 mg per day in developed countries. These compounds have antioxidative, anti-inflammatory, antihypertensive, lipolytic and anti-cancer effects. Emerging evidence suggests that kaempferol exhibits nutraceutical properties since it

---

* mena@mbox1.ufsc.br.a
† leila.zanatta@gmail.com.
‡ marisafrederico1@hotmail.com.
§ moacir.pizzolatti@gmail.com.
‖ angelacampos@gmail.com.

produces physiological benefits and is able to combat disease processes, among them those concerned to diabetes mellitus. Several results have demonstrated that kaempferol can influence glucose metabolism acting as antihyperglycemic and especially as hypoglycemic agent through *in vitro* and *in vivo* approaches. Kaempferol aglycone and/or glycosides stimulate glucose uptake and glycogen synthesis in skeletal muscle, inhibit intestinal glucose absorption and act *per se* as insulin-mimetic, probably, by influencing the pleiotropic mechanisms of insulin signaling. Thus, these nutraceutical compounds have attracted the interest of both academia and industry as possible therapy options. This review cover the frontiers of the development of chemistry and biological properties of kaempferol (aglycone/glycosides), especially those associated with glucose homeostasis and carbohydrate metabolism able to ameliorate diabetes status. Moreover, updates the debate about the development of new approaches to study the interactions of nutraceutical compounds with target-sites of action combining *in vivo, ex vivo* and *in vitro* pharmacological studies. Also, the perspectives of controlled delivering of these substances in multiple and specific biological locations useful to diabetes mellitus therapy are highlighted.

Keywords: Kaempferol, Kaempferitrin, Kaempferol glycosides, Nutraceuticals, Glucose homeostasis, Diabetes

Abbreviations

ACC	Acetyl-CoA carboxylase
ATP	*Adenosine* triphosphate
AUC	Area under the curve
Bcl-2	B-cell lymphoma 2
cAMP	Cyclic adenosine monophosphate
4CL	4-coumarate CoA-lygase
C4H	Cinnamate 4-hydroxylase
CREB	cAMP-responsive element binding protein
CBG	Cytosolic β-glucosidase
CHI	Chalcone isomerase
CHS	Chalcone synthase
COSY	Correlation spectroscopy
DDS	Drugs delivery systems
DEPT	Distortionless enhancement by polarization transfer
2D HMBC	Two-dimensional heteronuclear multiple bond correlation
2D HMQC	Two-dimensional_heteronuclear multiple-quantum correlation
EtOH	Ethanol
F3H	Flavanone-3-dioxygenase
FLS	Flavonol synthase
FL7RT	Flavonol-3-*O*-glycoside:7-*O*-rhamnosyltransferase
GSK-3	Glycogen synthase kinase 3
GTs	Glycosyltransferases
HRMS	High-resolution mass spectrometry
i.p.	Intraperitoneal
K_m	Michaelis constant
MAPK	Mitogen-activated protein kinase

MEK	Mitogen-activated *protein kinase kinase*
MRP	Multi-drug resistance protein
NMR	Nuclear magnetic resonance
PAL	L-phenylalanine ammonia-lyase
PDX-1	Pancreatic and duodenal homeobox 1
Pgp	P-glycoprotein
PI-3K	Phosphatidylinositol 3-kinase
PKA	Protein kinase A
PKC	Protein kinase C
P.o.	Per oral
PP1	Protein phosphatase 1
PPARγ	Peroxisome proliferator-activated receptor gamma
SNEDDS	Self-nanoemulsifying drugs delivery systems
SGLT	Na^+/glucose cotransporter or symporter
siRNA	Small interfering ribonucleic acid
TAL	L-tyrosine ammonia-lyase
UDP	Uridine diphosphate glucose
UGT78D1	UDP-rhamnose:flavonol-3-*O*-rhanmosyltransferase
VO(IV)	Vanadium (IV)

Introduction

1. General Aspects

Kaempferol (3,4',5,7-tetrahydroxyflavone) is among the 5,000 naturally occurring flavonoids characterized from various plants. Phytochemicals containing natural phenolic compounds have demonstrated a wide capability to ameliorate or prevent chronic diseases (Kang *et al.*, 2010; Zhang and Liu 2011; Calderón-Montaño *et al.*, 2011). Among the abundant flavonoids in food and in nutraceuticals, kaempferol, is an active biological substance with potential health effects (Zhang and Liu, 2011; Weng and Yen, 2012).

For flavonols, as well as to any drug, factors as absorption, clearance rate, solubility, permeability and transporter systems on enterocytes (influx and/or efflux) determine the bioavailability of these compounds and their physiological role. Moreover, the small intestine and liver metabolism as well as the disposition of kaempferol have been yield good insights about pharmacokinetics of kaempferol. Barve *et al.* (2009) reported the comparative small intestine and hepatic metabolism of kaempferol and examined the bioavailability and gastro-intestinal first-pass effects. It was demonstrated that kaempferol undergoes low to moderate absorption and the bioavailability (around 2%), probably, may be due to the extensive first-pass metabolism at phase I oxidative metabolism and by phase II glucuronidation in the intestine as well as in the liver.

On one hand, the type of metabolite of kaempferol, quercetin as an oxidative product or methyl, sulfate or glucuronide conjugates that appear from extensive biotransformation is crucial for its biological activity (Williamson and Manach, 2005; Cazarolli *et al.*, 2008a). On the other hand, kaempferol glycosylated forms also exist and attract interest due to their multiple sites of action and exhibition of different biological activities (Cazarolli *et al.*,

2008b); for an excellent review see Calderón-Montaño *et al.* (2011); Weng and Yen (2012). This structural glycosylated variability consists of one or more rhamnose or glucose moieties attached at different positions in aglycone sctructure that result in increased solubility and bioavailability compared with aglycone moiety. The glycosilation of a phenolic hydroxyl group of a flavonoid aglycone can occur through the hemiacetal bound (*O*-glycosides) or straight attachment to the C-1 of the sugar unit *via* a carbon-carbon bond (*C*-glycosides). The most common monosaccharides, such as D-glucose and L-rhamnose, and other less frequent glucorhamnose, galactose, xylose and arabinose, or a combinations of these (di- or trisaccharides), can bind to hydroxyl groups or directly to a carbon atom at different positions on the flavonoid aglycone (Harborne and Williams, 2000; Aherne and O´Brien, 2002; Erlund, 2004). The *O*-glycosilation of the flavonoid molecule occurs most frequently at the C-3 and/or C-7 positions and for *C*-glycoside flavonoids the preferential glycosilation sites are at positions C-6 and C-8 (Cazarolli *et al.*, 2008a). Although the real difference between hepatic and intestinal disposition of kaempferol (aglycone or glycosylated forms) exists, it is largely unknown. However, the identification of some glycosides in human plasma revealed that the absorption of glycosides from the diet occurs (Paganga and Rice-Evans, 1997; Calderón-Montaño *et al.*, 2011).

It has been demonstrated that flavonols glycosides from food intake, are hydrolyzed by lactase phlorizin hydrolase and aglycone diffuses into the enterocytes and produces sulphates, glucuronides, ethylated as well as a mix of conjugates by consequence of phase-II (Barrington *et al.*, 2009; Day, 2000). In general, the conjugation of flavonoids alters significantly the chemical properties and determines either absorption or efflux (return to intestinal lumen). Even that the flavonoids absorption and metabolism seem to be complex, some reports in the literature are consistent. Barrington *et al.* (2009) showed that in differentiated Caco2 cell kaempferol and galangin are rapidly conjugated to glucuronide and sulphates. So, these conjugates are then effluxed out of the cells to both the apical (luminal) and basolateral (serosal) sides, with glucuronides and sulphates preferentially effluxes to the basolateral and apical compartments, respectively. Furthermore, it has been described a multitude of active transporters engaged on intestinal efflux pump, among them, multi-drug resistance protein- 1 and 2 (MRP-1; MRP-2) (Vaidyanathan and Walle, 2003; Walle *et al.*, 1999) and atypical P-glycoproteins (Pgp) (Zhang and Morris, 2003). In addition, the cytosolic β-glucosidase (CBG), with broad specificity, is in abundance in small intestine, liver and kidney of mammals and requires active transport of hydrophilic glucosides into the cells (Németh *et al.*, 2003). Also, CBG acts on flavonoid and isoflavone glycosides according to the position and the structure of the sugar moiety attached to the flavonoid aglycone (Lambert *et al.*, 1999).

The glycoside moiety increases the hydrophilicity and reduces the possibility of passive transport. The sodium-glucose transporters (and/or galactose/mannose), also known as Na^+/glucose cotransporters or symporters (SGLTs: SGLT1-6 expressed in the human body as well as other species), are also known to transport a variety of glycosides albeit with a lower affinity than glucose and different apparent affinity for inhibitors phlorizin and phloretin (Diez-Sampedro *et al.*, 2000; Wright *et al.*, 2011). Increased evidences have been accumulated regarding to the crucial role of epithelial brush border membrane transporters, sodium dependent glucose transporter 1 (SGLT-1) in the small intestine (Gee *et al.*, 2000; Chang *et al.*, 2005; See an excellent review by Wright *et al.*, 2011). Concerned to kaempferol, Rodríguez *et al.* (2010), reported a significant effect of kaempferol 3-*O*-α-rhamnoside as a competitive inhibitor of intestinal SGLT-1 cotransporter and an additive inhibitory effect on intestinal glucose absorption when combined with phlorizin.

From several studies, it has been characterized that the natural compound dihydrochalcone glucoside (1-[2-β-D-glucopyranosyloxy)-4,6-dihydroxyphenyl]-3-(4-hydroxyphenyl)-1-propanone, phlorizin, is a competitive inhibitor of SGLTs (Ehrenkranz *et al.*, 2005). Based on flexible structure of phlorizin, it makes predictions about the binding site of this uncertain phlorizin structure to enhance selectivity for SGLT-2 over SGLT-1 and also, phlorizin has a greater than fourfold higher affinity for human SGLT-2 than SGLT-1 (Katsuno *et al.*, 2007; Hummel *et al.*, 2011). However, as much SGLT-1 (most abundant in intestine) as SGLT-2 (most abundant in kidney) exhibit a significant physiological role to glucose homeostasis (Kasahara *et al.*, 2001; Katsuno *et al.*, 2007). Taking it in mind, nutraceutical and pharmaceutical companies have been attracted by natural and modified compounds that are able to act selectively in multiple potential targets to diabetes therapy, among them, kaempferol aglycone and its glycosides and metabolites.

2. Kaempferitrin: Structure, Properties, Synthesis and Biosynthesis

The flavonoid kaempferitrin (kaenferol-3,7-*O*-(α)-L-dirhamnoside) is a bisglycosylated natural product derivative from 4',5,7-trihydroxyflavonol (kaempferol), which was first isolated from *Indigo arrecia* (Fabaceae) by Perkin in 1907 (Perkin, 1907).

Kaempferol (3,5,7-trihydroxy-2-(4-hydroxyphenyl)-4H-1-benzopyran-4-one) and its glycosyl derivatives including coumaroyl, caffeoyl, cinnamoyl, hydroxybenzoyl, sinapoyl, galoyl, malonyl, acetyl, methyl and sulfate groups, are widely distributed in plant kingdom being found in vegetables, fruits, food plants and in plants used in traditional medicine (Calderón-Montaño *et al.*, 2011). However, kaempferitrin has been isolated and identified in a few plants species belonging to the division Magnoliophyta, although it has been identified only once in the fern species *Asplenium trichomanes*, family Aspleniaceae (Imperato, 1979).

The highest occurrence of kaempferitrin in Magnoliophyta has been in the family Fabaceae where it has been found in *Indigo arrecta* (Perkin, 1907), *Lespedeza cyrtobotrya* Miquel (Hattori, 1951), *Lespedeza virgate* Murray DC. (Chen *et al.*, 2008). *Bauhinia forficata* Link (Pizzolatti *et al.*, 2003), *Bauhinia acuminate* (Vishnu Prasad *et al.*, 2009), *Pterogyne nitens* (Regasini *et al.*, 2008) and *Lotus corniculatus* var. São Gabriel (Koelzer *et al.*, 2009). It has also been identified in *Trichosanthus cucumeroides* Maxim (Cucurbitaceae) Nakaoki, 1957), *Momordica grosvenori* Swingle (Curcubitaceae) (Yang *et al.*, 2008), *Epimedzum acuminatum* Franch (Berberidaceae) (Hu *et al.*, 1992), *Vancouveria hexandra* (Berberidaceae) (Yamamoto *et al.*, 1993), *Hedyotis verticillata* (Rubiaceae) (Hamzah *et al.*, 1994), *Uncaria guianensis* (Rubiaceae) (Valente *et al.*, 2009), *Celastrus orbiculatus* Thunberg (Celastraceae) (Mototake and Shimokoriyama, 1949), Ligustrum *neilgherense* var. obovata C.B.Cl (Bignoniaceae) (Joshi and Gawad, 1977), *Jusficia spicigera* Schlectend (Acanthaceae) (Euler and Alam, 1982), *Chenopodium murale* L. (Amaranthaceae) (Gohar and Elmazar, 1997), *Weigela subsessillis* (Diervillaceae) (Won et al., 2004), *Cinnamomum osmophloeum* Kaneh (Lauraceae) (Fang et al., 2005), *Ipomoea pes-caprae* Linn. (Convolvulaceae) (Wang *et al.*, 2008), *Sedum dendroideum* Moc & Sesse (Crassulaceae) (De

Melo *et al.*, 2009), *Cardamine leucanta* (Brassicaceae) (Yang *et al.*, 2010), *Arabidopsis thaliana* (Brassicaceae) (Veit and Pauli, 1999), *Tagetes erecta* (Asteraceae) (Zhang and Zhang, 2010), *Hibiscus cannabinus L* (Malvaceae) (Rho *et al.*, 2010), *Bryophyllum pinnatum* Lank. Oken (Crassulaceae) (Tatsimo *et al.*, 2012) and *Geranium thunbergii* Sieb. et Zucc. TG (Geraniaceae) (Choi *et al.*, 2012). Kaempferitrin was also obtained from long-grain rice (*Oryza sativa*) fermented with the endophytic fungus *Annulohypoxylon boveri* var. microspore (Cheng *et al.*, 2012).

Structure and Properties

Indeed kaempferitrin is a yellowish straw compound with a molecular weight of the 578 g/mol ($C_{27}H_{30}O_{14}$) and have a diphenylpropane structure containing two molecules of rhamnose α-linked at C-3 and C-7 positions (Figure 1). Among the physical properties of kaempferitrin, the crystal dimorphism can be observed on the recrystallization process: recrystallization from 90% EtOH give isoform I as yellowish needles with 3.5 mols of H_2O of crystallization with melting point 234 °C; recrystallization from H_2O gives the isoform II as thin prisms with melting point 193 °C. Thus, isoform I was identified as Lespedin and isoform II as kaempferitrin (Hasegawa, 1940).

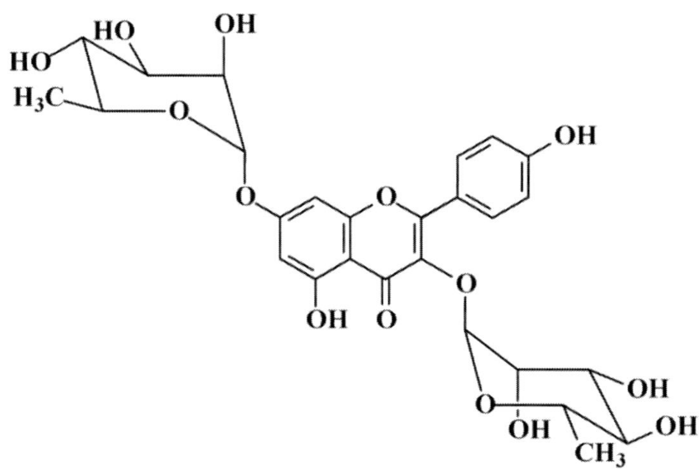

Figure 1. Kaempferitrin structure.

The true molecular structure of the kaempferitrin can be unambiguously confirmed through HRMS and detailed NMR spectroscopic analysis. The H^1 NMR and COSY spectrum (Figure 2) show clarity the hydrogen atoms for A-ring (2 H meta correlated) and B-ring (4 H ortho correlated) for flavonol moiety and the hydrogen atoms for two unit of the rhamnose. In the C^{13} NMR/DEPT (Figure 3) spectrum can be distinguish all carbon atoms for flavonol moiety as well as for two rhamnose units according to assigned in the spectrum of the figure 3 with the aid of the HMQC 2D spectrum (Figure 3). Finally, the J^2 and J^3 heteronuclear correlations observed in the HMBC 2D spectrum (Figure 4) was decisive for determination of the position and stereochemistry of the glycoside linkage as 3,7-*O*-(α)-L-dirhamnoside.

Biosynthesis

The recent advances in the identification of genes for plant enzyme biosynthesis, structure, function and regulatory factors, become the biosynthesis of flavonoids which have a C6-C3-C6 structure, one of the best known pathways in plant secondary metabolism (Winkel-Shirley, 2001; 2002). The flavonoid biosynthesis pathway represents the convergence of two major phenylpropanoid and polyketide biosynthetic pathways for building the A-ring (C6-bulck from phenylpropanol) and B-ring (C6-bulck from polyketide). This system requires two key enzymes Chalcone Synthase (CHS) (EC 2.3.1.74) and Chalcone Isomerase (CHI) (EC 5.5.1.6). CHS catalyzes the Claisen-like condensation of 4-coumaroil-CoA (building block C6-C3) with three molecules of malonyl-CoA (building block C6) resulting in the formation of the naringenin chalcone (C6-C3-C6), the first committed step leading phenylpropanoids pathway to biosynthesis of the flavonoids. Following, by action of the enzyme chalcone isomerase (CHI) occurs the C3 ring closure through Michael-like intramolecular nucleophilic addition to give the flavonoid naringenin, the central precursor flavonoid in this pathway. Then, the kaempferol biosynthesis is completed in two subsequent steps including the hydroxylation of the naringenin at C-3 position by enzyme flavanone-3-dioxygenase (F3H) (EC 1.14.11.9) to give dihydrokaempferol intermediate which finally undergoing dehydrogenation catalyzed by enzyme flavonol synthase (FLS) (EC 1.14.11.23) at C2 and C3 position to produce kaempferol (Figure 5).

Kaempferol as well as many others flavonoids aglycone are widely distributed in nature owing to the fact that in their biosynthesis there are involvement of a set of enzymes fairly common in the plant kingdom. Likewise the biosynthesis of the flavonoids glycosylated requires additional enzymes that are widespread in the plant kingdom. Enzymes that catalyze the glycoside formation, known as Glycosyltransferases (UGTs), have been intensively studied. These enzymes transfer the nucleotide-diphosphate-actived sugars (sugar donor) to a small molecular weight compounds as substrate acceptor, such as flavonoids. The activated sugar form is usually UDP-sugar, e.g. UDP-rhamnose that requires an appropriate enzyme glycosyltransferase to catalyze the transfer of rhamnose from uridinediphosphate for kaempferol. However a few flavonoids glycosylated such as kaempferol rhamnosides requires most specific enzymatic system and they can only be synthesized by plant species that have such genetic information. Studies carried out by Jones *et al.* (2003), Vogt and Jones (2000) and Wang *et al.* 2010 provided important contributions to the proposition of a model for biosynthesis of the kampferitrin. Jones *et al.* (2003) and Vogt and Jones (2000) have identified two glycosyltransferases involved in flavonol glycoside biosynthesis in *Arabidopsis thaliana* coded as UGT78D1 and UGT73C6 that appear to be flavonol-specific enzymes. Since UGT78D1 catalyzed the 3-*O*-rhamnosylation of kaempferol and quercetin and UGT73C6 catalyzed the 7-*O*-glucosylation both kaempferol-3-*O*-rhamnoside and quercetin-3-*O*-rhamnoside, should be classified as UDP-rhamnose:flavonol-3-*O*-rhanmosyltransferase and UDP-glucose:flavonol-3-*O*-glycoside-7-*O*-glucosyl transferase respectively. Furthermore, the Jones' group results indicate that 7-*O*-glucosylation and possibly also 7-*O*-rhamnosylation could occur after 3-*O*-glycosylation thereby allowing propose the biosynthetic scheme for kaempferitrin depicted in figure 5 (Figure 5). So the biosynthesis of the kaempfetrin is achieved from kaempferol in two steps: first the enzyme flavonol-3-*O*-rhamnosyltransferase (UGT78D1) catalyzes the transfer of rhamnose from UDP-rhamnose to the 3-OH position of kaempferol and then the supposed enzyme flavonol-3-*O*-glycoside:7-*O*-

rhamnosyltransferase (FL7RT) catalyzes the transfer of a second moiety rhamnose from UDP-rhamnose to 7-OH position of the afzelin (kaempferol-3-*O*-rhamnosyde) (Figure 5). Furthermore, whereas GTs involved in secondary metabolism of plants might follow a S_N2 mechanism whose transfer of the rhamnose from the α linkage UDP-rhamnose to kaempferol would result in glycoside with β configuration. Actually GTs are regioespecific and stereospecific enzymes, thus being able to catalyze the glycosylation with two possible stereochemical outcomes resulting in either inversion or retention of the anomeric configuration of the transferred sugar (Ren *et al.*, 2012; Gómez *et al.*, 2012). Glycosylation catalyzed by inverting GTs certainly occur through a single S_N2 reaction with displacement UDP living group by OH nucleophile of the flavonol (Figure 6a). On the other hand, retaining GTs enzymes have been proposed to work by different mechanisms (double displacement and/or internal return mechanisms) depending of the actual structure of the active site. The double displacement mechanism that involves two subsequent S_N2 reactions, going through a β-linkage covalent glycosyl-enzyme intermediate. In this template-driven mechanistic, a nucleophylic amino acid residue appropriately positioned witting the enzyme active center attack the face-β anomeric center of the UDP-glycoside forming the β-linkage covalent glycosyl-enzyme intermediate. In the second step, the enzyme is displaced by an OH nucleophylic group of the flavonol acceptor giving the glycoside with net overall retention of configuration (Figure 6b). The internal return (S_{Ni}) mechanism is proposed when lack a good catalytic nucleophile residue witting the active center of the retaining enzyme. Then due an interaction between the nucleophilic hydroxyl group of the flavonol and UDP living group, the nucleophilic attack of the anomeric center occurs on the same side as living group involving an stabilized oxycarbenium ion that on one face by the enzyme with retention of configuration of the glycoside formed (Figure 6c) (Breton *et al.*, 2012). Worth noting that a full understanding of the retaining glycosyltransferases must be investigated in each case.

Synthesis of the Kaempferitrin

Glycosylated flavonoids have drawn attention of the scientific community to synthetic preparation of this group of natural products. The major challenge seems to be in the bis-*O*-glycosylation of flavonoids which require a tedious protecting groups manipulation to distinguish any hydroxyl groups before every glycosylation (Du *et al.*, 2003; Liu *et al.*, 2011). Briefly for kaempferitrin two large strategies have been discussed as selective glycosylation of the kaempferol commercially available (or easily synthesized) and previous synthesis of the suitably protected 3,7-dihydroxyflavone for further glycosylation. The classical procedure for kaempferol synthesis commenced with claisen-Schimidt condensation between 2,4,6-trihydroxyacetophenone and *p*-hydroxybenzaldehyde to give 2',4,4',6'-tetrahydroxychalcone which is then treated with alkaline hydrogen peroxide by Algar-Flyn and Oyamada reaction to produce kaempferol (Britton *et al.*, 2012) (Figure 7).

The first synthesis of kaempferitrin was conceived by Urgaonkar and Shaw (2007). The Shaw' strategies relies on the formation of protected kaempferitrin upon bis-*O*-glycosylation of a suitably protected 3,7-dihydroxyflavone. This intermediate was prepared in six steps involving fist the Houben-Hoesch reaction of phloroglucinol following selective protection of two out of three hydroxyl groups which chloromethyl methyl ether, benzoylation of the thirty HO phenolic to give an ortho-benzoyl α-methoxy acetophenone the key compound for cyclization to 3-methoxy-4'-benzyl-5-hydroxy-7-methoxymethylflavona which being tosylated gives the full protected kaempferol. Thus the kaempferitrin synthesis is achieved by protecting group's manipulation and glycosylation method with tri-*O*-acetylrhaminose bromide as showed in Figure 8.

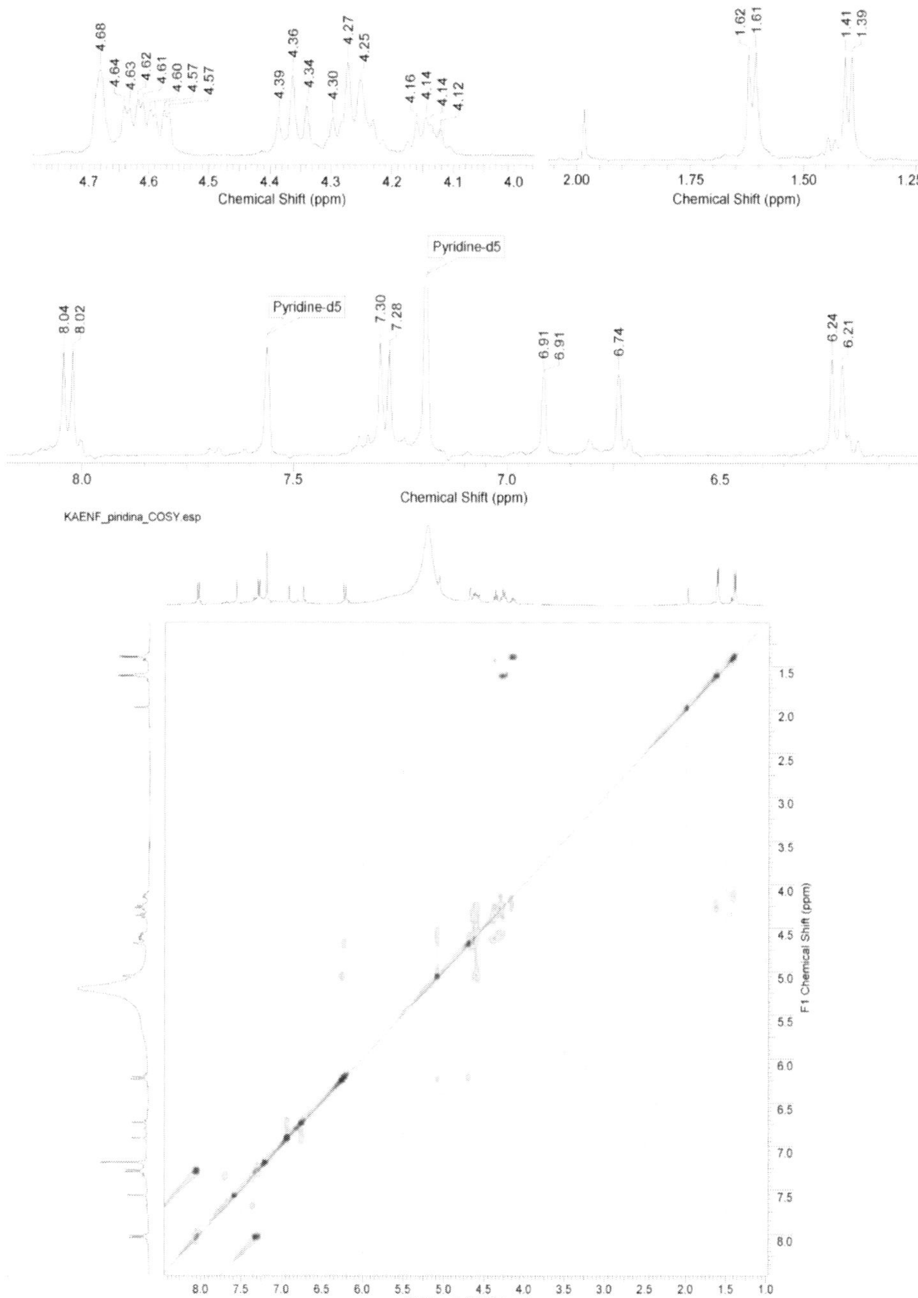

Figure 2. H^1 NMR and COSY spectrum of the kaempferitrin.

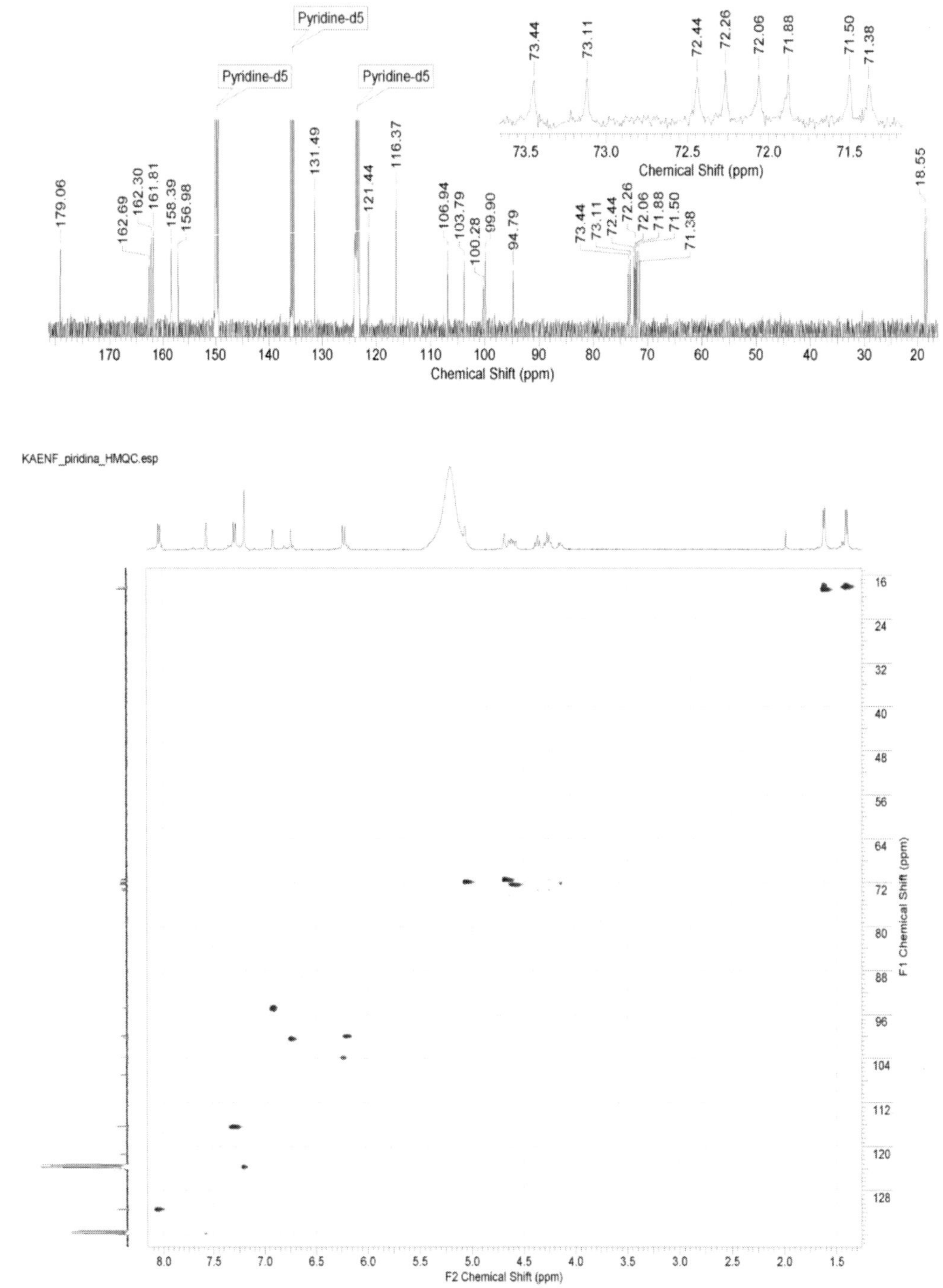

Figure 3. C^{13} NMR and HMQC spectrum of the kaempferitrin.

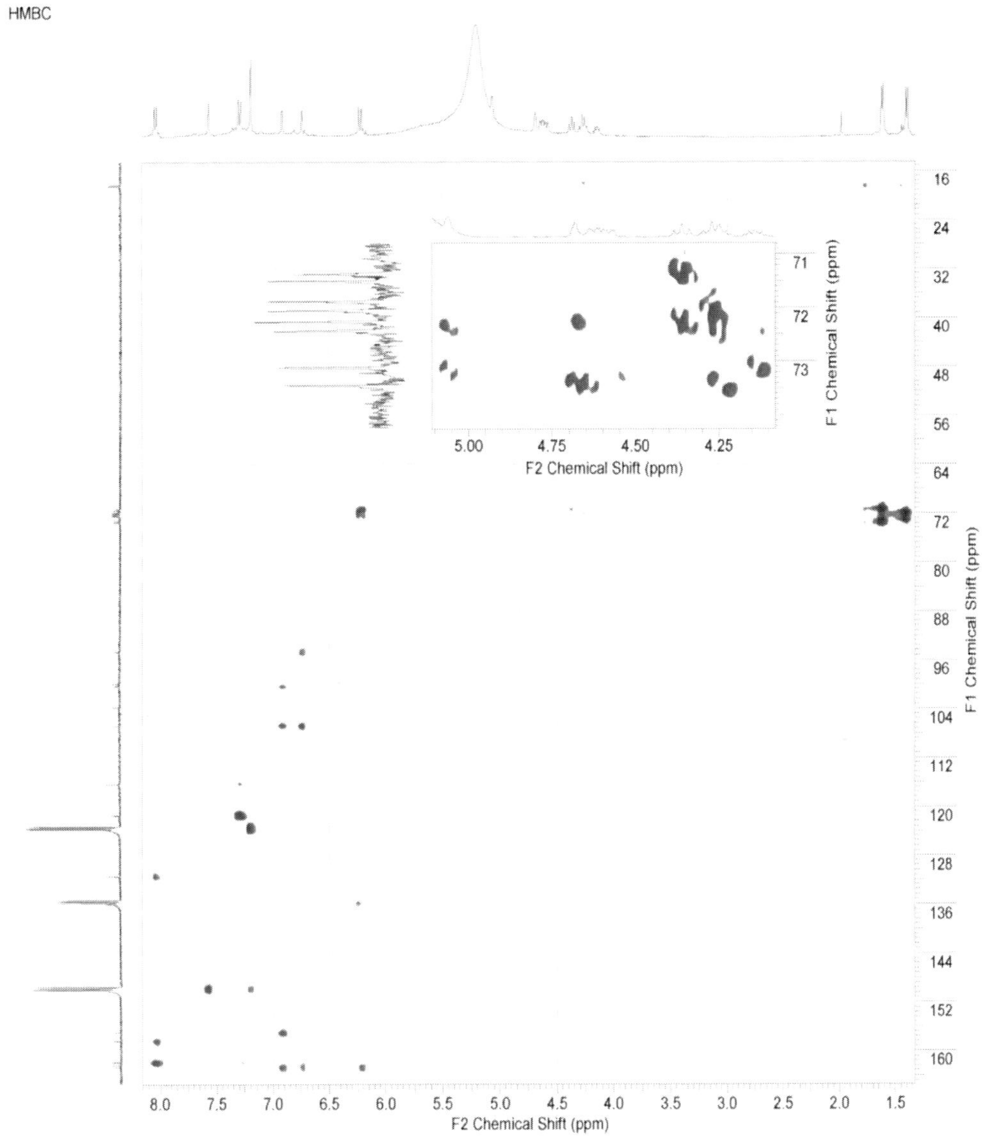

Figure 4. HMBC spectrum of the kaempferitrin.

The approach to synthesize kaempferitrin described by Yang *et al.* (2012) stand out by a selective glycosylation of the 3-OH of 5,4'-di-*O*-acetylkaempferol by glycosylation method with glycosyl ortho-alkynylbenzoates and subsequent glycosylation of the remaining 7-OH with glycosyltrifluorocetimidates. Preparation of the required 5,4'-di-*O*-acetylkaempferol is achieved in three steps by 3,7-benzylation and 4',5-acethylation of the kempferol followed removal of the two benzyl groups by hydrogenolysis which catalytic hydrogenation (Figure 9).

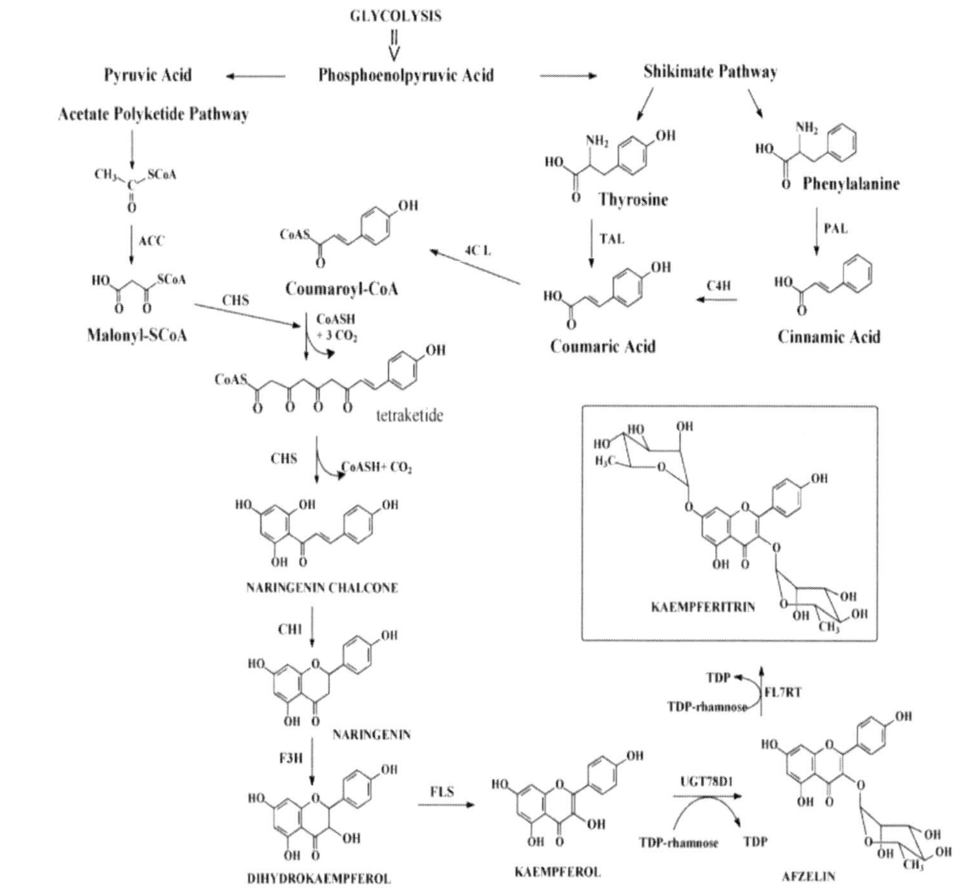

Figure 5. Overview *schematic for biosynthesis of the kaempferol and kaempferitrin.* ACC = acetyl-CoA carboxylase; TAL = L-tyrosine ammonia-lyase; PAL = L-phenylalanine ammonia-lyase; C4H = cinnamate 4-hydroxylase; 4CL = 4-coumarate CoA-lygase; CHS = chalcone synthase; CHI = chalcone isomerase; F3H = flavanone-3-dioxygenase; FLS = flavonol synthase; UGT78D1= UDP-rhamnose:flavonol-3-*O*-rhanmosyltransferase; FL7RT = flavonol-3-*O*-glycoside:7-*O*-rhamnosyltransferase.

3. Biological Activities of Kaempferol Aglycone and Glycosides: Different Targets to Modulate Glucose Homeostasis

Glycemia

Dietary polyphenols, such as kaempferol and kaempferitrin, can effectively prevent many diseases and protect the human health (Calderón-Montaño *et al.*, 2011; Weng and Yen, 2012). These both natural phytochemicals appear in vegetative foods and nutraceuticals and have been extensively studied not only *in vivo* and *in vitro* but also *in situ* approach (Cazarolli *et al.*, 2006; Cazarolli *et al.*, 2008a; Rodríguez *et al.*, 2010). The pharmacological activities, including antioxidant (De Sousa *et al.*, 2004), antiviral (Lyu *et al.*, 2005), anti-inflammatory (Parveen *et al.*, 2007), antiallergic (Hirose *et al.*, 2009), antidiabetic (Pereira *et al.*, 2011; Calderón-Montaño *et al.*, 2011) and anticancer properties (Barve *et al.*, 2009; Weng and Yen, 2012) are widely described. However, the mechanism of action of these flavonols in specific targets able to regulate glucose homeostasis is not yet well understood. Taking in mind that kaempferol and kaempferitrin, the major active nutraceutical ingredients in plants with effective influence on glucose metabolism (Cazarolli *et al.*, 2008a,b; Tapas *et al.*, 2008), some reports were selected to be discussed herein in order to better comprehend the current knowledge.

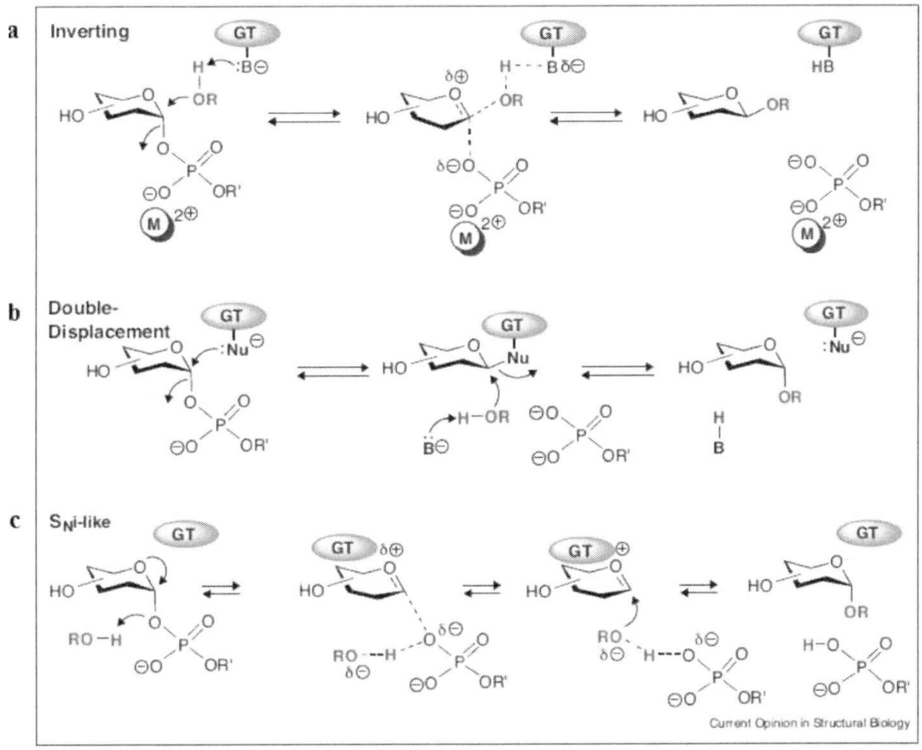

Figure 6. Reaction *mechanisms proposed for inverting and retaining glycosyltransferases. (From C. Breton, S. Fournel-Gigleux and M. M. Palcic, Cur. Op. Structural Biology 2012, p. 544, figure 2. Elsevier. Used by permission).*

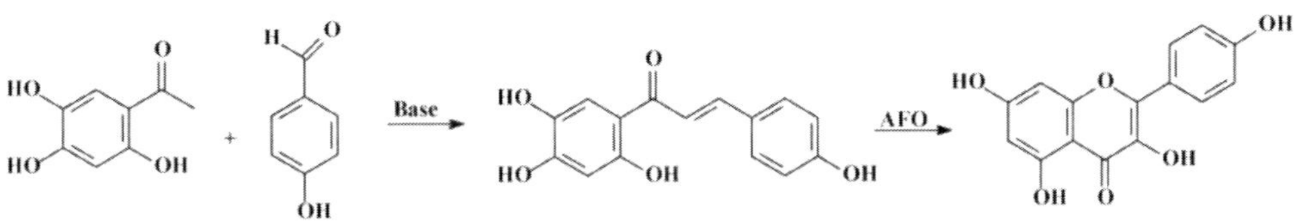

Figure 7. Synthesis of kaempferol by Claisen-Schimitdt condensation following Algar-Flyn-Oyamada reaction.

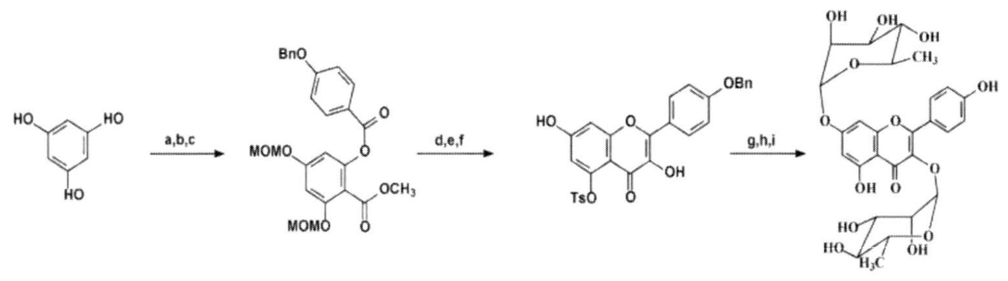

Figure 8. Shaw'synthesis of kaempferitrin. a) CH_3OCH_2CN, $ZnCl_2$, HCl; b) MOMCl, K_2CO_3, ACETONE, 23 ^0C; c) EDCl, DMAP, CH_2Cl_2; d) K_2CO_3, pyridine, reflux; e) p-TosCl, K_2CO_3, CH_3CN,60 ^0C; f) A_LBr_3, CH_3CN, 0 ^0C, 1.25 M HCl in MeOH, 70 ^0C; g) tri-*O*-acetylrhamnose bromide, Ag_20,CH_2Cl_2, 40 ^0C; h) H_2, Pd/C, EtOAc; i) K_2CO_3, MeOH, 70 ^0C.

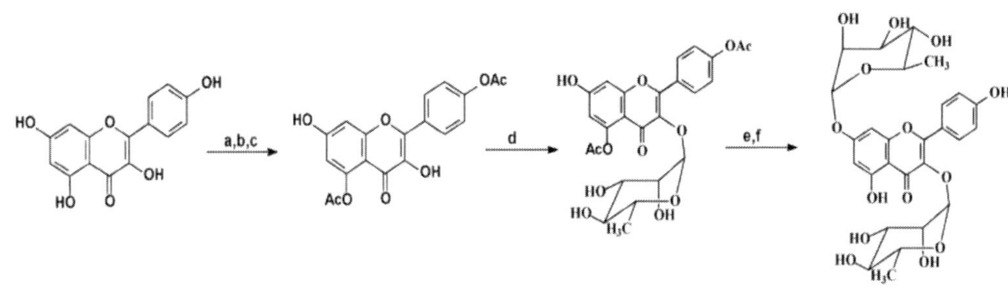

Figure 9. Synthesis of kaempferitrin proposed by Weizhun Yang *et al.* (2012): a) BnBr, K_2CO_3; b) Ac_2O, pyridine, DMAP; c) H_2, 10% Pd/C; d) rhamnosyl ortho-alkynylbenzoate, PPh_3AuNTf_2, CH_2Cl_2, 4A MS; e) rhamnosyltrifluorocetimidate, BF_3EtO_2, CH_2Cl_2, 4A MS; f) K_2CO_3, MeOH, THF.

It has been reported antidiabetic activity of several flavonoids (Cazarolli *et al.*, 2008b; Tapas *et al.*, 2008) (Tabe 1). The oral treatment in euglycemic and diabetic rats with kaempferol-3,7-*O*-(α)-dirhamnoside (kaempferitrin) isolated from *Bauhinia forficate* leaves, induced an acute hypoglycemic activity (from zero to 6 h). However, a slight serum glucose lowering was observed in hyperglycemic rats overloaded with glucose (4g/kg; p.o), before the compound treatment (Silva *et al.*, 2002; De Sousa *et al.*, 2004; Jorge *et al.*, 2004). Following-up some reports that sought to develop vanadium-based drugs for diabetes therapy (Shechter *et al.*, 2003; Thompson *et al.*, 2004) and considering the hypoglycemic property of vanadium compounds, kaempferol-3,7-*O*-(α)-dirhamnoside complex with vanadium (IV; VO(IV)) was assayed to evaluate glycemia. As much free vanadium, insulin and kaempferol as vanadium-based kaempferitrin complex (kaempferitrin-VO(IV)), administrated by intraperitoneal rout in diabetic rats, exhibited hypoglycemic effect (glucose lowering was around 60%) for at least 24 h (Cazarolli *et al.*, 2006). Also, another glycosylated form of kaempferol (kaempferol-3-neohesperidoside) structurally very similar to kaempferitrin and naturally occurring in *Cyathea phalerata* stalk demonstrated hypoglycemic effect in both oral and intraperitoneal treatments in diabetic rats. Additionally, the complex kaempferol-3-neohesperidoside–VO (IV) presented an effect on plasma glucose levels from diabetic rats (reduction of approximately 70%) that persisted for up to 24 h (Cazarolli *et al.*, 2006). Furthermore, recently it was described that an acetylated kaempferol glycoside, 3-*O*-[6-*O*-acetyl]-β-D-glucopyranosyl(1→4)-α-L-rhamnopyranoside (Multiflorin A), isolated from peach leaf, exhibits a dose-dependent anti-hyperglycemic effect on mice submitted to oral glucose-loading test (Shirosaki *et al.*, 2012a).

Although the increased contribution about the biotransformation products of kaempferol on biological activities as well as the real differences between hepatic and intestinal disposition of aglycone and their glycosylated forms (Barve *et al.*, 2009; Li *et al.*, 2012), some results show different effect and mechanism of action to a series of kaempferol (aglycone or glycosylated) through distinct experimental models (Jorge *et al.*, 2004; Cazarolli *et al.*, 2006; Rodríguez *et al.*, 2010). For example, the glycosylated forms of kaempferol demonstrated antidiabetic activity but the kaempferol aglycone did not have significant effect on blood glucose levels (Cazarolli *et al.*, 2006; Zang *et al.*, 2011) indicating that glycosyl residues are necessary for biological activity (Yamasaki *et al.*, 2011).

Moreover, it was proposed that acetylated sugar moiety is responsible by suppressive effect on glucose absorption in mice since the deacetylated form of Multiflorin A (kaempferol 3-*O*-β-D-glucopyranosyl-(1 →4)-α-L-rhamnopyranoside: Multiflorin B) did not present activity (Shirosaki *et al.*, 2012a).

Insofar as different glycosylated forms of kaempferol exhibited an anti-hyperglycemic and/or hypoglycemic effect and, in a whole those data point a potential insulin-secretagogue and/or insulin-mimetic effect to these compounds, *in vivo* and *in vitro* studies have been performed in order to establish the mechanism responsible for glycemia homeostasis. The targets of action studied and discussed below are muscle, liver, adipose tissue, pancreas and intestine.

Protein Synthesis

Few reports concerning kaempferol and kaempferitrin on protein synthesis in physiological sites involved on glucose metabolism were described. Zhang and Liu (2011) demonstrated that kaempferol improves chronic hyperglycemia-impaired pancreatic β-cell viability and insulin secretory function. The cytoprotective effect of kaempferol on cultured clonal β-cells and pancreatic human islets was promoted by inhibition of cellular apoptosis and reduced caspase-3 activity in a high and chronic glucose exposition. Also, kaempferol improved the expression of anti-apoptotic proteins Akt and B-cell lymphoma 2 (Bcl-2) and restored intracellular cyclic adenosine monophosphate (cAMP) and *adenosine* triphosphate (ATP) production. In addition, these cytoprotective effects of kaempferol were associated with improved insulin secretory function and synthesis in β-cells and human islets. Also, Zhang *et al.* (2012) showed that kaempferol promoted viability, inhibited apoptosis and reduced caspase-3 activity in INS-1E cells and human islets chronically exposed to palmitate. In addition, kaempferol prevented the lipotoxicity-induced down-regulation of anti-apoptotic proteins Akt and Bcl-2. The cytoprotective effects of kaempferol also were associated with improved insulin secretion, synthesis, and pancreatic and duodenal homeobox-1 (PDX-1) expression. Kaempferol restored cAMP production, protein kinase A (PKA) activation, cAMP-responsive element binding protein (CREB) phosphorylation and its regulated transcriptional activity in β-cells, all of which were significantly diminished by chronic hyperlipidemia. Disruption of CREB expression by transfection of CREB small interfering ribonucleic acid (siRNA) in INS-1E cells or adenoviral transfer of dominant-negative forms of CREB in human islets ablated kaempferol protection of β-cell apoptosis and dysfunction caused by palmitate. Incubation of INS-1E cells or human islets with kaempferol induced PDX-1 expression. This effect of kaempferol on PDX-1 expression was not shared by a host of structurally related flavonoid compounds. PDX-1 gene knockdown reduced kaempferol-stimulated cAMP generation and CREB activation in INS-1E cells. These findings demonstrate that kaempferol is a novel survivor factor for pancreatic β-cells via upregulating the PDX-1/cAMP/PKA/CREB signaling cascade. The data point a pivot role of kaempferol on β-cells, one of the most significant tissues to the glucose homeostasis.

On the other hand, although the efficient action of kaempferitrin on glycemia and the insulin-mimetic effect demonstrated to this flavonoid, ^{14}C-leucine incorporation into proteins in rat soleus muscle was not modified by oral treatment with kaempferitrin either in euglycemic or in diabetic rats (Jorge *et al.*, 2004). In a whole, it seems that the glucose uptake in soleus muscle does not depend on active protein synthesis. However, further studies needs to be carried out with different protocols in order to clarify the mechanism of action kaempferitrin on glucose uptake in skeletal muscle.

Glucose Uptake

The glucose uptake in insulin-dependent tissues is one of the most prompt effects of insulin to regulate glucose homeostasis. Based on insulin-mimetic properties of sundry flavonoids, studies have uncovered that kaempferitrin increases the ^{14}C-Deoxy-D-glucose, as a nonmetabolic glucose substitute, in soleus muscle in euglycemic rats. Kaempferitrin was found to have an acute effect on serum glucose lowering in diabetic rats and to stimulate the glucose uptake percentile, as efficiently as insulin in soleus muscle (Jorge *et al.*, 2004). Likewise, kaempferol-3-neoespheridoside increased glucose uptake in skeletal muscle at 1 and 100 nM (35 and 21%, respectively). This flavonoid at 1 nM also proved to be as effective as insulin (7 nM) in stimulatory effect of glucose uptake (Zanatta *et al.*, 2008). Additionally, both glycosilated forms (kaempferitrin and kaempferol-3-neohesperidoside) presented hypoglycemic effect in alloxan-induced diabetic rats (De Sousa *et al.*, 2004; Cazarolli *et al.*, 2006) therefore; the effect of kaempferol-3-neohesperidoside probably results from the ability of these dietary compounds to increase glucose transport into tissues such as skeletal muscle and adipocyte. Taking it in mind, studies were performed in order to highlight the signaling pathways on glucose uptake triggered by glycosilated forms of kaempferol. Zanatta *et al.* (2008) demonstrated that kaempferol-3-neoesperidoside induces the glucose uptake on skeletal muscle via phosphatidylinositol 3-kinase (PI-3K) and protein kinase C (PKC) activation but independently of mitogen-activated *protein kinase kinase (*MEK) pathways and active protein synthesis. Study performed by Yamasaki *et al.*, (2011) confirmed the stimulatory effect of kaempferol-3-neoesperidoside on glucose uptake and suggested that the substructure responsible for insulin-mimetic events on muscle is neohesperidose since kaempferol and its 3-glucoside showed negligible effect.

Another glycosilated form of kaempferol that enhanced glucose uptake in C2C12 muscle cell line was kaempferol-3-*O*-(6″-caffeoylglucoside) (Muhammad *et al.*, 2012). This compound isolated from *Sarracenia purpurea* L. increased glucose uptake compared with control group. Nevertheless, not all glycosylated forms of kaempferol showed positive effect on glucose metabolism. Vishnu Prasad *et al.* (2009) found that kaempferitrin inhibits GLUT-4 translocation stimulated by insulin and consequently glucose uptake in differentiated 3T3-

L1 cells. These data point a tissue-specific effect since kaempferitrin stimulates glucose uptake in skeletal muscle as described by Jorge *et al.* (2004).

Despite the evidences that kaempferitrin inhibits glucose uptake though interference on insulin signaling pathways in 3T3-L1 cells another study reports that both glycosilated forms, kaempferol 3-*O*-*β*-D-glucopyranosyl-(1 → 4)-*α*-L-rhamnopyranosyl-7-*O*-*α*-L-rhamnopyrano side (CO-1) and kaempferol 3-*O*-*β*-D-apiofuranosyl-(1 → 2)-*α*-L-arabinofuranosyl-7-*O*-*α*-L-rhamnopyranoside, i*solated from Cinnamomum osmophloeum* increased the level of insulin receptor-beta (IR*β)* phosphorylation as well as the GLUT-4 translocation in 3T3-L1 adipocytes albeit they did not show the glucose uptake (Lee *et al*., 2009). In a whole, further studies are needed in order to better understand the intracellular pathways triggered by kaempferol and kaempferitrin on glucose uptake in insulin-dependent tissues.

Although kaempferol aglycone is not able to reduce glycemia neither in diabetic nor hyperglycemic rats, studies using HepG2 cells (human liver hepatocellular carcinome cell line) improved the basal glucose uptake. The effect induced by kaempferol was stronger than that of rosiglitazone, which has been used as an antidiabetic drug (Chen *et al*., 2010). Moreover, kaempferol improved glucose uptake in adipocytes stimulated by insulin although the glucose uptake was not changed by flavonoid in the absence of insulin (Fang *et al*., 2008). The authors proposed that kaempferol acts at multiple targets to improve glucose uptake in adipocytes, among them, as partial agonist of peroxisome proliferator-activated receptor gamma (PPARγ). The study showed that kaempferol can compete with rosiglitazone in a dose-dependent manner suggesting that it binds to the same pocket of PPARγ as rosiglitazone. Agreeing to these data Christensen *et al.* (2010) demonstrated that kaempferol elicits the activation of PPARγ, in contrast to glycoside forms that are unable to activate PPARγ.

On the other hand, Nomura *et al.* (2008) reported that kaempferol inhibits insulin-stimulated glucose uptake in mouse bone marrow-derived stromal cell lines (MC3T3-G2/PA6) differentiated into mature adipose cells. In addition, it was showed that the inhibitory mechanism of action involves inhibition of insulin-stimulated insulin receptor phosphorylation, Akt activation and GLUT-4 translocation to plasma membrane. Also, the structure-activity relationship seems to be relevant since the effect of different classes of flavonoids, the position of hydroxyl substitution (4'- or 3',4'-OH) in the B-ring as well as C2-C3 double bond determine inhibitory activity on glucose uptake.

Accordingly, stimulating glucose uptake into peripheral tissues may be responsible for glycemia homeostasis. The findings discussed herein present these flavonoids as promising antidiabetic drugs. However, it is still necessary to define some missing steps in these flavonoids-signaling pathways and elucidate the mechanism in order to provide new insights in the role of kaempferol and its glycosides to diabetes treatment. Further studies are also required concerning safety (assessment of toxicity) and human trial to develop potential anti-diabetic remedies of choice.

Glycogen Content and Synthesis

It is clear that insulin reduces serum glucose levels by increasing glucose uptake, storage of carbohydrates, synthesis of lipids and proteins in skeletal and adipose tissue (Khan and Pessin, 2002; Martin-Gronert and Ozanne, 2012), stimulates glycogen synthase (Syed and

Khandelwal, 2000), inhibits gluconeogenesis, among other key steps in energy metabolism (Saltiel and Kahn, 2001). Most of the glucose that enters on muscle fibers in response to insulin is converted into glycogen (Roach, 2002). Based on that, it has been proposed that the influence of kaempferol glycosides on glucose homeostasis is mediated by an insulin-mimetic effect. Hereupon, Zanatta *et al.* (2008) reported that kaempferol-3-neohesperidoside, which presented hypoglycemic effect, also increased glycogen content in soleus muscle from hyperglycemic rats. These data were lather confirmed by Cazarolli *et al.* (2009) who demonstrated the stimulatory effect of kaempferol 3-neohesperidoside on D -[^{14}C (U)] -glucose incorporation into glycogen in rat soleus muscle. The authors supported their hypothesis about insulin-mimetic role of kaempferol 3-neohesperidoside in glucose homeostasis through pharmacological studies. From those data it was showed beyond a prompt stimulatory effect of the compound, the involvement of PI-3K/GSK-3 (glycogen synthase kinase 3) pathway and mitogen-activated protein kinase (MAPK)/ protein phosphatase 1 (PP1) pathway on glycogen synthesis in rat soleus muscle. The similarities in the involvement of protein kinases and phosphatases in the stimulation of glycogen deposition by kaempferol 3-neohesperidoside and by insulin constitute strong evidence for the insulin-mimetic role of kaempferol 3-neohesperidoside in glucose homeostasis. All together, it points that kaempferol and its different forms of glycosides have potential role to develop antidiabetic drugs.

Glucose Intestinal Absorption

Glucose intestinal absorption is one of the factors that increase glycemia. The regulation of postprandial glycemia through the digestion and/or absorption of carbohydrates can be effective to avoid hyperglycemia and prevent diabetes. In this context, manifold studies have reported significant inhibitory activity of kaempferol aglycone and glycosides on disaccharidases and intestinal glucose transporter.

It was showed significant inhibitory effect of kaempferitrin, in different concentrations, on duodenal maltase and sucrase activity after *in vitro* incubation although the specific activity of lactase was unaltered by it (Pereira *et al.*, 2011). Another kaempferol glycoside, kaempferol-3-*O*-rutinoside, also demonstrated a potent inhibitory effect against α-glucosidase on *in vitro* studies. Kaempferol-3-*O*-rutinoside was 8-times more potent than acarbose, an antidiabetic drug that act on intestinal glucose absorption. Besides this newsworthy outcome it exhibited a synergistic effect with non glycosilated kaempferol (Habtemariam, 2011).

Likewise, kaempferol aglycone on *in vitro* assays using rat duodenum inhibited maltase, but not sucrase and lactase activities at any concentration studied (Pereira *et al.*, 2011). On the other hand, Matsui *et al.* (2002) previously demonstrated that kaempferol aglycone inhibits both maltase and sucrase *in vitro*. Wang *et al.* (2010) also find out that kaempferol, isolated from guava leaves, exhibits important inhibitory activities against rat intestinal α-glucosidases (sucrase and maltase) and α-amylase thus indicating that sugar binding to kaempferol is not essential for its inhibitory activity on disaccharidases.

Beyond kaempferol, it was verified the inhibitory activity of other 2-substituted-phenyl-chromen-4-one flavonoids like kaempferol. From these data, it was deduced that "large conjugated-system" skeleton formed by A-ring, B-ring and C-ring can be necessary for these compounds to act upon α-glucosidase and α-amylase. Accordingly, it was concluded that the free 3-position hydroxyl group is a key functional group for inhibits α-glucosidase and α-amylase and this inhibitory activity increases with the increase of the number of hydroxyl groups on the C-ring (Wang *et al.*, 2010).

Concerning intestinal glucose absorption, Shirosaki *et al.* (2012a, b) found in murine intestine that a natural acetylated kaempferol glycoside, multiflorin A, has potent inhibitory activity against glucose absorption. Likewise, Rodriguez *et al.* (2010) demonstrated in isolated intestinal segment, that kaempferol 3-*O*-α-rhamnoside inhibits intestinal glucose absorption although a different glycosylated form; kaempferol 3-*O*-α-(2″-galloyl) rhamnoside did not alter glucose absorption. It was discussed that, probably, the presence of the galloyl group bound to rhamnosyl moiety annuls the inhibitory effect of kaempferol 3-*O*-α-rhamnoside on glucose intestinal absorption and the presence of an H (kaempferol 3-*O*-α-rhamnoside) instead of an OH in the 3′ position of the B ring is determinant for the biological activity. Regarding the mechanism of action of kaempferol 3-*O*-α-rhamnoside on glucose intestinal absorption the results indicate this flavonoid as a competitive inhibitor of intestinal SGLT-1 since it demonstrated an additive effect to that of phlorizin, a classic inhibitor of SGLT-1, and it significantly increased the K_m of glucose intestinal absorption.

A variety of inhibitors of α-glucosidases are studied as potential therapeutic drugs to treat diabetes and an important drug commercially available is acarbose. Maybe some of the kaempferol forms discussed above can become drugs but before that it is necessary to know the mechanism of action of these flavonoids beyond of their possible toxicity. Among some essential characteristics concerning the structure and the activity relationship of α-glucosidase inhibitors structurally different are: sugar (substrate)-mimic structures, the ability to form hydrogen bonds or ionic bonds with nucleophilically catalyzing residues, transition-state-like structures, the ability to make ionic and hydrophobic interactions at sites other than the active site, and the ability to form covalent bond with enzymes through an epoxy or aziridine group (Moorthy *et al.*, 2012). Taking this data together, natural, synthetic or semi-synthetic new compounds have been studied in order to characterize specific site of action on intestine in order to develop efficient drugs able to regulate carbohydrates absorption.

4. Perspective of Technical Approach for the Delivered Control of Bioactive Flavonoids at Specific Target

Flavonoids, plant secondary metabolites, are natural phenolic compounds widely found in chocolates, cereals, beverages, fresh and dried fruits and vegetables. These compounds have been reported as able to promote beneficial health effects, especially regarding the prevention of chronic human diseases (Scalbert *et al.*, 2005; Arts and Hollman, 2005; Pandey and Rizvi, 2009). However, the bioavailability of each flavonoid depends on the actual amount of each nutrient that is digested and absorbed as well as its biotransformation by intestinal mucosa and liver. Moreover, it is unknown the relation between the quantity of polyphenols intake and their bioavailability in human body (DuPont *et al.*, 2004; Barve *et al.*, 2009).

Table 1. Effects of kaempferol aglycone and glycosides on glucose homeostasis and carbohydrate metabolism

Kaempferol form	Treatment	Concentration	Experimental model	Tissue target	Effect	Reference
Kaempferol	Oral	0.0297 and 0.0594 mmol/kg	Diabetic rats induced by alloxan and genetically type 2 diabetic *KK-A*^{y} mice	Blood	No effect on glycemia	Cazarolli *et al.*, 2006 Zang *et al.*, 2011
Kaempferol-3,7-*O*-(α)-L-dirhamnoside (kaempferitrin)	Oral	200 mg/kg	Euglycemic rats	Blood	Reduces glycemia	De Sousa *et al.*, 2004
Kaempferitrin	Oral	50, 100, 200 mg/kg	Diabetic rats	Blood	Reduces glycemia	De Sousa *et al.*, 2004
Kaempferitrin	i.p. and oral	0.0294 mmol/kg	Diabetic rats induced by alloxan	Blood	Reduces glycemia	Cazarolli *et al.*, 2006
Kaempferitrin - VO(IV))	i.p.	0.0294 mmol/kg	Diabetic rats induced by alloxan	Blood	Reduces glycemia	Cazarolli *et al.*, 2006
Kaempferol-3-neohesperidoside	i.p. and oral	0.0286 mmol/kg	Diabetic rats induced by alloxan	Blood	Reduces glycemia	Cazarolli *et al.*, 2006
Kaempferol-3-neohesperidoside–VO(IV)	i.p.	0.0286 mmol/kg	Diabetic rats induced by alloxan	Blood	Reduces glycemia	Cazarolli *et al.*, 2006
3-*O*-[6-*O*-acetyl]-β-D-glucopyranosyl(1→4)-α-L-rhamnopyranoside (Multiflorin A)	Oral	0.7 and 5.6 mg/kg	Glucose-loaded mice	Blood	Reduces glycemia	Shirosaki *et al.*, 2012a
Kaempferol	*In vitro* assay	10 μM	INS-1E beta-cells and human islets chronically exposed to high glucose concentration	Pancreas	Improved insulin secretion	Zhang and Liu, 2011
Kaempferol	*In vitro* assay	10 μM	INS-1E cells and human islets chronically exposed to palmitate	Pancreas	Improved insulin secretion	Zhang *et al.*, 2012
Kaempferitrin	Oral	100 mg/kg	Euglycemic and diabetic rats induced by alloxan	Soleus muscle	No changes on protein synthesis	Jorge *et al.*, 2004
Kaempferitrin	*In vitro* assay	2 and 104 mM	Euglycemic rats	Soleus muscle	Increases glucose uptake	Jorge *et al.*, 2004
Kaempferol-3-neohesperidoside	*In vitro* assay	1 and 100 nM	Euglycemic rats	Soleus muscle	Increases glucose uptake	Zanatta *et al.*, 2008
Kaempferol	*In vitro* assay	10 and 100 μM	HepG2 human cell line	Liver	Increases glucose uptake	Chen *et al.*, 2010
Kaempferol	*In vitro* assay	10 μM	MC3T3-G2/PA6 mouse cells line	Adipocytes	Inhibits insulin-stimulated glucose uptake	Nomura *et al.*, 2008
Kaempferol	*In vitro* assay	20 and 50 μM	3T3-L1 cell culture	Adipocytes	Increases insulin-stimulated glucose uptake	Fang *et al.*, 2008
Kaempferol-3-*O*-(6″-caffeoylglucoside)	*In vitro* assay	50 μM	Mouse myoblast *cell line (C2C12)*	Myoblast	Increases glucose uptake	Muhammad *et al.*, 2012
Kaempferitrin	*In vitro*	0.1, 1,10 and	3T3-L1 cell culture	Adipocytes	Inhibits	Vishnu

Kaempferol form	Treatment	Concentration	Experimental model	Tissue target	Effect	Reference
	assay	20 μM			insulin-stimulated glucose uptake	Prasad *et al.*, 2009
Kaempferol-3-neohesperidoside	Oral	100 mg/kg	Hyperglycemic rats	Soleus muscle	Increases glycogen content	Zanatta *et al.*, 2008
Kaempferol-3-neohesperidoside	*In vitro* assay	1 μM	Euglycemic rats	Soleus muscle	Glycogen synthesis	Cazarolli *et al.*, 2009
Kaempferol	*In vitro* assay	3.125 – 400 μM	Euglycemic rats	Duodenum	Inhibits maltase activity	Pereira *et al.*, 2011
Kaempferol	*In vitro* assay	0.1 mg/mL	Euglycemic rats	Intestine	Inhibits maltase and sucrase activity	Matsui *et al.*, 2002
Kaempferol	*In vitro* assay	5.2 μM* for sucrase 5.6 μM* for maltase 5.3 μM* for amylase	Porcine pancreatic α-amylase and rat intestinal sucrase and maltase	Intestine and pancreas	Inhibits α-glucosidase and α-amylase	Wang *et al.*, 2010
Kaempferitrin	*In vitro* assay	3.125 – 25 μM	Euglycemic rats	Duodenum	Inhibits maltase and sucrase activity	Pereira *et al.*, 2011
Kaempferol 3-*O*-α-rhamnoside	*In situ* assay	5 mM	Euglycemic rats	Isolated intestinal segments	Inhibits intestinal glucose absorption	Rodriguez *et al.*, 2010
Kaempferol 3-*O*-α-(2″-galloyl) rhamnoside	*In situ* assay	5 mM	Euglycemic rats	Isolated intestinal segments	No effect on glucose absorption	Rodriguez *et al.*, 2010

* = IC_{50}; i.p. = intraperitoneal.

In the last decade, there has been an increasing interest in nutraceutical compounds since their potential health benefits and inherent biological activities in human metabolism are widely reported (Cazarolli *et al.,* 2008a; Pandey and Rizvi, 2009; Calderón-Montaño *et al.,* 2011). However, the success of drug therapies based on these compounds depends on the pharmacodynamics and dose-response relationship of each molecule. Especially regarding this issue, the dose-response of the drug in a specific physiological site of action allows the determination of safe and hazardous dosages and therefore defining the safety and efficacy of treatment (Kaparissides *et al.,* 2006).

For a variety of active substances, their inclusion in conventional solid dosage forms is enough to obtain the desired therapeutic effects. On the other hand, most natural origin active ingredients present specific biopharmaceutical characteristics that compromise their absorption and therefore the effectiveness of the treatment when included in this type of formulation. Usually only a small proportion of flavonoids remain available after oral administration. This occurs mainly due to their low solubility/permeability in the gastrointestinal tract and chemical instability during processing and storage (temperature, humidity and exposure to light) and after administration (pH, enzymes, and other nutrients) (Bell, 2001). As consequence, only a low concentration of flavonoids reaches specific therapeutic targets, which can result in treatment failure, toxicity and serious adverse effects.

The development of new therapeutic systems consists in a promising strategy for the improvement of the biopharmaceutical profile of drugs. Mainly those molecules with poor water solubility and low bioavailability can benefit from their inclusion in these systems. Sustained release dosage forms offer numerous advantages compared to conventional pharmaceutical formulations thereby improving therapeutic efficacy, delivering the therapeutic agent to a focused site for a prolonged period with appropriate amounts, reducing the toxicity and increasing the bioavailability of the active fraction in the diseased area (Grund *et al.,* 2011; Granada *et al.,* 2007; Alagusundaram *et al.,* 2009).

Drug controlled release systems (drug delivery systems-DDS) may be classified according to their structure as reservoir (e.g.: microcapsules, nanocapsules) or matrix systems (e.g.: microspheres, nanospheres) (William et al., 2010). Reservoir-type drug delivery systems are vesicular systems in which the drug is confined to a cavity surrounded by a unique polymer membrane, while matrix-type DDS are compact particles, formed from a polymer network in which the active substance is distributed in its solid or molecular state. In these systems the drug is dissolved, entrapped, encapsulated or attached and the method of preparation is what defines which type of system is obtained (Soppimath *et al.,* 2001). Depending mainly on its solubility, the drug may be released by gradual diffusion through the matrix system. Moreover, several stimuli such as pH changes, ionic strength, temperature or any other chemical or physical alteration are able to induce matrix degradation allowing drug release by polymer erosion (Chien and Lin, 2002).

The microparticulate and nanoparticulate systems have particle sizes greater or less than 1 μm, respectively, and that size difference entails real differences at many levels as levels from formulation to *in vivo* usage (Kohane, 2006). Microparticles are systems with excellent ability to sustained drug release that may be suitable in the treatment of chronic diseases. Besides the benefit of drug control release, nanoparticles opened new perspectives in targeted drug delivery (Olivier, 2005).

For kaempferol, a successful nanoparticulate system improved the inhibition of cancer cell viability indicating a potential use of these systems in the therapy to inhibit cell proliferation (Luo *et al.*, 2012). In another study, persimmon leaf extract *Diospyros kaki* was incorporated into a self-nanoemulsifying DDS to compare its *in vitro* dissolution and relative bioavailability with commercially available tablets of Naoxinqing. Compared with the commercial tablets, the area under curve (AUC) of both quercetin and kaempferol, which are representative active flavonoids of persimmon leaf extract, were increased by 1.5-fold and 1.6-fold, respectively, following oral administration of persimmon leaf extract-loaded self-nanoemulsifying DDS in fasting beagle dogs. The authors suggested that this system is a promising DDS for increasing the oral bioavailability of this extract (Li *et al.*, 2011). Also, Chao *et al.* (2012) demonstrate that submicron emulsion system may be a promising vehicle for topical kaempferol application.

Concerning kaempferol and its glycosilated forms, the nanotechnology field provides new and modern drug delivery systems of growing interest due to the promise of increasing their bioavailability in human body to treat glucose unbalance and diabetes. It is anticipated that the effectual and valuable relevance of nanocarriers applied to the development of natural products and herbal medicines will increase the importance of developing new drug delivery systems (Yadav *et al.*, 2011). In the near future, flavonoids nanoparticulate formulations may lead to breakthroughs in diabetes therapy by delivering the drug in a specific target tissue with reduced dose and superior effect when compared with conventional solid dosage forms.

Acknowledgments

Studies in the authors' laboratory were supported by Conselho Nacional de Desenvolvimento e Tecnológico (CNPq), Coordenação de Pessoal de Nível Superior (CAPES), Fundação de Amparo à Pesquisa do Estado de Santa Catarina (FAPESC) and Financiadora de Estudo e Projetos (FINEP).

References

Aherne, SA; O´Brien, NM. Dietary Flavonols: chemistry, food content, and metabolism. *Nutrition*, 2002 18, 75-81.

Alagusundaram, M; Madhu, SCC; Umashankari, K; Attuluri, VBLC; Ramkanth, S. Microspheres as a novel drug delivery system - a review. *International Journal of ChemTech Research*, 2009 1 3, 526-534.

Arts, ICW; Hollman, PCH. Polyphenols and disease risk in epidemiologic studies. *American Journal of Clinical Nutrition*, 2005 81, 317-325.

Barrington, R; Williamson, G; Bennett, RN; Davis, BD; Brodbelt, JS; Kroon, PA. Absorption, conjugation and efflux of the flavonoids, kaempferol and galangin, using the intestinal Caco-2/TC7 cell model. *Journal of Functional Foods,* 2009 1 1, 74-87.

Barve, A; Chen, C; Hebbar, V; Desiderio, J; Saw, CLL; Kong, A-N. Metabolism, oral bioavailability and pharmacokinetics of chemopreventivek in rats. *Biopharmaceutics & Drug Disposition,* 2009 30, 356-365.

Breton, C; Fournel-Gigleux, S; Palcic, MM. Recent structures, evolution and mechanisms of glycosyltransferases. *Current Opinion in Structural Biology,* 2012 22, 540-549.

Britton, RG; Horner-Glister, E; Pomenya, OA; Smith, EE; Denton, R; Jenkins, PR; Steward, WP; Brown, K; Gescher, A; Sale, S. Synthesis and biological evaluation of novel flavonols as potential anti-prostate cancer agents. *European Journal of Medicinal Chemistry,* 2012 54, 952-958.

Calderón-Montaño, JM; Burgos-Morón, E; Pérez-Guerrero, C; López-Lázaro, M. A review on the dietary flavonoid kaempferol. *Mini-Reviews in Medicinal Chemistry*, 2011 11 4, 298-344.

Cazarolli, LH; Folador, P; Pizzolatti, MG; Mena Barreto Silva, FR. Signaling pathways of kaempferol-3-neohesperidoside in glycogen synthesis in rat soleus muscle. *Biochimie*, 2009 91 7, 843-849.

Cazarolli, LH; Zanatta, L. Alberton, EH; Figueiredo, MSRB; Folador, P; Damazio, RG; Pizzolatti, MG; Silva, FRMB. Flavonoids: cellular and molecular mechanism of action in glucose homeostasis. *Mini-Reviews in Medicinal Chemistry,* 2008b 8 10, 1032-1038.

Cazarolli, LH; Zanatta, L; Alberton, EH; Figueiredo, MSRB; Folador, P; Damazio, RG; Pizzolatti, MG; Silva, FRMB. Flavonoids: prospective drug candidates. *Mini - Reviews in Medicinal Chemistry*, 2008a 13, 1429-1440.

Cazarolli, LH; Zanatta, L; Jorge, AP; Horst, H; De Sousa, E; Woehl, VM; Pizzolatti, MG; Szpoganicz, B; Silva, FRMB. Follow-up studies on glycosylated flavonoids and their complexes with vanadium: their anti-hyperglycemic potential role in diabetes. *Chemico-Biological Interactions*, 2006 163, 177-191.

Chang, Q; Zuo, Z; Chow, MSS; Ho, WKK. Difference in absorption of the two structurally similar flavonoid glycosides, hyperoside and isoquercitrin in rats. *European Journal of Pharmaceutics and Biopharmaceutics,* 2005 59, 549-555.

Chao, Y; Huang, CT; Fu, LT; Huang, YB; Tsai, YH; Wu, PC. The effect of submicron emulsion systems on transdermal delivery of kaempferol. *Chemical & Pharmaceutical Bulletin (Tokyo*), 2012 60 9, 1171-1175.

Chen, QC; Zhang, WY; Jin, W; Lee, IS; Min, BS; Jung, HJ; Na, M; Lee, S; Bae, K. Flavonoids and isoflavonoids from Sophorae Flos improve glucose uptake *in vitro. Planta Medica*, 2010 76 1, 79-81.

Chen, Y; Deng, HZ; Liang, L. Contents of flavonoids in the extract of *Lespedeza virgate. Nan fang yi ke da xue xue bao*, 2008 28 5, 858-860.

Cheng, M-J; Wu, M-D; Hsieh, S-Y; Su, Y-S; Chen, I-S; Yuan,G-F. Secondary metabolites from the endophytic fungus *Annulohypoxylon boveri* var.*microspora* BCRC 34012. *Chemistry of Natural Compounds*, 2011 47 4, 536-540.

Chien, YW; Lin, S. Optimisation of treatment by applying programmable rate-controlled drug delivery technology. *Clinical Pharmacokinet*ics, 2002 41 15 1267-1300.

Choi, SJ; Kim, JK; Jang, JM; Shin, KH; Lim, SS. Rapid identification of the α-glucosidase inhibitory compounds from Thunberg's Geranium (Geranium thunbergii Sieb. et Zucc.). *Food Science Biotechnology,* 2012 21 4, 987-996.

Christensen, KB; Petersen, RK; Kristiansen, K; Christensen, LP. Identification of bioactive compounds from flowers of black elder (*Sambucus nigra* L.) that activate the human peroxisome proliferator activated receptor (PPAR)γ. *Phytotherapy Research*, 2010 24, S129-S132.

Day, AJ. *Uptake and metabolism of dietary flavonoid glycosides*. Norwich - UK: University of East Anglia (Institute of Food Research); 2000.

De Melo, GO; Malvar, DC; Vanderlinde, FA; Rocha, FF; Pires, PA; Costa, EA; de Matos, LG; Kaiser, CR; Costa, SS. Antinociceptive and anti-inflammatory kaempferol glycosides from *Sedum dendroideum. Journal of Ethnopharmacology,* 2009 124, 228-232.

De Sousa, E; Zanatta, L; Seifriz, I; Creczynski-Pasa, TB; Pizzolatti, MG; Szpoganicz, B; Silva, FRMB. Hypoglycemic effect and antioxidant potential of kaempferol-3,7-*O*-(r)-dirhamnoside from *Bauhinia forficata* leaves. *Journal Natural Products,* 2004 67, 829-832.

Diez-Sampedro, A; Lostao, MP; Wright, EM; Hirayama, BA. Glycoside binding and translocation in Na^+-dependent glucose cotransporters: comparison of SGLT1 and SGLT3. *Journal of Membrane Biology,* 2000 176, 111-117.

Du, Y; Wei, G; Linhardt, RJ. The first total synthesis of calabricoside A. *Tetrahedron Letters,* 2003 44, 6887-6890.

Dupont, MS; Day, AJ; Bennett, RN; Mellom, FA; Kroon, PA. Absorption of kaempferol from endive, a source of kaempferol-3-glucuronide in humans. *European Journal of Clinical Nutrition,* 2004 58 6, 947-954.

Ehrenkranz, JR; Lewis, NG; Kahn, CR; Roth, J. Phlorizin: a review. *Diabetes/ Metabolism Research and Reviews,* 2005 21, 31-38.

Erlund, I. Review of the flavonoids quercetin, hesperetin, and naringenin. Dietary sources, bioactivities, bioavailability and epidemiology. *Nutrition Research,* 2004 24, 851-874.

Euler, KL; Alam, M. Isolation of kaempferitrin from *Justicia spicigera. Journal Natural Products*, 1982 45 2, 220-221.

Fang, S-H; Rao, YK; Tzeng, Y-M. Inhibitory effects of flavonol glycosides from Cinnamomum osmophloeum on inflammatory mediators in LPS/IFN-γ-activated murine macrophages. *Bioorganic and Medicinal Chemistry,* 2005 13 7, 2381-2388.

Fang, XK; Gao, J; Zhu, DN. Kaempferol and quercetin isolated from *Euonymus alatus* improve glucose uptake of 3T3-L1 cells without adipogenesis activity. *Life Sciencies,* 2008 82 11-12, 615-622.

Gee, JM; DuPont, MS; Day, AJ; Plumb, GW; Williamson, G; Johnson, IT. Intestinal transport of quercetin glycosides in rats involves both deglycosylation and interaction with the hexose transport pathway. *Journal of Nutrition,* 2000 130, 2765-2771.

Gohar, AA; Elmazar, MMA. Isolation of hypotensive flavonoids from Chenopodium species growing in Egypt. *Phytotherapy Research,* 1997 11 8, 564-567.

Gómez, H; Polyak, I; Thiel, W; Lluch, JM; Masgrau, L. Retaining glycosyltransferase mechanism studied by QM/MM Methods: Lipopolysaccharyl-α-1,4-galactosyltransferase C transfers α-galactose via an oxocarbenium ion-like transition state. *Journal of the American Chemical Society,* 2012 134, 4743-4752.

Granada, A; Nemen, D; Dora, CL; Neckel, GL; Lemossenna, E. O emprego de sistemas de liberação como estratégia para melhorar as propriedades terapêuticas de fármacos de origem natural: o exemplo da camptotecina e seus derivados. *Revista de Ciências Farmacêuticas Básica e Aplicada,* 2007 28 2, 129-139.

Grund, S; Bauer, M; Fischer, D. Polymers in drug delivery- state of the art and future trends. *Advanced Engineering Materials*, 2011, 9999.

Habtemariam, S. A-glucosidase inhibitory activity of kaempferol-3-*O*-rutinoside. *Natural Products Communication*, 2011 6 2, 201-203.

Hamzah, AS; Lajis, NH; Sargent, MV. Kaempferitrin from the leaves of *Hedyotis verticillata* and its biological activity. *Planta medica*, 1994 60 4, 388-389.

Harborne, JB; Williams, CA. Advances in flavonoid research since 1992. *Phytochemistry,* 2000 55, 481-504.

Hasegawa, M. Lespedin, a birhamnoside of kaempferol. *Acta Phytochimica,* 1940 11, 299-308.

Hattori, S. Identity of lespedin with kaempferitrin. *Nature,* 1951 168 4279, 788-788.

Hirose, E; Matsushima, M; Takagi, K; Ota, Y; Ishigami, K; Hirayama, T; Hayashi, Y; Nakamura, T; Hashimoto, N; Imaizumi, K; Baba, K; Hasegawa, Y; Kawabe, T. Involvement of heme oxygenase-1 in kaempferol-induced anti-allergic actions in RBL-2H3 cells. *Inflammation*, 2009 32, 99-108.

Hu, BH; Zhou, LD; Liu, YL. New tetrasaccharide flavonol glycoside from *Epimedium acuminatum. Journal Natural Products*, 1992 55 5, 672-675.

Hummel, C; Lu, C; Loo, DDF; Hirayama, BA; Voss, AA; Wright, EM. Glucose transport by human renal Na^+/D-glucose cotransporters SGLT1 and SGLT2. *American Journal of Physiology - Cell Physiology*, 2011 300, C14-C21.

Imperato, F. New Kaempferol 3,7-Diglycosides and kaempferitrin in the fern Asplenium-Trichomanes. *Experientia,* 1979 35 9 1134-1135.

Jones, P; Messner, B; Nakajima, J-I; Schaffner, AR; Saito, K. UGT73C6 and UGT78D1, Glycosyltransferases involved in flavonol glycoside biosynthesis in Arabidopsis thaliana. *The Journal of Biological Chemistry*, 2003 278 45, 43910-43918.

Jorge, AP; Horst, H; de Sousa, E; Pizzolatti, MG; Silva, FR. Insulinomimetic effects of kaempferitrin on glycaemia and on ^{14}C-glucose uptake in rat soleus muscle. *Chemico-Biological Interactions*, 2004 149 2-3, 89-96.

Joshi, BS; Gawad, DH. Flavones of *Pajanelia multijuga* P.DC. and *Ligustrum neilgherense* var. obovata *C.B.Cl. Proceedings of the Indian Academy of Sciences - Section A*, 1977 86 1, 41-44.

Kang, JW; Kim, JH; Song, K; Kim, SH; Yoon, JH; Kim, KS. Kaempferol and quercetin, components of Ginkgo biloba extract (EGb 761), induce caspase-3-dependent apoptosis in oral cavity cancer cells. *Phytotherapy Research,* 2010 24 (Suppl 1), S77-S82.

Kaparissides, C; Alexandridou, S; Kotti, K; Chaitidou, S. Recent advances in novel drug delivery systems. *Journal of Nanotechynolog Online*, 2006 2, 1-11.

Kasahara, M; Maeda, M; Hayashi, S; Mori, Y; Abe, T. A missense mutation in the Na^+/glucose cotransporter gene SGLT1 in a patient with congenital glucose-galactose malabsorption: normal trafficking but inactivation of the mutant protein. *Biochima et Biophysica Acta,* 2001 1536, 141-147.

Katsuno, K; Fujimori,Y; Takemura, Y; Hiratochi, M; Itoh, F; Komatsu, Y; Fujikura, H; Isaji, M. Sergliflozin, a novel selective inhibitor of low-affinity sodium glucose cotransporter (SGLT2), validates the critical role of SGLT2 in renal glucose reabsorption and modulates plasma glucose level. *Journal of Pharmacology and Experimental Therapeutics*, 2007 320, 323-330.

Khan, AH; Pessin, JE. Insulin regulation of glucose uptake: a complex interplay of intracellular signalling pathways. *Diabetologia*, 2002 45 11, 1475-1483.

Koelzer, J; Pereira, DA; Dalmarco, JB; Pizzolatti, MG; Fröde, TS. Evaluation of the anti-inflammatory efficacy of *Lotus corniculatus. Food Chemistry*, 2009 117 3, 444-450.

Kohane, DS. Microparticles and nanoparticles for drug delivery. *Biotechnology and Bioengineering,* 2007 96, 203-209.

Lambert, N; Kroon, PA; Faulds, CB; Plumb, GW; McLauchlan, WR; Day, AJ; Williamson, G. Purification of cytosolic betaglucosidase from pig liver and its reactivity towards flavonoid glycosides. *Biochimica et Biophysica Acta,* 1999 1435, 110-116.

Lee, MJ; Rao, YK; Chen, K; Lee, YC; Tzeng, YM. Effect of flavonol glycosides from *Cinnamomum osmophloeum* leaves on adiponectin secretion and phosphorylation of insulin receptor-beta in 3T3-L1 adipocytes. *Journal Ethnopharmacology* 2009 126 1, 79-85.

Li, G; Zeng, X; Xie, Y; Cai, Z; Moore, JC; Yuan, X; Cheng, Z; Ji, G. Pharmacokinetic properties of isorhamnetin, kaempferol and quercetin after oral gavage of total flavones of Hippophae rhamnoides L. in rats using a UPLC-MS method. *Fitoterapia,* 2012 83, 182-191.

Li, W; Yi, S; Wang, Z; Chen, S; Xin, S; Xie, J; Zhao, C. Self-nanoemulsifying drug delivery system of persimmon leaf extract: Optimization and bioavailability studies. *International Journal of Pharmacy*, 2011 420 1, 161-171.

Liu, Q; Li, W; Guo, T; Li, D; Fan, Z; Yan, S. Efficient synthesis of a bisglycosyl kaempferol from *Fagonia taeckholmiana. Chemistry Letters,* 2011 40, 324-325.

Luo, H; Jiang, B; Li, B; Li, Z; Jiang, BH; Chen, YC. Kaempferol nanoparticles achieve strong and selective inhibition of ovarian cancer cell viability. *International Journal of Nanomedicine*, 2012 7 3951-3959.

Lyu, SY; Rhim, JY; Park, WB. Antiherpetic activities of flavonoids against herpes simplex virus type 1 (HSV-1) and type 2 (HSV-2) *in vitro*. *Archives of Pharmacol Research,* 2005 28, 1293-1301.

Martin-Gronert, MS; Ozanne, SE. Metabolic programming of insulin action and secretion. *Diabetes, Obesity and Metabolism*, 2012 Suppl 3, 29-39.

Matsui, T; Kobayashi, M; Hayashida, S; Matsumoto, K. Luteolin, a flavone, does not suppress postprandial glucose absorption through an inhibition of alpha-glucosidase action. *Bioscience Biotechnology & Biochemistry,* 2002 66 3, 689-692.

Moorthy, NSHN; Ramos, MJ; Fernandes, P.A.Studies on α-Glucosidase inhibitors development: magic molecules for the treatment of carbohydrate mediated diseases. *Mini-Reviews in Medicinal Chemistry*, 2012 12, 713-720.

Mototake, K; Shimokoriyama, M. A flavone glycoside of *Celastrus orbiculata*. *Acta Phytochimica*, 1949 15, 229-321.

Muhammad, A; Guerrero-Analco, JA; Martineau, LC; Musallam, L; Madiraju, P; Nachar, A; Saleem, A; Haddad, PS; Arnason, JT. Antidiabetic compounds from *Sarracenia purpurea* used traditionally by the Eeyou Istchee Cree First Nation. *Journal Natural Product,* 2012 75, 1284-1288.

Nakaoki, M. *Journal of the Pharmaceutical Society of Japan,* 1957 77, 108.

Németh, K; Plumb, GW; Berrin, JG; Juge, N; Jacob, R; Naim, HY; Williamson, G; Swallow, DM; Kroon, PA. Deglycosylation by small intestinal epithelial cell beta-glucosidases is a critical step in the absorption and metabolism of dietary flavonoid glycosides in humans. *European Journal of Nutrition,* 2003 42, 29-42.

Nomura, M; Takahashi, T; Nagata, N; Tsutsumi, K; Kobayashi, S; Akiba, T; Yokogawa, K; Moritani, S; Miyamoto, K. Inhibitory mechanisms of flavonoids on insulin-stimulated glucose uptake in MC3T3-G2/PA6 adipose cells. *Biological & Pharmaceutical Bulletin,* 2008 31 1403-1409.

Olivier, JC. Drug Transport to Brain with Targeted Nanoparticles. *NeuroRx*, 2005 2, 108–119.

Paganga, G; Rice-Evans, CA. The identification of flavonoids as glycosides in human plasma. *FEBS Lett*ers 1997 401 1,78-82.

Pandey, KB; Rizvi, SI. Plant polyphenols as dietary antioxidants in human health and disease. *Oxidative Medicine and Cellular Longevity*, 2009 2 5, 270-278.

Parveen, Z; Deng, Y; Saeed, MK; Dai, R; Ahamad, W; Yu,YH. Antiinflammatory and analgesic activities of Thesium chinense Turcz extracts and its major flavonoids, kaempferol and kaempferol-3-*O*-glucoside. *Yakugaku Zasshi*, 2007 127, 1275-1279.

Pereira, DF; Cazarolli, LH; Lavado, C; Mengatto, V; Figueiredo, MSRB; Guedes, A; Pizzolatti, MG; Silva, FRMB. Effects of flavonoids on α-glucosidase activity: Potential targets for glucose homeostasis. *Nutrition*, 2011 27 11-12, 1161-1167.

Perkin, AG. *Journal of the Chemical Society*, 1907 91, 433.

Pizzolatti, MG; Cunha Jr., A; Szpoganicz, B; de Sousa, E. Flavonóides glicosilados das folhas e flores de *Bauhinia Forficata* (Leguminosae). *Química Nova*, 2003 26 4, 466-469.

Regasini, LO; Vellosa, JCR; Silva, DHS; Furlan, M; de Oliveira, OMM; Khalil, NM; Brunetti, IL; Young, MCM; Barreiro, EJ; Bolzani, VS. Flavonols from *Pterogyne nitens* and their evaluation as myeloperoxidase inhibitors. *Phytochemistry,* 2008 69, 1739-1744.

Ren, G; Hou, J; Fang, Q; Sun, H; Liu, X; Zhang, L; Wang, PG. Synthesis of flavonol 3-*O*-glycoside by UGT78D1. *Glycoconjugate Journal,* 2012 29, 425-432.

Rho, HS; Ahn, SM; Lee, BC; Kim, MK; Ghimeray, AK; Jin, CW; Cho, DH. Changes in flavonoid content and tyrosinase inhibitory activity in kenaf leaf extract after far-infrared treatment. *Bioorganic & Medicinal Chemistry Letters,* 2010 20, 7534-7536.

Roach, PJ. Glycogen and its metabolism, *Current Molecular Medicine*, 2002 02, 101-120.

Rodriguez, P; González-Mujica, F; Bermúdez, J; Hasegawa, M. Inhibition of glucose intestinal absorption by kaempferol 3-*O*-α-rhamnoside purified from *Bauhinia megalandra* leaves. *Fitoterapia,* 2010 81, 1220-1223.

Saltiel, AR; Kahn, CR.Insulin signalling and the regulation of glucose and lipid metabolism. *Nature,* 2001 414 6865, 799-806.

Scalbert, A; Manach, C; Morand, C; Remesy, C. Dietary polyphenols and the prevention of diseases. *Critical Reviews in Food Science and Nutrition*, 2005 45, 287-306.

Shechter, Y; Goldwaser, I; Mironchik, M; Fridkin, M; Gefel, D. Historic perspective and recent developments on the insulin-like actions of vanadium; toward developing vanadium-based drugs for diabetes. *Coordination Chemistry Reviews,* 2003 237, 3-11.

Shirosaki, M; Goto, Y; Hirooka, S; Masuda,H; Koyama, T; Yazawa, K. Peach leaf contains multiflorin a as a potent inhibitor of glucose absorption in the small intestine in mice. *Biological & Pharmaceutical Bulletin,* 2012a 35 8, 1264-1268.

Shirosaki, M; Koyama, T; Yazawa, K. Supressive effect of peach leaf extract on glucose absorption from the small intestine of mice. *Bioscience, Biotechnology and Biochemistry*, 2012b 76, 89-94.

Silva, FRMB; Szpoganicz, B; Pizzolatti, MG; Willrich, MAV; De Sousa, E. Acute effect of *Bauhinia forficata* on serum glucose levels in normal and alloxan-induced diabetic rats. *Journal of Ethnopharmacology,* 2002 83, 33-37.

Soppimath, KS; Aminabhavi, TM; Kulkarni, AR; Rudzinski, WE. Biodegradable polymeric nanoparticles as drug delivery devices. *Journal of Controlled Release*, 2001 70, 1–20.

Syed, NA; Khandelwal, RL. Reciprocal regulation of glycogenphosphorylase and glycogen synthase by insulin involving phosphatidylinositol-3 kinase and protein phosphatase-1 in HepG2 cells. *Molecular and Cellular Biochemistry*, 2000 211, 123-136.

Tapas, AR; Sakarkar, DM; Kakde, RB. Flavonoids as nutraceuticals: a review. *Tropical Journal of Pharmaceutical Research,* 2008 7 3, 1089-1099.

Tatsimo, SJ; Tamokou, Jde D; Havyarimana, L; Csupor, D; Forgo, P; Hohmann, J; Kuiate, JR; Tane, P. Antimicrobial and antioxidant activity of kaempferol rhamnoside derivatives from *Bryophyllum pinnatum. BMC Research Notes*, 2012 5, 158.

Thompson, KH; Chiles, J; Yuen, VG; Tse, J; McNeill, JH; Orvig, C. Comparison of anti-hyperglycemic effect amongst vanadium, molybdenum and other metal maltol complexes. *Journal of Inorganic Biochemistry,* 2004 98, 683-690.

Urgaonkar, S; Shaw, JT. Synthesis of kaempferitrin. *The Journal of Organic Chemistry*, 2007 72, 4582-4585.

Vaidyanathan, JB; Walle, T. Cellular uptake and efflux of the tea flavonoid (-) epicatechin-3-gallate in the human intestinal cell line Caco-2. *Journal of Pharmacology and Experimental Therapies,* 2003 307, 745-752.

Valente, LMM; Bizarri, CHB; Liechocki, S; Barboza, RS; da Paixão, D; Almeida, MBS; Benevides, PJC; Magalhães, A; Siani, AC. Kaempferitrin from *Uncaria guianensis* (Rubiaceae) and its potential as a chemical marker for the species. *Journal of the Brazilian Chemical Society*, 2009 20 6, 1041-1045.

Veit, M; Pauli, GF. Major flavonoids from *Arabidopsis thaliana* leaves. *Journal Natural Products*, 1999 62, 1301-1303.

Vishnu Prasad, CN; Suma Mohan, S; Banerji, A; Gopalakrishnapillai, A. Kaempferitrin inhibits GLUT4 translocation and glucose uptake in 3T3-L1 adipocytes. *Biochemical and Biophysical Research Communications,* 2009 380 1, 39-43.

Vogt, T; Jones, P. Glycosyltransferases in plant natural product synthesis: characterization of a supergene family. *Trends in Plant Science,* 2000 5 9, 380-386.

Walle, UK; Galijatovic, A; Walle, T. Transport of the flavonoid chrysin and its conjugated metabolites by the human intestinal cell line Caco-2. *Biochemical Pharmacology,* 1999 58, 431-438.

Wang, H; Dub, YJ; Song, HC. α-Glucosidase and α-amylase inhibitory activities of guava leaves. *Food Chemistry,* 2010 123, 6-13.

Wang, Q-J; Lou, Z-P; Wang, Y-S; He, L._Study on chemical constituents from Lpomoea pes-caprae (Linn.) sweet III. *Chinese Pharmaceutical Journal,* 2008 43 1, 20-22.

Yang, W; Sun, J; Yang, Z; Han, W; Zhang, WD; Yu, B. Efficient synthesis of kaempferol 3,7-O-bisglycosides via successive. *Tetrahedron Letters* 2012 53, 2773–2776.

Weng, C-J; Yen, G-C. Flavonoids, a ubiquitous dietary phenolic subclass, exert extensive in vitro anti-invasive and in vivo anti-metastatic activities. *Cancer Metastasis Review,* 2012 31, 323-351.

William, B; Liechty, WB; David, R; Kryscio, DR; Brandon, V; Slaughter, BV; Peppas NA. Polymers for Drug Delivery Systems. *Annual Review of Chemical and Biomolecular Engineering,* 2010 1, 149-173.

Williamson, G; Manach, C. Bioavailability and bioefficacy of polyphenols in humans II. Review of 93 intervention studies. *American Journal of Clinical Nutrition,* 2005 81 (Suppl 1), 243S-255S.

Winkel-Shirley, B. Biosynthesis of flavonoids and effects of stress. *Current Opinion on Plant Biology*, 2002 5, 218-223.

Winkel-Shirley, B. Flavonoid biosynthesis. A colorful model for genetics, biochemistry, cell biology, and biotechnology. *Plant Physiology*, 2001 126, 485-493.

Won, HM; Kwon, YS; Lee, JH; Kim, CM. Chemical constituents of the leaves of *Weigela subsesillis*. *Korean Journal of Pharmacognosy,* 2004 35 1, 1-5.

Wright, EM; Loo, DDF; Hirayama, BA. Biology of human sodium glucose transporters. *Physiological Reviews,* 2011 91, 733-794.

Yadav, D; Suri, S; Choudhary, AA; Sikender, M; Hemant; Beg, NM; Garg, V; Ahmad, A; Asif, M.. Novel approach: Herbal remedies and natural products in pharmaceutical science as nano drug delivery systems. *International Journal of Pharmacy and Technology,* 2011 3, 3092-3116.

Yamamoto, H; Yan, K; Ieda, K; Tanaka, T; Iinuma, M; Mizuno, M. Flavonol glycosides production cultures of *Vancouveria hexandra* in cell suspension. *Phytochemistry*, 1993 33 4, 841-846.

Yamasaki, K; Hishiki, R; Kato, E; Kawabata, J. Study of kaempferol glycoside as an insulin mimic reveals glycon to be the key active structure. *ACS Medicinal Chemistry Letter,* 2011 2, 17-21.

Yang, F; Su, Y-F; Bi, Y-P; Xu, J; Zhu, Z-Q; Tu, G-Z ; Gao, X-M. Three new kaempferol glycosides from *Cardamine leucantha*. *Helvetica Chimica Acta,* 2010 93 3, 536-541.

Yang, W; Sun, J; Yang, Z; Han, W; Zhang, W-D; Yu, B. Efficient synthesis of kaempferol 3,7-*O*-bisglycosides via successive. *Tetrahedron Letters*, 2012 53, 2773-2776.

Yang, X-W; Zhang, J-Y; Qian, Z-M._New natural saponins from fruits of *Momordica grosvenori. Chinese Traditional and Herbal Drugs,* 2008 39 6, 810-813.

Zanatta, L; Rosso, A; Folador, P; Figueiredo, MSRB; Pizzolatti, MG; Leite, LD; Silva, FRMB. Insulinomimetic Effect of kaempferol 3-neohesperidoside on the rat soleus muscle. *Journal Natural Products,* 2008 71, 532-535.

Zang, Y; Sato, H; Igarashi, K. Anti-diabetic effects of a kaempferol glycoside-rich fraction from unripe soybean (Edamame, Glycine max L. Merrill. 'Jindai') leaves on KK-A(y) mice. *Bioscience Biotechnology & Biochemistry,* 2011 75 9, 1677-1684.

Zhang, S; Morris, ME. Effects of flavonoids biochanin A, morin, phloretin, and silymarin on P-glycoprotein-mediated transport. *The Journal of Pharmaceutical and Experimental Therapeutics,* 2003 304, 1258-1267.

Zhang, Y; Liu, D. Flavonol kaempferol improves chronic hyperglycemia-impaired pancreatic beta-cell viability and insulin secretory function. *European Journal of Pharmacology,* 2011 670, 325-332.

Zhang, Y; Zhen, W; Maechler, P; Liu, D. Small molecule kaempferol modulates PDX-1 protein expression and subsequently promotes pancreatic β-cell survival and function via CREB. *The Journal of Nutritional Biochemistry*, 2012.

Zhang, Y; Zhang, TT. Studies on the chemical constituents from the stem and leaves of *Tagetes erecta. Zhong Yao Cai*, 2010 33 9, 1412-1414.

In: Kaempferol
Editors: Garion Villers and Yves Fougere
ISBN: 978-1-62618-515-9

Chapter 3

Distribution of Kaempferol Glycosides and their Function in Plants

Tsukasa Iwashina[1,*] and Yoshinori Murai[2]
[1]Department of Botany, National Museum of Nature and Scieince, Tsukuba, Japan
[2]Department of Chemistry, Hiyoshi Campus, Keio University, Yokohama, Japan

Abstract

About 350 kaempferol glycosides are found in plant kingdom. Their glycosides are distributed to most plant species from mosses and ferns to vascular plants. Though the flavonoids involving kaempferol were considered to be waste products of plant metabolism in the early days of 20th century, they now may be to serve as many functions and activities in land plants. Copigmentation with the anthocyanins was well-known as a function of the kaempferol glycosides in the flowers, especially blue and purple flowers. Additionally, it is known that kaempferol, which can hardly be seen by human eyes, also act as the important pollinator attractants in addition to visible anthocyanins. Recently, other functions, e.g., feeding attractants (kaempferol 3-*O*-xylosyl-(1→2)-galactoside), allelopathy (kaempferol itself), phytoalexins (kaempferol 3-*O*-robinobioside-7-*O*-rhamnoside) of naturally occurring kaempferol, were reported by some authors. In this review, we describe the distribution of kaempferol glycosides and their functions and activities in plants.

Introduction

Kaempferol and quercetin glycosides are most popular flavonoids in plant kingdom. Kaempferol was first isolated in plants as the aglycone of a glycoside in the flowers of *Delphinium consolida* (Gripenberg, 1962). As the glycosides, kaempferol 3-*O*-sophoroside

* E-mail: iwashina@kahaku.go.jp.

and then 3-*O*-rutinoside have been found in the fruits of *Sophora japonica* by Rabaté and Dussy in 1938 (Harborne and Williams, 1975) and whole plants of *Calystegia japonica* by Hukuti in 1939 (Hukuti, 1939).

About 350 glycosides are now reported from various plants (Williams, 2006). Though kaempferol occurs as 3-*O*-mono- or 3-*O*-diglycosides in most cases, it is also present as 7-, 4′-, 3,7- and rarely 5-, 3, 4′-, 7,4′-*O*-glycosides and so on. Recently, many acylated kaempferol glycosides were identified by NMR and MS spectra. In this review, we describe kaempferol glycosides and their distribution and functions, such as copigmentation with anthocyanins in flowers, pollinator and feeding attractants, allelopathy and phytoalexins, in plants.

Distribution of Kaempferol in Plants

Bryophytes

Kaempferol glycosides are widespreadly distributed in plant kingdom. In the bryophytes, their occurrence is relatively rare and kaempferol 3-*O*-rhamnosylglucoside-7-*O*-rhamnoside, 3,7-di-*O*-rhamnoside and 3-*O*-rhamnosylglucoside were isolated from the Hepaticae, *Hymenophyton leptopodum* (Markham et al., 1976). Kaempferol glycoside was found in *Takakia* species, which are thought to be among the most primitive of extent liverworts, and identified as kaempferol 3-*O*-glucoside-7-*O*-xyloside, together with flavone glycosides, takakin 8-*O*-glucuronide and 8-*O*-xylosylglucoside, luteolin 6-*C*-arabinoside-8-*C*-pentoside and tricetin *C*-glycosides (Markham and Porter, 1979). However, they are not reported from the algae and fungi.

Pteridophytes

In the ferns, kaempferol glycosides are common. Its 3-*O*-glucoside, 3-*O*-glucuronide and 3-*O*-rutinoside were found in many *Adiantum* species (Adiantaceae) with quercetin glycosides (Hasegawa and Akabori, 1968; Akabori and Hasegawa, 1969; Cooper-Driver and Swain, 1977; Iwashina et al., 1995a). *Asplenium* species (Aspleniaceae) synthesize many kaempferol glycosides as major flavonoids. Thirty-three *Asplenium* species have been surveyed for flavonoids (Iwashina and Matsumoto, 2011). Of their species, kaempferol glycosides were found in 21 species as major compounds except for flavone *O*-glycoside containing species, *A. boreale*, *A. normale*, *A. shimurae*, *A. oligophlebium* (Iwashina et al., 1990; Iwashina and Matsumoto, 1994) and *A. belangeri* (Umikalsom et al., 1994), and xanthone, mangiferin containing species, *A. adiantum-nigrum*, *A. bolearicum* (Richardson and Lorenz-Liburnau, 1982), *A. bradley*, *A. montanum*, *A. pinnatifida* and *A. stotleri* (Smith and Harborne, 1971). In this genus, rare acylated kaempferol glycosides have been isolated from a few species, i.e., 3-*O*-rhamnoside-7-*O*-[(6′′-feruloylglucosyl)-(1→3)-rhamnoside] from *A. prolongatum* (Mizuno et al., 1990a), 3-*O*-[glucosyl-(1→3)-(2′′-caffeoylglucoside)]-7-*O*-rhamnoside from *A. scolopendrium* (Mizuno et al., 1990b) and 3-*O*-[(6′′-*E*-

caffeoylglucosyl)-(1→3)-glucoside]-7-*O*-glucoside from *A. ruta-muraria* (Iwashina et al., unpublished data).

In the genus *Dryopteris* (Dryopteridaceae), rare kaempferol 5-*O*-glucoside and 3,5-di-*O*-glucoside were isolated from *D. dickinsii* (Fuchino et al., 1997). The occurrence of 5-*O*-glycosides and 3,5-di-*O*-glycosides of flavonols are comparatively unusual in plants, since the 5-hydroxyl group forms hydrogen bonding with the adjacent 4-carbonyl group (Harborene, 1967). Four rare acylated glycosides, 3-*O*-(4′′-acetylrhamnoside)-7-*O*-rhamnoside, 3-*O*-(2′′,3′′-di-acetylrhamnoside)-7-*O*-rhamnoside, 3-*O*-(2′′,4′′-di-acetylrhamnoside)-7-*O*-rhamnoside and 3-*O*-(3′′,4′′-di-acetylrhamnoside)-7-*O*-rhamnoside, were found in the rhizomes of *Dryopteris crassirhizoma* (Min et al., 2001). Other rare acylated glycosides, 3-*O*-(3′′-*E*-*p*-coumaroyl-6′′-*Z*-*p*-coumaroylglucoside), 3-*O*-(3′′-*Z*-*p*-coumaroyl-6′′-*E*-*p*-coumaroylglucoside), 3-*O*-(3′′-*E*-*p*-coumaroyl-6′′-*Z*-feruloylglucoside), 3-*O*-(3′′-*Z*-*p*-coumaroyl-6′′-*E*-feruloylglucoside), 3-*O*-(3′′-*E*-*p*-coumaroylglucoside) and 3-*O*-(3′′-*E*-*p*-coumaroyl)-[6′′-*E*-(4-*O*-(1-(4-hydroxy-3-methyl-phenyl)-1,3-dihydroxy-isoprenyl)-feruloylglucoside)], were isolated from the fronds of *Stenochlaena palustris* (Pteridaceae) (Liu et al., 1999). Various kaempferol glycosides have been reported from fern allies, *Equisetum* species (Saleh et al., 1972; Saleh, 1975) including rare 3-*O*-rutinoside-7-*O*-sophoroside (Veit et al., 1993).

Gymnosperms

Kaempferol glycosides are also popular compounds in the gymnosperms. In *Ginkgo biloba* (Ginkgoaceae), many acylated and non-acylated glycosides were reported from the leaves. Rare glycosides, 3-*O*-[(6′′-*p*-coumaroylglucosyl)-(1→2)-rhamnoside]-7-*O*-glucoside and 3-O-[(6′′-*p*-coumaroylglucosyl)-(1→2)-rhamnoside] (Tang et al., 2001), 3-*O*-[6′′-(7′′′′-glucosyl-*p*-coumaroyl)-glucosyl]-(1→2)-rhamnoside (Hasler et al., 1992), 3-*O*-glucosyl-(1→2)-rhamnoside (Markham et al., 1992; Hasler et al., 1992), 3-*O*-(6′′-*p*-coumaroylglucosyl)-(1→4)-rhamnoside (Nasr et al., 1986), together with common 3-*O*-glucoside, 3-*O*-rhamnoside and 3-*O*-rutinoside (Yoshitama, 1997). A kaempferol glycoside, 3-*O*-(6′′-acetylglucoside), was found in the needles of *Picea abies* (Pinaceae) (Slimestad et al., 1995).

Dicotyledons

Kaempferol glycosides are detected from all over flowering plants, dicotyledons and monocotyledons. In comparative primitive family Ranunculaceae, acylated and non-acylated kaempferol polyglycosides are frequently found. 3-*O*-Glucosyl-(1→3)-glucoside-7-*O*-glucoside was found in the seeds of *Nigella sativa* (Merfort et al., 1997). Kaempferol 3-*O*-sophoroside-7-*O*-glucoside and 3-*O*-sophoroside-7-*O*-rhamnoside were detected in *Callianthemum alatavicum* in the Tien Shan Mountains, and *C. hondoense* in Japanese endemic alpine plant, respectively (Asakawa et al., 2010). Recently, a new tetragycoside, 3-*O*-{rhamnosyl-(1→2)-[rhamnosyl-(1→6)-glucoside]}-7-*O*-glucoside, was isolated from the flowers of *Clematis* cultivar (Sakaguchi et al., 2012), and may be distribute to almost *Clematis* cultivars (Sakaguchi et al., unpublished data). Acylated polyglycosides were found in other Ranunculaceae genera, i.e., 3-*O*-{[6′′-*E*-caffeoylglucosyl]-(1→2)-glucoside}-7-*O*-

rhamnoside and 3-*O*-[(6′′-*E*-*p*-coumaroylglucosyl)-(1→2)-glucoside]-7-*O*-rhamnoside from the flowers of *Aconitum napellus* subsp. *tauricum* (Fico et al., 2001b), 3-*O*-glucoside-7-*O*-[6′′-*E*-caffeoylglucosyl-(1→3)-rhamnoside] and 3-*O*-glucoside-7-*O*-[(6′′-*E*-*p*-coumaroylglucosyl-(1→3)-rhamnoside] from the flowers of *Aconitum napellus* subsp. *neomontanum* (Fico et al., 2001a), 3-*O*-(6′′-acetylglucoside)-7-*O*-rhamnoside from the flowers of *Delphinium fontanum* (Özden et al., 1998).

Kaempferol polyglycosides are frequently reported from the Brassicaceae. Ten acylated and non-acylated kaempferol polyglycosides, 7-*O*-glucosyl-(1→3)-[glucosyl-(1→6)-glucoside], 3-*O*-sophoroside-7-*O*-glucoside, 3-*O*-sophoroside-7-*O*-[glucosyl-(1→4)-glucoside], 3-*O*-[(2′′-*E*-feruloylglucosyl)-(1→2)-glucoside]-7-*O*-glucoside, 3-*O*-[(2′′-*E*-feruloylglucosyl)-(1→2)-glucoside]-7-*O*-[glucosyl-(1→4)-glucoside], 3-*O*-[(2′′-*E*-caffeoylglucosyl)-(1→2)-glucoside]-7-*O*-glucoside, 3-*O*-[(2′′-*E*-caffeoylglucosyl)-(1→2)-glucoside]-7-*O*-[glucosyl-(1→4)-glucoside], 3-*O*-[(2′′-*E*-*p*-coumaroylglucosyl)-(1→2)-glucoside]-7-*O*-glucoside, 3-*O*-[(2′′-*E*-sinapoylglucosyl)-(1→2)-glucoside]-7-*O*-glucoside and 3-*O*-[(2′′-*E*-sinapoylglucosyl)-(1→2)-glucoside]-7-*O*-[glucosyl-(1→4)-glucoside], were isolated from two *Brassica* species, *B. juncea* and cabbage (*B. oleracea*) (Nielsen et al., 1993, 1998; Kim et al., 2002). Other polyglycosides, kaempferol 3-*O*-neohesperidoside-4′-*O*-glucoside and 3-*O*-neohesperidoside-7,4′-di-*O*-glucoside and 7-*O*-(6′′-caffeoylglucosyl)-(1→4)-rhamnoside, were found in other Brassicaceae species, *Pseuderucaria clavata* (Sharaf et al., 1997) and *Rorippa indica* (Lin and Kuo, 1995), respectively. Recently, two polyglycosylated kaempferols were isolated from the flowers of *Aubrieta* ×*cultorum* and identified as kaempferol 3-*O*-[glucosyl-(1→2)-(6′′-*E*-sinapoylglucoside)]-7-*O*-[glucosyl-(1→4)-(6′′-*E*-sinapoylglucoside)] and 3-*O*-[glucosyl-(1→2)-(6′′-*E*-sinapoylglucoside)] (Tatsuzawa et al., 2012). An acylated kaempferol glycoside, 3-*O*-(6′′-*E*-feruloylglucoside)-7-*O*-cellobioside-4′-*O*-glucoside was isolated from the flowers of *Heliophila coronopifolia* (Saito et al., 2011).

The Leguminosae is the very abundant family for flavonoids, especially isoflavonoids. Rare kaempferol glycosides were also reported from many species, i.e., 7-*O*-alloside, 3-*O*-arabinoside-7-*O*-rhamnoside, 3,7-di-*O*-arabinoside, 3-*O*-rhamnoside-7-*O*-arabinoside, 3-*O*-[rhamnosyl-(1→2)-galactoside]-7-*O*-arabinofuranoside and 3-*O*-[rhamnosyl-(1→6)-galactoside]-7-*O*-arabinofuranoside from *Indigofera hebepetala* (Hasan et al., 1994, 1996), 7-*O*-glucosyl-(1→4)-xyloside from the seeds of *Cratalaria labrnifolia* (Yadava and Singh, 1993), 3-*O*-rhamnosyl-(1→2)-rhamnoside from the flowers of *Cassia hirsuta* (Rao et al., 1999), 3-*O*-rhamnosyl-(1→2)-glucosyl-(1→6)-galactoside from *Cassia marginata* (Chauhan et al., 2002), 3-*O*-rhamnosyl-(1→6)-glucosyl-(1→6)-galactoside from the leaves of *Albizia lebbeck* (El-Mousallamy, 1998), 3-*O*-[glucosyl-(1→2)-galactoside]-7-*O*-glucoside and 3-*O*-[glucosyl-(1→2)-(6′′-acetylgalactoside)]-7-*O*-glucoside from *Trigonella foenum-graecum* (Han et al., 2001), 3-*O*-rhamnosyl-(1→2)-rhamnosyl-(1→6)-galactoside and 3-*O*-(4′′-acetylrhamnosyl)-(1→2)-rhamnosyl-(1→6)-galactoside from *Galega officinalis* (Champavier et al., 2000), 3-*O*-rhamnosyl-(1→2)-[rhamnosyl-(1→6)-glucoside] and 3-*O*-rhamnosyl-(1→2)-(6′′-malonylglucoside) from the flowers of *Clitoria ternatea* (Kazuma et al., 2003), 3-*O*-robinobioside-7-*O*-rhamnoside, 3-*O*-rutinoside-7-*O*-glucuronide and 3-*O*-isorhamninoside-7-*O*-rhamnoside from *Vigna*, *Phaseolus* and *Macroptilium* spp. (Zallocchi and Pomilio, 1994), and 3-*O*-[rhamnosyl-(1→6)-glucosyl-(1→2)-glucoside]-7-*O*-rhamnoside from the seeds of *Sophora japonica* (Wang et al., 2003). The genus *Astragalus* also synthesizes many kaempferol polyglycosides. 3-*O*-[Rhamnosyl-(1→2)-galactoside]-7-*O*-rhamnoside and 3-*O*-rhamnosyl-(1→6)-[rhamnosyl-(1→2)-galactoside]-7-*O*-rhamnoside were isolated from the rare species, *Astragalus shikokianus* (Yahara et al., 2000), and 3-*O*-rhamnosyl-(1→3)-[rhamnosyl-(1→4)-galactoside] was reported from *A. tana* (Alaniya and Chkadua, 2000). Six rare glycosides, kaempferol 3-*O*-xylosyl-(1→3)-rhamnosyl-(1→6)-galactoside, 3-*O*-[xylosyl-(1→3)-rhamnosyl]-(1→6)-[rhamnosyl-(1→2)-(3′′-*E*-*p*-coumaroylgalactoside)], 3-*O*-[xylosyl-(1→3)-rhamnosyl]-(1→6)-[rhamnosyl-(1→2)-(4′′-*E*-*p*-coumaroylgalactoside)], 3-*O*-[xylosyl-(1→3)-rhamnosyl]-(1→6)-[rhamnosyl-(1→2)-(3′′-*E*-feruloylgalactoside)] and 3-*O*-[xylosyl-(1→3)-rhamnosyl]-(1→6)-[rhamnosyl-(1→2)-(4′′-*E*-feruloylgalactoside)], were isolated from the leaves of *Astragalus caprinus* (Semmar et al., 2002a, 2002b).

Glycine max, which is one of the important crops in the world, also synthesizes kaempferol glycosides. However, though kaempferol polyglycosides, e.g., 3-*O*-glucosyl-(1→2)-[rhamnosyl-(1→6)-galactoside], 3-*O*-rhamnosyl-(1→4)-[glucosyl-(1→6)-galactoside] and 3-*O*-glucosyl-(1→2)-[rhamnosyl-(1→6)-glucoside] and so on are found in the leaves (Murai et al., 2013; Buttery and Buzzell, 1975), the major flavonol in the flowers of all cultivars surveyed is kaempferol 3-*O*-gentiobioside, together with minor kaempferol 3-*O*-rutinoside, 3-*O*-glucoside and 7-*O*-glucoside (Iwashina et al., 2007).

As kaempferol glycosides of other dicotyledon families, three acylated glycosides, 3-*O*-glucosyl-(1→6)-[rhamnosyl-(1→3)]-(2′′-*E*-*p*-coumaroyl)-glucoside, 3-*O*-{glucosyl-(1→6)-[rhamnosyl-(1→3)]-(2′′-*E*-*p*-coumaroyl)-glucoside}-7-*O*-[rhamnosyl-(1→3)-rhamnosyl-(1→3)-(4′′-*Z*-*p*-coumaroylrhamnoside)] and 3-*O*-{rhamnosyl-(1→6)-[rhamnosyl-(1→3)]-(2′′-*E*-*p*-coumaroylglucoside)}-7-*O*-[rhamnosyl-(1→3)-rhamnosyl-(4′′-*E*-*p*-coumaroylrhamnoside)], were reported from *Planchonia grandis* (Lecythidaceae) (Crublet et al., 2003).

From green tea (*Camellia sinensis*), kaempferol 3-*O*-rhamnosyl-(1→3)-(4′′-acetylrhamnosyl)-(1→6)-glucoside and 3-*O*-rhamnosyl-(1→3)-rhamnosyl-(1→6)-galactoside are reported, together with various catechins (Lakenbrink et al., 2000). Recently, a rare tetraglycoside, 3-*O*-[glucosyl-(1→3)-rhamnosyl]-(1→6)-[xylosyl-(1→2)-glucoside] was isolated from the flowers of *Camellia uraku* with 3-*O*-glucosyl-(1→3)-rhamnosyl-(1→6)-glucoside (Tanikawa et al., 2011).

Monocotyledons

In many monocotyledon families such as the Araceae and Poaceae, major flavonoids are flavone *O*- and *C*-glycosides. However, in some genera of the Liliaceae sensu lato, flavonol glycosides are major compounds. In the genus *Allium*, many kaempferol polyglycosides, e.g., 3-*O*-glucoside-7-*O*-sophoroside, 3-*O*-glucosyl-(1→4)-rhamnosyl-(1→6)-glucoside and 3-*O*-neohesperidoside-7-*O*-glucoside, were reported from *Allium neapolitanum* with common glycosides (Carotenuto et al., 1997), 3-*O*-sophoroside-7-*O*-glucuronide from *A. cepa* (Urushibara et al., 1992), 3-*O*-(2′′-*E*-feluroylgalactosyl)-(1→4)-glucoside and 3-*O*-(2′′-*E*-feluroylglucosyl)-(1→6)-glucoside from *A. porrum* (Fattorusso et al., 2001), and 3-*O*-neohesperidoside-7-*O*-(2′′-*E*-*p*-coumaroylglucoside), 3-*O*-neohesperidoside-7-*O*-[glucosyl-(1→3)-(2′′-*E*-*p*-coumaroylglucoside)] and 3-*O*-neohesperidoside-7-*O*-(2′′-*E*-feruloyl-glucoside) from *A. ursianum* (Carotenuto et al., 1996).

Though major flavonoids of many Iridaceous genera, especially in the genus *Iris*, are *C*-glycosylflavones (Iwashina and Ootani, 1998), those of *Gladiolus* and *Crocus* species are flavonol *O*-glycosides.

Four kaempferol glycosides, 3-*O*-sophoroside, 3-*O*-rutinoside, 3-*O*-glucoside and 3-*O*-glucosyl-(1→2)-rhamnoside were identified as major flavonoids from the flowers of *Gladiolus* cultivars (Takemura et al., 2005; Takemura, 2008). Kaempferol 3-*O*-[glucosyl-(1→2)-rhamnoside]-7-*O*-(6′′-acetylglucoside), 3-*O*-sophoroside-7-*O*-glucoside, 3-*O*-glucosyl-(1→2)-[glucosyl-(1→3)-rhamnoside], 3-*O*-[glucosyl-(1→2)-rhamnoside]-7-*O*-glucoside, 7-*O*-sophoroside and 3,4′-di-*O*-glucoside were isolated from the stigmas of *Crocus* species and cultivars (Nørbæk and Kondo, 1999; Nørbæk et al., 1999; Straubinger et al., 1997).

Flavonol *C*-glycosides are very rare flavonoids. Their occurrence is limited in a few species in nature (Iwashina et al., 2005a). Kaempferol 6-*C*-glucoside was first isolated from the heartwoods of *Zelkova* spp. (Ulmaceae, dicotyledon) (Hayashi et al., 1987). Thereafter, the compound was found in the leaves of Japanese endemic species, *Japonolirion osense* (Petrosaviaceae, Liliaceae sensu lato), together with a new flavonoid, kaempferol 6-*C*-glucoside-3-*O*-glucoside, which is a compound of flavonol *C*-glycoside-*O*-glycoside in nature together with another quercetin 6-*C*-glucoside-3-*O*-glucoside (Iwashina et al., 2005a).

By development of various spectral methods such as NMR and MS, and chromatography for isolation, the chemical structures of the comparatively complicated flavonols were determined, so that many kaempferol polyglycosides are recently found. However, the plant species which were not surveyed for flavonoids are very much, especially wild plants. In the future, many kaempferol glycosides, especially acylated and/or polyglycosylated ones, may be isolate.

Function of Kaempferol Glycosides in Plants

Flower Color

As described above, many kaempferol glycosides have been isolated from the plant species (Tables 1 and 2). However, the information of their function in plants is meager.

Copigmentation in flowers is well known as the function of kaempferol glycosides. In almost purple and blue flowers, major pigments are anthocyanins such as cyanidin, delphinidin, petunidin and malvidin. However, the colors of their anthocyanins are red to red purple.

Their pigments are co-existent with certain colorless substances such as flavones, flavonols or organic acids, so that their compounds are bathochromically shifted the colors of anthocyanins to purple and bluish colors.

Kaempferol glycosides are important copigment substances in the flowers of some plants. In bluish flower cultivar of *Gladiolus*, it was shown that kaempferol 3-*O*-rutinoside and 3-*O*-glucosyl-(1→2)-rhamnoside act as the copigment substances to malvidin 3,5-di-*O*-glucoside. However, another major glycoside, kaempferol 3-*O*-sophoroside did not function as the copigment (Takemura et al., 2005; Takemura, 2008).

Many Himalayan blue poppy, *Meconopsis* spp. put blue or deep purple flowers. Their blue flowers were reported to be colored by co-existence of anthocyanin, kaempferol 3-*O*-gentiobioside and 3-*O*-glucosyl-(1→6)-galactoside, and metals, Fe and Mg (Takeda et al., 1996; Tanaka et al., 2001; Yoshida et al., 2006).

Table 1. Kaempferol glycosides detected until 1989

Monoglycosides

3-*O*-galactoside, 3-*O*-glucoside, 3-*O*-glucuronide, 3-*O*-(6´´-ethylglucuronide), 3-*O*-rhamnoside, 5-*O*-rhamnoside, 7-*O*-arabinoside, 7-*O*-glucoside, 7-*O*-xyloside, 4´-*O*-glucoside, 4´-*O*-rhamnoside

Diglycosides

3-*O*-apiosyl-(1→2)-galactoside, 3-*O*-apiosyl-(1→2)-glucoside, 3-*O*-arabinosyl-(1→6)-galactoside, 3-*O*-gentiobioside, 3-*O*-glucosyl-(1→4)-rhamnoside, 3-*O*-glucosyl-(1→6)-galactoside, 3-*O*-rhamnosyl-(1→2)-galactoside, 3-*O*-rhamnosyl-(1→3)-glucoside, 3-*O*-rutinoside, 3-*O*-xylosyl-(1→2)-galactoside, 3-*O*-xylosyl-(1→2)-glucoside, 3-*O*-xylosyl-(1→2)-rhamnoside, 7-*O*-galactosyl-(1→4)-rhamnoside, 7-*O*-neohesperidoside, 3,5-di-*O*-galactoside, 3,7-di-*O*-glucoside, 3-*O*-glucoside-7-*O*-galactoside, 3-*O*-glucoside-7-*O*-glucuronide, 3-*O*-glucuronide-7-*O*-glucoside, 3-*O*-rhamnoside-7-*O*-galacturonide, 3-*O*-rhamnoside-7-*O*-xyloside, 3-*O*-xyloside-7-*O*-glucoside, 3-*O*-xyloside-7-*O*-rhamnoside, 3,4´-di-*O*-xyloside, 3-*O*-galactoside-4´-*O*-glucoside, 3-*O*-rhamnoside-4´-*O*-arabinoside, 3-*O*-rhamnoside-4´-*O*-xyloside, 7,4´-di-*O*-glucoside, 7,4´-di-*O*-rhamnoside

Triglycosides

3-*O*-apiosyl-(1→2)-[rhamnosyl-(1→6)-galactoside], 3-*O*-galactosyl-(1→2)-[rhamnosyl-(1→6)-glucoside], 3-*O*-glucosyl-(1→4)-arabinofuranosyl-(1→2)-arabinoside, 3-*O*-glucosyl-(1→2)-glucosyl-(1→6)-glucoside, 3-*O*-glucosyl-(1→2)-[glucosyl-(1→6)-glucoside], 3-*O*-glucosyl-(1→4)-glucosyl-(1→6)-glucoside, 3-*O*-glucosyl-(1→2)-[rhamnosyl-(1→6)-galactoside], 3-*O*-glucosyl-(1→3)-rhamnosyl-(1→6)-galactoside, 3-*O*-glucosyl-(1→2)-[rhamnosyl-(1→6)-glucoside], 3-*O*-glucosyl-(1→3)-rhamnosyl-(1→6)-glucoside, 3-O-rhamnosyl-(1→2)-glucosyl-(1→3)-glucoside, 3-*O*-rhamnosyl-(1→2)-[glucosyl-(1→6)-glucoside], 3-*O*-rhamnosyl-(1→2)-rhamnosyl-(1→6)-glucoside, 3-*O*-rhamnosyl-(1→2)-[rhamnosyl-(1→6)-glucoside], 3-*O*-rhamnosyl-(1→4)-rhamnosyl-(1→6)-glucoside, 3-*O*-sophorotrioside, 3-*O*-xylosyl-(1→2)-[rhamnosyl-(1→6)-glucoside], 4´-*O*-rhamnosyl-(1→2)-[rhamnosyl-(1→6)-galactoside], 3-*O*-[apiofuranosyl-(1→2)-arabinofuranoside]-7-*O*-rhamnoside,

Table 1. (Continued)

3-*O*-gentiobioside-7-*O*-glucuronide,
3-*O*-gentiobioside-7-*O*-rhamnoside,
3-*O*-glucoside-7-*O*-gentiobioside,
3-*O*-laminaribioside-7-*O*-rhamnoside,
3-*O*-lathyroside-7-*O*-rhamnoside,
3-*O*-[rhamnosyl-(1→2)-galactoside]-7-*O*-glucoside,
3-*O*-[rhamnosyl-(1→4)-rhamnoside]-7-*O*-rhamnoside,
3-*O*-robinobioside-7-*O*-glucoside,
3-*O*-rutinoside-7-*O*-galactoside,
3-*O*-rutinoside-7-*O*-rhamnoside,
3-*O*-sambubioside-7-*O*-glucoside,
3-*O*-rutinoside-4´-*O*-glucoside,
3-*O*-sophoroside-4´-*O*-glucoside,
3-*O*-glucoside-7,4´-di-*O*-rhamnoside

Tetraglycosides
3-*O*-rhamnosyl-(1→6)-[glucosyl-(1→2)-glucoside]-7-*O*-glucoside,
3-*O*-sophorotrioside-7-*O*-glucoside,
3-*O*-sophorotrioside-7-*O*-rhamnoside
3-*O*-gentiobioside-7,4´-di-*O*-glucoside,
3-*O*-neohesperidoside-7,4´-di-*O*-glucoside

***C*-Glycosides**
6-*C*-glucoside,
8-*C*-glucoside

Acylated glycosides
3-*O*-(2´´-acetylarabinosyl)-(1→6)-galactoside,
3-*O*-(4´´-acetyl-6´´-*p*-coumaroylglucoside),
3-*O*-(6´´-acetylglucosyl)-(1→3)-galactoside,
3-*O*-(2´´-acetylrhamnoside),
3-*O*-(3´´-acetylrhamnoside),
3-*O*-(4´´-acetylrhamnoside),
3-*O*-(6´´-caffeoylglucosyl)-(1→2)-galactoside,
3-*O*-(*Z*-cinnamoylglucoside),
3-*Z*/*E*-*p*-coumarate,
3-*O*-(6´´-*p*-coumaroylgalactoside),
3-*O*-(2´´,3´´-di-acetyl-4´´-*p*-coumaroylrhamnoside),
3-*O*-(2´´,4´´-di-*p*-coumaroylglucoside),
3-*O*-glucosyl-(1→3)-(4´´-acetylrhamnosyl)-(1→6)-galactoside,
3-*O*-(6´´-malonylgalactoside),
3-*O*-(6´´-malonylglucoside),
3-*O*-rhamnosyl-(1→4)-(3´´-acetylrhamnosyl)-(1→6)-galactoside,

3-*O*-rhamnosyl-(1→6)-(4´´-*E*-*p*-coumaroylglucoside),
3-*O*-rhamnosyl-(1→6)-(4´´-*Z*-*p*-coumaroylglucoside),
3-*O*-(6´´-succinylglucoside),
7-*O*-(6´´-succinylglucoside),
3-*O*-(3´´-acetylarabinofuranoside)-7-*O*-glucoside,
3-*O*-(6´´-acetylgalactoside)-7-*O*-rhamnoside,
3-*O*-(6´´-acetylglucoside)-7-*O*-glucoside,
3-*O*-(4´´-acetylrhamnoside)-7-*O*-rhamnoside,
3-*O*-(2´´-caffeoylglucoside)-7-*O*-rhamnoside,
3-*O*-[glucosyl-(1→3)-(4´´-caffeoylglucoside)]-7-*O*-rhamnoside,
3-*O*-[rhamnosyl-(1→6)-(4´´-*E*-*p*-coumaroylgalactoside)]-7-*O*-rhamnoside,
3-*O*-[rhamnosyl-(1→6)-(4´´-*Z*-*p*-coumaroylgalactoside)]-7-*O*-rhamnoside,
3-*O*-sophoroside-7-*O*-(2´´-feluroylglucoside),
3-*O*-[xylosyl-(1→3)-(4´´-acetylrhamnoside)]-7-*O*-rhamnoside,
3-*O*-(2´´-feluroylglucoside)-7,4´-di-*O*-glucoside,
3,4´-di-*O*-glucoside-7-*O*-(2´´-feluroylglucoside)

Sulphated glycosides

3-*O*-[glucosyl-(1→6)-(6´´-sulphatoglucoside)],
3-sulphate,
3-*O*-(3´´-sulphatoglucoside),
3-*O*-(6´´-sulphatoglucoside),
7-sulphate,
3,7-di-sulphate,
3-*O*-glucuronide-7-sulphate,
3,7,4´-tri-sulphate

Table 2. Kaempferol glycosides reported between 1990 and 2013

Monoglycosides

3-*O*-alloside

Glaucidium palmatum (fl, lv, Glaucidiaceae) (Iwashina and Ootani, 1990)

3-*O*-arabinofuranoside

Prunus spinosa (fl, lv, Rosaceae) (Olszewska and Wolbiś, 2001, 2002)

3-*O*-arabinoside

Rorippa indica (wp, Brassicaceae) (Lin and Kuo, 1995), *Persea americana* (lv, Lauraceae) (de Almeida et al., 1998), *Persicaria lapathifolia* (wp, Polygonaceae) (Park et al., 1999), *Warburgia ugandensis* (lv, Canellaceae) (Manguro et al., 2003)

3-*O*-xyloside

Prunus spinosa (fl, Rosaceae) (Olszewska and Wolbiś, 2001)

5-*O*-glucoside

Dryopteris dickinsii (fr, Dryopteridaceae) (Fuchino et al., 1997)

5-*O*-glucuronide

Leucanthemum vulgare (lv, Asteraceae) (Williams et al., 2001)

Table 2. (Continued)

7-*O*-alloside

Indigofera hebepetala (lv, Leguminosae) (Hasan et al., 1994)

7-*O*-rhamnoside

Cephalocereus senilis (wp, Cactaceae) (Liu et al., 1994), *Rorippa indica* (wp, Brassicaceae) (Lin and Kuo, 1995), *Sedum telephium* subsp. *maximum* (lv, Crassulaceae) (Mulinacci et al., 1995), *Rhodiala crenulata* (rt, Crassulaceae) (Du and Xie, 1995), *Indigofera hebepetala* (fl, Leguminosae) (Hasan et al., 1996), *Delphinium formosum* (fl, Ranunculaceae) (Özden et al., 1998), *Chenopodium murale* (ap, Chenopodiaceae) (El-Sayed et al., 1999), *Prunus spinosa* (fl, lv, Rosaceae) (Olszewska and Wolbiś, 2001, 2002)

Diglycosides

3-*O*-glucosyl-(1→2)-galactoside

Thevetia peruviana (lv, Apocynaceae) (Abe et al., 1995), *Trigonella foenum-graecum* (st, Leguminosae) (Han et al., 2001)

3-*O*-glucosyl-(1→2)-rhamnoside

Ginkgo biloba (lv, Ginkgoaceae) (Markham et al., 1992; Hasler et al., 1992), *Crocus speciosus*, *C. antalyensis* (sg, Iridaceae) (Nørbæk and Kondo, 1999)

3-*O*-neohesperidoside

Allium ursinum (wp, Alliaceae) (Carotenuto et al., 1996), *Allium neapolitanum* (wp, Alliaceae) (Carotenuto et al., 1997), *Crocus speciosus*, *C. antalyensis* (sg, Iridaceae) (Nørbæk and Kondo, 1999)

3-*O*-rhamnosyl-(1→2)-arabinofuranoside

Artabotrys hexapetalus (lv, Annonaceae) (Li et al., 1997)

3-*O*-rhamnosyl-(1→2)-rhamnoside

Cassia hirsuta (fl, Leguminosae) (Rao et al., 1999)

3-*O*-robinobioside

Astragalus tana (ap, Leguminosae) (Alaniya and Chkadua, 2000), *Silphium perfoliatum* (lv, Asteraceae) (El-Sayed et al., 2002), *Cassytha filiformis* (wp, Lauraceae) (Murai et al., 2008)

3-*O*-sambubioside

Triantha japonica (lv, Liliaceae) (Iwashina et al., 2013)

3-*O*-sophoroside

Petunia hybrid 'Mitchell' (lv, Solanaceae) (Bloor et al., 1998), *Crocus speciosus*, *C. antalyensis* (sg, Iridaceae) (Nørbæk and Kondo, 1999), *Triantha japonica* (lv, Liliaceae) (Iwashina et al., 2013)

3-*O*-[xylosyl-(1→2)-glucoside]

Warburgia ugandensis (lv, Canellaceae) (Manguro et al., 2003)

7-*O*-glucosyl-(1→3)-rhamnoside

Rhodiola crenulata (rt, Crassulaceae) (Du and Xie, 1995)

7-*O*-glucosyl-(1→4)-xyloside

Crotalaria laburnifolia (sd, Leguminosae) (Yadava and Singh, 1993)

7-*O*-sophoroside

Crocus sativus (sg, Iridaceae) (Straubinger et al., 1997)
3,5-di-*O*-glucoside
Dryopteris dickinsii (fr, Dryopteridaceae) (Fuchino et al., 1997)
3-*O*-arabinofuranoside-7-*O*-rhamnoside
Prunus spinosa (lv, Rosaceae) (Olszewska and Wolbiś, 2002)
3-*O*-arabinoside-7-*O*-rhamnoside
Indigofera hebepetala (lv, Leguminosae) (Hasan et al., 1994, 1996), *Asplenium trichomanes-ramosum* (= *A. viride*) (fr, Aspleniaceae) (Iwashina et al., 1995b)
3,7-di-*O*-arabinoside
Indigofera hebepetala (lv, Leguminosae) (Hasan et al., 1994),
3,7-di-*O*-rhamnoside
Sedum telephium subsp. *maximum* (lv, Crassulaceae) (Mulinacci et al., 1995), *Asplenium trichomanes-ramosum* (= *A. viride*) (fr, Aspleniaceae) (Iwashina et al., 1995b), *Siraitia grosvenori* (ft, Cucurbitaceae) (Si et al., 1994), *Chenopodium murale* (ap, Chenopodiaceae) (El-Sayed et al., 1999), *Prunus spinosa* (lv, Rosaceae) (Olszewska and Wolbiś, 2002)
3-*O*-galactoside-7-*O*-rhamnoside
Asarum canadense (lv, Aristolochiaceae) (Iwashina and Kitajima, 2000)
3-*O*-glucoside-7-*O*-arabinoside
Callianthemum hondoense (lv, Ranunculaceae) (Asakawa et al., 2010)
3-*O*-glucoside-7-*O*-rhamnoside
Asplenium trichomanes-ramosum (= *A. viride*) (fr, Aspleniaceae) (Iwashina et al., 1995b), *Rorippa indica* (wp, Brassicaceae) (Lin and Kuo, 1995), *Delphinium formosum* (fl, Ranunculaceae) (Özden et al., 1998), *Chenopodium murale* (ap, Chenopodiaceae) (Gohar et al., 2000), *Callianthemum hondoense* (lv, Ranunculaceae) (Asakawa et al., 2010)
3-*O*-glucoside-7-*O*-xyloside
Callianthemum hondoense (lv, Ranunculaceae) (Asakawa et al., 2010)
3-*O*-rhamnoside-7-*O*-arabinoside
Indigofera hebepetala (lv, Leguminosae) (Hasan et al., 1994, 1996)
3-*O*-rhamnoside-7-*O*-glucoside
Chenopodium murale (ap, Chenopodiaceae) (El-Sayed et al., 1999)
3,4´-di-*O*-glucoside
Crocus speciosus, *C. antalyensis* (sg, Iridaceae) (Nørbæk and Kondo, 1999)
3-*O*-rhamnoside-4´-*O*-galactoside
Warburgia ugandensis (lv, Canellaceae) (Manguro et al., 2003)
7-*O*-rhamnoside-4´-*O*-glucoside
Pteridium aquilinum (ap, Dennstaedtiaceae) (Imperato, 1998)

Triglycosides
3-*O*-glucosyl-(1→2)-galactosyl-(1→2)-glucoside
Nigella sativa (sd, Ranunculaceae) (Merfort et al., 1997)
3-*O*-glucosyl-(1→2)-[glucosyl-(1→6)-galactoside]
Tofieldia nuda (lv, Liliaceae) (Iwashina et al., 2013)
3-*O*-glucosyl-(1→2)-[glucosyl-(1→6)-glucoside]
Triantha japonica (lv, Liliaceae) (Iwashina et al., 2013)

Table 2. (Continued)

3-O-glucosyl-(1→2)-[glucosyl-(1→3)-rhamnoside]
Crocus speciosus, *C. antalyensis* (sg, Iridaceae) (Nørbæk and Kondo, 1999)
3-*O*-glucosyl-(1→2)-[rhamnosyl-(1→6)-galactoside]
Glycine max (lv, Leguminosae) (Murai et al., 2013)
3-*O*-glucosyl-(1→2)-rhamnosyl-(1→6)-glucoside
Glycine max (lv, Leguminosae) (Murai et al., 2013)
3-*O*-glucosyl-(1→3)-rhamnosyl-(1→6)-glucoside
Camellia uraku (fl, Theaceae) (Tanikawa et al., 2011)
3-*O*-glucosyl-(1→4)-rhamnosyl-(1→6)-glucoside
Allium neapolitanum (wp, Alliaceae) (Carotenuto et al., 1997)
3-*O*-isorhamninoside
Vigna spp., *Phaseolus lunatus*, *Macroptilium lathyroides* (ap, Leguminosae) (Zallocchi and Pomilio, 1994)
3-*O*-rhamnosyl-(1→2)-glucosyl-(1→6)-galactoside
Cassia marginata (st, Leguminosae) (Chauhan et al., 2002)
3-*O*-rhamnosyl-(1→4)-[glucosyl-(1→6)-galactoside]
Glycine max (lv, Leguminosae) (Murai et al., 2013)
3-*O*-rhamnosyl-(1→6)-glucosyl-(1→6)-galactoside
Albizia lebbeck (lv, Leguminosae) (El-Mousallamy, 1998)
3-*O*-rhamnosyl-(1→2)-[glucosyl-(1→3)-glucoside]
Impatiens balsamina (fl, Balsaminaceae) (Fukumoto et al., 1994)
3-*O*-rhamnosyl-(1→2)-[rhamnosyl-(1→6)-galactoside]
Galega officinalis (ap, Leguminosae) (Champavier et al., 2000)
3-*O*-rhamnosyl-(1→3)-[rhamnosyl-(1→4)-galactoside]
Astragalus tana (ap, Leguminosae) (Alaniya and Chkadua, 2000)
3-*O*-rhamnosyl-(1→3)-rhamnosyl-(1→6)-galactoside
Camellia sinensis (green tea, Theaceae) (Lakenbrink et al., 2000)
3-*O*-rhamnosyl-(1→4)-[rhamnosyl-(1→6)-galactoside]
Glycine max (lv, Leguminosae) (Murai et al., 2013)
3-*O*-rhamnosyl-(1→2)-[rhamnosyl-(1→6)-glucoside]
Nicotiana spp. (fl, Solanaceae) (Snook et al., 1992), *Elaeagnus bockii* (lv, Elaeagnaceae) (Cao et al., 2001), *Clitoria ternatea* (fl, Leguminosae) (Kazuma et al., 2003)
3-*O*-rhamnosyl-(1→3)-rhamnosyl-(1→6)-glucoside
Camellia sinensis (green tea, Theaceae) (Lakenbrink et al., 2000)
3-*O*-xylosyl-(1→2)-[glucosyl-(1→6)-glucoside]
Triantha japonica (lv, Liliaceae) (Iwashina et al., 2013)
3-*O*-xylosyl-(1→6)-glucosyl-(1→2)-rhamnoside
Helicia nilagirica (lv, Proteaceae) (Wu et al., 2002)
3-*O*-xylosyl-(1→3)-rhamnosyl-(1→6)-galactoside
Astragalus caprinus (lv, Leguminosae) (Semmar et al., 2002a)
7-*O*-glucosyl-(1→3)-[glucosyl-(1→6)-glucoside]
Brassica juncea (lv, Brassicaceae) (Kim et al., 2002)

4′-*O*-rhamnosyl-(1→3)-rhamnosyl-(1→6)-galactoside
Rhamnus thymifolius (ft, Rhamnaceae) (Satake et al., 1993)
3-*O*-apiofuranoside-7-*O*-robinobioside
Silphium perfoliatum (lv, Asteraceae) (El-Sayed et al., 2002)
3-*O*-[apiosyl-(1→4)-rhamnoside]-7-*O*-rhamnoside
Chenopodium murale (ap, Chenopodiaceae) (Gohar et al., 2000)
3-*O*-gentiobioside-7-*O*-glucoside
Asarum spp. (lv, Aristolochiaceae) (Iwashina et al., 2005b)
3-*O*-glucoside-7-*O*-sophoroside
Allium neapolitanum (wp, Alliaceae) (Carotenuto et al., 1997)
3-*O*-[glucosyl-(1→4)-galactoside]-7-*O*-arabinofuranoside
Corchorus depressus (wp, Tiliaceae) (Zahid et al., 2002)
3-*O*-[glucosyl-(1→6)-galactoside]-7-*O*-arabinofuranoside
Corchorus depressus (wp, Tiliaceae) (Zahid et al., 2002)
3-*O*-[glucosyl-(1→2)-galactoside]-7-*O*-glucoside
Nicotiana spp. (fl, Solanaceae) (Snook et al., 1992), *Trigonella foenum-graecum* (st, Leguminosae) (Han et al., 2001)
3-*O*-[glucosyl-(1→3)-glucoside]-7-*O*-glucoside
Asplenium altajense (fr, Aspleniaceae) (Iwashina et al., unpublished data)
3-*O*-[glucosyl-(1→2)-rhamnoside]-7-*O*-glucoside
Crocus chrysanthus-biflorus cultivars 'Eye-catcher', 'Spring Pearl' (sg, Iridaceae) Nørbæk et al., 1999), *Mentha lavandulacea* (ap, Lamiaceae) (El-Desoky et al., 2001), *Reseda muricata* (lv, Resedaceae) (El-Sayed et al., 2001)
3-*O*-neohesperidoside-7-*O*-glucoside
Allium neapolitanum (wp, Alliaceae) (Carotenuto et al., 1997)
3-*O*-neohesperidoside-7-*O*-rhamnoside
Sedum telephium subsp. *maximum* (lv, Crassulaceae) (Mulinacci et al., 1995)
3-*O*-rhamnoside-7-*O*-[glucosyl-(1→2)-rhamnoside]
Siraitia grosvenori (ft, Cucurbitaceae) (Si et al., 1994)
3-*O*-rhamnoside-7-*O*-rutinoside
Asarum caulescens (lv, Aristolochiaceae) (Iwashina et al., 2005b)
3-*O*-rhamnoside-7-*O*-[xylosyl-(1→2)-rhamnoside]
Chenopodium murale (ap, Chenopodiaceae) (El-Sayed et al., 1999)
3-*O*-[rhamnosyl-(1→2)-galactoside]-7-*O*-arabinofuranoside
Indigofera hebepetala (fl, Leguminosae) (Hasan et al., 1996)
3-*O*-[rhamnosyl-(1→6)-galactoside]-7-*O*-arabinofuranoside
Indigofera hebepetala (fl, Leguminosae) (Hasan et al., 1996)
3-*O*-[rhamnosyl-(1→2)-galactoside]-7-*O*-rhamnoside
Astragalus shikokianus (ap, Leguminosae) (Yahara et al., 2000)
3-*O*-[rhamnosyl-(1→6)-galactoside]-7-*O*-rhamnoside
Cephalocereus senilis (wp, Cactaceae) (Liu et al., 1994)
3-*O*-[xylosyl-(1→2)-rhamnoside]-7-*O*-rhamnoside
Kalanchoe streptantha (lv, Crassulaceae) (Costa et al., 1996)
3-*O*-[xylosyl-(1→4)-rhamnoside]-7-*O*-rhamnoside
Chenopodium murale (ap, Chenopodiaceae) (Gohar et al., 2000)
3-*O*-robinobioside-7-*O*-rhamnoside

Table 2. (Continued)

Vigna spp., *Phaseolus* spp., *Macroptilium* spp. (ap, Leguminosae) (Zallocchi and Pomilio, 1994)

3-*O*-rutinoside-7-*O*-glucoside

Asarum takaoi (lv, Aristolochiaceae) (Iwashina et al., 2005b), *Clematis* cultivar (fl, Ranunculaceae) (Sakaguchi et al., 2012)

3-*O*-rutinoside-7-*O*-glucuronide

Vigna spp., *Macroptilium* spp. (ap, Leguminosae) (Zallocchi and Pomilio, 1994)

3-*O*-sophoroside-7-*O*-arabinoside

Callianthemum alatavicum (lv, Ranunculaceae) (Asakawa et al., 2010)

3-*O*-sophoroside-7-*O*-glucoside

Nicotiana spp. (fl, Solanaceae) (Snook et al., 1992), *Brassica oleracea* (cabbage), *B. napus*, *B. juncea*(lv, Brassicaceae) (Nielsen et al., 1993; Olsson et al., 1998; Kim et al., 2002)

3-*O*-sophoroside-7-*O*-glucuronide

Allium cepa (lv, Alliaceae) (Urushibara et al., 1992)

3-*O*-sophoroside-7-*O*-rhamnoside

Callianthemum hondoense (lv, Ranunculaceae) (Asakawa et al., 2010)

3-*O*-sophoroside-7-*O*-glucoside

Crocus sativus (sg, Iridaceae) (Straubinger et al., 1997)

3-*O*-gentiobioside-4´-*O*-glucoside

Asplenium incisum (fr, Aspleniaceae) (Iwashina et al., 2000)

3-*O*-neohesperidoside-4´-*O*-glucoside

Pseuderucaria clavata (fl, Brassicaeae) (Sharaf et al., 1997)

3-*O*-rhamnoside-7,4´-di-*O*-galactoside

Warburgia ugandensis (lv, Canellaceae) (Manguro et al., 2003)

3,7,4´-tri-*O*-glucoside

Warburgia ugandensis (lv, Canellaceae) (Manguro et al., 2003)

Tetraglycosides

3-*O*-[glucosyl-(1→3)-rhamnosyl]-(1→2)-[rhamnosyl-(1→6)-galactoside]

Maytenus aquifolium (lv, Celastraceae) (Sannomiya et al., 1998)

3-*O*-[glucosyl-(1→3)-rhamnosyl]-(1→6)-[xylosyl-(1→2)-glucoside]

Camellia uraku (fl, Theaceae) (Tanikawa et al., 2011)

3-*O*-[xylosyl-(1→3)-rhamnosyl]-(1→6)-[rhamnosyl-(1→2)-galactoside]

Astragalus caprinus (lv, Leguminosae) (Semmar et al., 2002b)

3-*O*-{glucosyl-(1→2)-[rhamnosyl-(1→6)-galactoside]}-7-*O*-rhamnoside

Cephalocereus senilis (wp, Cactaceae) (Liu et al., 1994)

3-*O*-isorhamninoside-7-*O*-rhamnoside

Vigna spp. (ap, Leguminosae) (Zallocchi and Pomilio, 1994)

3-*O*-[rhamnosyl-(1→6)-glucosyl-(1→2)-glucoside]-7-*O*-rhamnoside

Sophora japonica (sd, Leguminosae) (Wang, et al., 2003)

3-*O*-{rhamnosyl-(1→6)-[rhamnosyl-(1→2)-galactoside]}-7-*O*-rhamnoside

Astragalus shikokianus (ap, Leguminosae) (Yahara et al., 2000)

3-*O*-{rhamnosyl-(1→2)-[rhamnosyl-(1→6)-glucoside]}-7-*O*-glucoside
Clematis cultivar (fl, Ranunculaceae) (Sakaguchi et al., 2012)
3-*O*-rutinoside-7-*O*-sophoroside
Equisetum palustre (wp, Equisetaceae) (Veit et al., 1993)
3-*O*-sophoroside-7-*O*-[glucosyl-(1→4)-glucoside]
Brassica oleracea (cabbage) (lv, Brassicaceae) (Nielsen et al., 1998)
3-*O*-neohesperidoside-7,4´-di-*O*-glucoside
Pseuderucaria clavata (fl, Brassicaceae) (Sharaf et al., 1997)

C-Glycoside
6-*C*-glucoside-3-*O*-glucoside
Japonolirion osense (lv, Petrosaviaceae) (Iwashina et al., 2005a)

Acylated glycosides
3-*O*-(3´´-acetylarabinosyl)-(1→6)-glucoside
Thalictrum atriplex (ap, Ranunculaceae) (Guangyao et al., 2000)
3-*O*-(3´´-acetyl-6´´-*p*-coumaroylglucoside)
Anaphalis aureo-punctata (wp, Asteraceae) (Wu et al., 2003)
3-*O*-(6´´-acetylglucoside)
Picea abies (lv, Pinaceae) (Slimestad et al., 1995)
3-*O*-(4´´-acetylrhamnosyl)-(1→2)-[glucosyl-(1→6)-glucoside]
Ammi majus (ap, Apiaceae) (Singab, 1998)
3-*O*-(4´´-acetylrhamnosyl)-(1→2)-[rhamnosyl-(1→6)-galactoside]
Galega officinalis (ap, Leguminosae) (Champavier et al., 2000)
3-*O*-(6´´-caffeoylglucoside)
Pteridium aquilinum (ap, Dennstaedtiaceae) (Imperato and Minutiello, 1997)
3-*O*-(2´´-caffeoylglucosyl)-(1→2)-glucoside
Petunia hybrid 'Mitchell' (lv, Solanaceae) (Bloor et al., 1998)
3-*O*-(2´´-*E*-*p*-coumaroyl-6´´-acetylglucoside)
Quercus dentata (lv, Fagaceae) (Zhou et al., 2001)
3-*O*-(2´´-*E*-*p*-coumaroylarabinofuranoside)
Prunus spinosa (fl, Rosaceae) (Olszewska and Wolbiś, 2001)
3-*O*-(2´´-*E*-*p*-coumaroyl-6´´-*E*-*p*-coumaroylglucoside)
Quercus canariensis (lv, Fagaceae) (Romussi et al., 1992)
3-*O*-(3´´-*E*-*p*-coumaroyl-6´´-*Z*-*p*-coumaroylglucoside)
Stenochlaena palustris (fr, Pteridaeae) (Liu et al., 1999)
3-*O*-(3´´-*Z*-*p*-coumaroyl-6´´-*E*-*p*-coumaroylglucoside)
Stenochlaena palustris (fr, Pteridaeae) (Liu et al., 1999)
3-*O*-(3´´-*E*-*p*-coumaroyl-6´´-*E*-feruloylglucoside)
Stenochlaena palustris (fr, Pteridaeae) (Liu et al., 1999)
3-*O*-(3´´-*Z*-*p*-coumaroyl-6´´-*E*-feruloylglucoside)
Stenochlaena palustris (fr, Pteridaeae) (Liu et al., 1999)
3-*O*-(2´´-*E*-*p*-coumaroylglucoside)
Eringium campestre (ap, Umbelliferae) (Hohmann et al., 1997), *Allium porrum* (bl, Alliaceae) (Fattorusso et al., 2001)
3-*O*-(2´´-*Z*-*p*-coumaroylglucoside)

Table 2. (Continued)

Eringium campestre (ap, Umbelliferae) (Hohmann et al., 1997)
3-*O*-(3´´-*E*-*p*-coumaroylglucoside)
Stenochlaena palustris (fr, Pteridaeae) (Liu et al., 1999)
3-*O*-(4´´-*E*-*p*-coumaroylglucoside)
Elaeagnus bockii (lv, Elaeagnaceae) (Cao et al., 2001)
3-*O*-(6´´-*E*-*p*-coumaroylglucoside)
Platanus acerifolia (bd, Platanaceae) (Kaouadji et al., 1993), *Platanus orientalis* (bd, Platanaceae) (Mitrokotsa et al., 1993), *Stenochlaena palustris* (fr, Pteridaeae) (Liu et al., 1999), *Elaeagnus bockii* (lv, Elaeagnaceae) (Cao et al., 2001), *Quercus dentata* (lv, Fagaceae) (Zhou et al., 2001)
3-*O*-[(6´´-*p*-coumaroylglucosyl)-(1→2)-rhamnoside]
Ginkgo bioba (lv, Ginkgoaceae) (Tang et al., 2001)
3-*O*-(3´´-*E*-*p*-coumaroyl)-[6´´-*E*-(4-*O*-(1-(4-hydroxy-3-methyl-phenyl)-1,3-dihydroxy-isoprenyl)-feruloyl)]-glucoside
Stenochlaena palustris (fr, Pteridaeae) (Liu et al., 1999)
3-*O*-(2´´-*E*-*p*-coumaroylrhamnoside)
Platanus acerifolia (bd, Platanaceae) (Kaouadji et al., 1993), *Platanus orientalis* (bd, Platanaceae) (Mitrokotsa et al., 1993), *Laurus nobilis* (lv, Lauraceae) (Fiorini et al., 1998)
3-*O*-(2´´-*Z*-*p*-coumaroylrhamnoside)
Platanus acerifolia (bd, Platanaceae) (Kaouadji et al., 1993)
3-*O*-(2´´,3´´-di-acetyl-4´´-*Z*-*p*-coumaroyl-6´´-*E*-*p*-coumaroylglucoside)
Quercus canariensis (lv, Fagaceae) (Romussi et al., 1992)
3-*O*-(2´´,6´´-di-*Z*-*p*-coumaroyl-3´´,4´´-di-acetylglucoside)
Quercus dentata (lv, Fagaceae) (Zhou et al., 2001)
3-*O*-(2´´,3´´-di-*E*-*p*-coumaroylglucoside)
Platanus acerifolia (bd, Platanaceae) (Kaouadji et al., 1993)
3-*O*-(2´´,6´´-di-*E*-*p*-coumaroylglucoside)
Quercus canariensis, Q. ilex, Q. dentata (lv, Fagaceae) (Romussi et al., 1992; Skaltsa et al., 1994; Zhou et al., 2001)
3-*O*-(3´´, 6´´-di-*E*-*p*-coumaroylglucoside)
Stenochlaena palustris (fr, Pteridaeae) (Liu et al., 1999)
3-*O*-(3´´,6´´-di-*Z*-*p*-coumaroylglucoside)
Stenochlaena palustris (fr, Pteridaeae) (Liu et al., 1999)
3-*O*-(2´´,3´´-di-*E*-*p*-coumaroylrhamnoside)
Platanus orientalis (bd, Platanaceae) (Mitrokotsa et al., 1993)
3-*O*-(2´´,4´´-di-*E*-*p*-coumaroylrhamnoside)
Pentachondra pulila (Iv, Epacidaceae) (Bloor, 1995), *Laurus nobilis* (Iv, Lauraceae) (Fiorini et al., 1998)
3-*O*-(2´´, 4´´-di-*Z*-*p*-coumaroylrhamnoside)
Laurus nobilis (lv, Lauraceae) (Fiorini et al., 1998)
3-*O*-(2´´, 4´´-di-*E*-*p*-coumaroylrhamnoside)
Laurus nobilis (lv, Lauraceae) (Fiorini et al., 1998)

3-*O*-(2´´,6´´-di-galloylglucoside)
Loropetalum chinense (lv, Hamamelidaceae) (Liu et al., 1997a, 1997b)
3-*O*-(5´´-feruloylapioside)
Pteridium aqulinum (ap, Dennstaedtiaceae) (Imperato, 1996)
3-*O*-(2´´-*E*-feruloylgalactosyl)-(1→4)-glucoside
Allium porrum (bl, Alliaceae) (Fattorusso et al., 2001)
3-*O*-(6´´-feruloylglucoside)
Polylepis incana (lv, Rosaceae) (Catalano et al., 1994)
3-*O*-(6´´-*E*-feruloylglucosyl)-(1→2)-galactoside
Hedyotis diffusa (ap, Rubiaceae) (Lu et al., 2000)
3-*O*-(2´´-feruloylglucosyl)-(1→2)-glucoside
Petunia hybrid 'Mitchell' (lv, Solanaceae) (Bloor et al., 1998)
3-*O*-(2´´-*E*-feruloylglucosyl)-(1→6)-glucoside
Allium porrum (bl, Alliaceae) (Fattorusso et al., 2001)
3-*O*-[2´´-*E*-feruloylglucosyl-(1→2)-glucoside]-7-*O*-glucoside
Brassica oleracea (cabbage), *B. juncea* (lv, Brassicaceae) (Nielsen et al., 1993; Kim et al., 2002)
3-*O*-[2´´-*E*-feruloylglucosyl-(1→2)-glucoside]-7-*O*-[glucosyl-(1→4)-glucoside]
Brassica oleracea (cabbage) (lv, Brassicaceae) (Nielsen et al., 1998)
3-*O*-(2´´-feruloylglucosyl)-(1→2)-(6´´-malonylglucoside)
Petunia hybrid 'Mitchell' (lv, Solanaceae) (Bloor et al., 1998)
3-*O*-(6´´-feruloylglucosyl)-(1→2)-[rhamnosyl-(1→6)-galactoside]
Brunfelsia grandiflora subsp. *grandiflora* (ap, Solanaceae) (Brunner et al., 2000)
3-*O*-(2´´-galloylarabinoside)
Eucalyptus rostrata (lv, Myrtaceae) (Okamura et al., 1993)
3-*O*-(6´´-galloylgalactoside)
Penphis acidula (lv, Lythraceae) (Masuda et al., 2001)
3-*O*-(2´´-galloylglucoside)
Loropetalum chinense (lv, Hamamelidaceae) (Liu et al., 1997a, 1997b)
3-*O*-(6´´-galloylglucoside)
Quercus canariensis (lv, Fagaceae) (Romussi et al., 1992), *Loropetalum chinense* (lv, Hamamelidaceae) (Liu et al., 1997a, 1997b), *Penphis acidula* (lv, Lythraceae) (Masuda et al., 2001)
3-*O*-[6´´-(7´´´´-glucosyl-*p*-coumaroyl)-glucosyl]-(1→2)-rhamnoside
Ginkgo biloba (lv, Ginkgoaceae) (Hasler et al., 1992)
3-*O*-[glucosyl-(1→3)-(6´´-*p*-coumaroylrhamnosyl)]-(1→6)-[rhamnosyl-(1→2)-galactoside]
Lysimachia capillipes (wp, Primulaceae) (Xie et al., 2002)
3-*O*-glucosyl-(1→2)-(6´´-*E*-sinapoylglucoside)
Aubrieta×*cultorum* (fl, Brassicaceae) (Tatsuzawa et al., 2012)
3-*O*-(6´´-*p*-hydroxybenzoylgalactoside)
Persicaria lapathifolia (wp, Polygonaceae) (Park et al., 1999)
3-*O*-[6´´-(3-hydroxy-3-methyl-glutaryl)-glucoside]
Citrus aurantifolia (cc, Rutaceae) (Berhow et al., 1994)

Table 2. (Continued)

3-O-rhamnosyl-(1→3)-(4´´-acetylrhamnosyl)-(1→6)-galactoside
Rhamnus thymifolius (ft, Rhamnaceae) (Satake et al., 1993)
3-*O*-rhamnosyl-(1→3)-(4´´-acetylrhamnosyl)-(1→6)-glucoside
Camellia sinensis (green tea, Theaceae) (Lakenbrink et al., 2000)
3-*O*-rhamnosyl-(1→6)-(2´´-*E*-*p*-coumaroylglucoside)
Alibertia sessilis (lv, Rubiaceae) (Olea et al., 1997)
3-*O*-rhamnosyl-(1→2)-(6´´-malonylglucoside)
Clitoria ternatea (fl, Leguminosae) (Kazuma et al., 2003)
3-*O*-(6´´-sinapoylglucosyl)-(1→2)-galactoside
Thevetia peruviana (lv, Apocynaceae) (Abe et al., 1995)
3-*O*-(6´´-sinapoylglucosyl)-(1→2)-[glucosyl-(1→4)-galactoside]
Thevetia peruviana (lv, Apocynaceae) (Abe et al., 1995)
3-*O*-(6´´-sinapoylglucosyl)-(1→2)-[(2´´-sinapoylglucosyl)-(1→4)-galactoside]
Thevetia peruviana (lv, Apocynaceae) (Abe et al., 1995)
3-*O*-(2´´,3´´,4´´-tri-acetylarabinosyl)-(1→6)-glucoside
Calluna vulgaris (fl, Ericaceae) (Simon et al., 1993)
3-*O*-(2´´,3´´,5´´-tri-acetylarabinosyl)-(1→6)-glucoside
Calluna vulgaris (fl, Ericaceae) (Allais et al., 1991)
3-*O*-[xylosyl-(1→3)-rhamnosyl]-(1→6)-[rhamnosyl-(1→2)-(3´´-*E*-*p*-coymaroyl-galactoside)]
Astragalus caprinus (lv, Leguminosae) (Semmar et al., 2002b)
3-*O*-[xylosyl-(1→3)-rhamnosyl]-(1→6)-[rhamnosyl-(1→2)-(4´´-*E*-*p*-coymaroyl-galactoside)]
Astragalus caprinus (lv, Leguminosae) (Semmar et al., 2002b)
3-*O*-[xylosyl-(1→3)-rhamnosyl]-(1→6)-[rhamnosyl-(1→2)-(3´´-*E*-feruloylgalactoside)
Astragalus caprinus (lv, Leguminosae) (Semmar et al., 2002a)
3-*O*-[xylosyl-(1→3)-rhamnosyl]-(1→6)-[rhamnosyl-(1→2)-(4´´-*E*-feruloylgalactoside)]
Astragalus caprinus (lv, Leguminosae) (Semmar et al., 2002a)
7-*O*-(6´´-*p*-coumaroylglucoside)
Buddleja coriacea (ap, Buddlejaceae) (Kubo and Yokokawa, 1992)
7-*O*-(6´´-caffeoylglucosyl)-(1→4)-rhamnoside
Rorippa indica (wp, Brassicaceae) (Lin and Kuo, 1995)
3-*O*-(6´´-acetylglucoside)-7-*O*-rhamnoside
Delphinium formosum (fl, Ranunculaceae) (Özden et al., 1998)
3-*O*-(4´´-acetylrhamnoside)-7-*O*-rhamnoside
Dryopteris crassirhizoma (rz, Dryopteridaceae) (Min et al., 2001)
3-*O*-apiofuranoside-7-*O*-[rhamnosyl-(1→6)-(2´´-*E*-caffeoylgalactoside)]
Silphium perfoliatum (lv, Asteraceae) (El-Sayed et al., 2002)
3-*O*-[(2´´-*E*-caffeoylglucosyl)-(1→2)-glucoside]-7-*O*-glucoside
Brassica oleracea (cabbage) (lv, Brassicaceae) (Nielsen et al., 1993)
3-*O*-[(6´´-*E*-caffeoylglucosyl)-(1→3)-glucoside]-7-*O*-glucoside
Asplenium ruta-muraria, *A. altajense* (fr, Aspleniaceae) (Iwashina et al., unpublished data)

3-*O*-[(2´´-*E*-caffeoylglucosyl)-(1→2)-glucoside]-7-*O*-[glucosyl-(1→4)-glucoside]
Brassica oleracea (cabbage) (lv, Brassicaceae) (Nielsen et al., 1998)
3-*O*-[(6´´-*E*-caffeoylglucosyl)-(1→2)-glucoside]-7-*O*-rhamnoside
Aconitum napellus subsp. *tauricum* (fl, Ranunculaceae) (Fico et al., 2001b)
3-*O*-(2´´-*E*-*p*-coumaroylarabinofuranoside)-7-*O*-rhamnoside
Prunus spinosa (lv, Rosaceae) (Olszewska and Wolbiś, 2002)
3-*O*-(6´´-*E*-*p*-coumaroylglucoside)-7-*O*-glucoside
Lotus polyphyllos (wp, Leguminosae) (El-Mousallamy et al., 2002)
3-*O*-[2´´-*E*-*p*-coumaroylglucosyl-(1→2)-glucoside]-7-*O*-glucoside
Brassica oleracea (cabbage) (lv, Brassicaceae) (Nielsen et al., 1993)
3-*O*-[(6´´-*E*-*p*-coumaloylglucosyl)-(1→2)-glucoside]-7-*O*-rhamnoside
Aconitum napellus subsp. *tauricum* (fl, Ranunculaceae) (Fico et al., 2001b)
3-*O*-[4´´-*p*-coumaroylglucosyl-(1→2)-rhamnoside]-7-*O*-glucoside
Mentha lavandulacea (ap, Lamiaceae) (El-Desoky et al., 2001)
3-*O*-[6´´-*p*-coumaroylglucosyl-(1→2)-rhamnoside]-7-*O*-glucoside
Ginkgo bioba (lv, Ginkgoaceae) (Tang et al., 2001), *Mentha lavandulacea* (ap, Lamiaceae) (El-Desoky et al., 2001)
3-*O*-(2´´-*p*-coumaroylrhamnoside)-7-*O*-rhamnoside
Cheilanthes fragrans (ap, Parkeriaceae) (Imperato, 1992)
3-*O*-(4´´,6´´-di-acetylglucoside)-7-*O*-rhamnoside
Delphinium formosum (fl, Ranunculaceae) (Özden et al., 1998)
3-*O*-(2´´,3´´-di-acetylrhamnoside)-7-*O*-rhamnoside
Dryopteris crassirhizoma (rz, Dryopteridaceae) (Min et al., 2001)
3-*O*-(2´´,4´´-di-acetylrhamnoside)-7-*O*-rhamnoside
Dryopteris crassirhizoma (rz, Dryopteridaceae) (Min et al., 2001)
3-*O*-(3´´,4´´-di-acetylrhamnoside)-7-*O*-rhamnoside
Dryopteris crassirhizoma (rz, Dryopteridaceae) (Min et al., 2001)
3-*O*-glucoside-7-*O*-[(6´´-*E*-caffeoylglucosyl)-(1→3)-rhamnoside]
Aconitum napellus subsp. *neomontanum* (fl, Ranunculaceae) (Fico et al., 2001a)
3-*O*-glucoside-7-*O*-[(6´´-*E*-*p*-coumaroylglucosyl)-(1→3)-rhamnoside]
Aconitum napellus subsp. *neomontanum* (fl, Ranunculaceae) (Fico et al., 2001a)
3-*O*-[glucosyl-(1→2)-(6´´-acetylgalactoside)]-7-*O*-glucoside
Trigonella foenum-graecum (st, Leguminosae) (Han et al., 2001)
3-*O*-[glucosyl-(1→3)-(2´´-caffeoylglucoside)]-7-*O*-rhamnoside
Asplenium scolopendrium (fr, Aspleniaceae) (Mizuno et al., 1990b)
3-*O*-[glucosyl-(1→2)-rhamnoside]-7-*O*-(6´´-acetylglucoside)
Crocus chrysanthus-biflorus 'Eye-cather', 'Spring Pearl' (sg, Iridaceae) Nørbæk et al., 1999)
3-*O*-[glucosyl-(1→2)-rhamnoside]-7-*O*-(6´´-*E*-*p*-coumaroylglucoside)
Reseda muricata (lv, Resedaceae) (El-Sayed et al., 2001)
3-*O*-[glucosyl-(1→2)-rhamnoside]-7-*O*-(6´´-malonylglucoside)
Crocus chrysanthus-biflorus 'Eye-cather', 'Spring Pearl' (sg, Iridaceae) Nørbæk et al., 1999)
3-*O*-{glucosyl-(1→6)-[rhamnosyl-(1→3)]-(2´´-*E*-*p*-coumaroylglucoside)}-7-*O*-[rhamnosyl-(1→3)-rhamnosyl-(1→3)-(4´´-*E*-*p*-coumaroylrhamnoside)]

Table 2. (Continued)

Planchonia grandis (lv, Lecythidaceae) (Crublet et al., 2003)
3-*O*-{glucosyl-(1→6)-[rhamnosyl-(1→3)]-(2´´-*E*-*p*-coumaroylglucoside)}-7-*O*-[rhamnosyl-(1→3)-rhamnosyl-(1→3)-(4´´-*Z*-*p*-coumaroylrhamnoside)]
Planchonia grandis (lv, Lecythidaceae) (Crublet et al., 2003)
3-*O*-[glucosyl-(1→2)-(6´´-*E*-sinapoylglucoside)]-7-*O*-[glucosyl-(1→4)-(6´´-*E*-sinapoylglucoside)
Aubrieta×*cultorum* (fl, Brassicaceae) (Tatsuzawa et al., 2012)
3-*O*-[6´´-(3-hydroxy-3-methyl-glutaryl)-glucoside]-7-*O*-glucoside
Citrus aurantifolia (cc, Rutaceae) (Berhow et al., 1994)
3-*O*-(6´´-malonylglucoside)-7-*O*-glucoside
Equisetum ×*litorale* (wp, Equisetaceae) (Veit et al., 1993)
3-*O*-neohesperidoside-7-*O*-(2´´-*E*-*p*-coumaroylglucoside)
Allium ursinum (wp, Alliaceae) (Carotenuto et al., 1996)
3-*O*-neohesperidoside-7-*O*-(2´´-*E*-feruloylglucoside)
Allium ursinum (wp, Alliaceae) (Carotenuto et al., 1996)
3-*O*-neohesperidoside-7-*O*-[glucosyl-(1→3)-(2´´-*E*-*p*-coumaroylglucoside)]
Allium ursinum (wp, Alliaceae) (Carotenuto et al., 1996)
3-*O*-rhamnoside-7-*O*-[6´´-feruloylglucosyl-(1→3)-rhamnoside]
Asplenium prolongatum (fr, Aspleniaceae) (Mizuno et al., 1990a)
3-*O*-{rhamnosyl-(1→6)-[rhamnosyl-(1→3)]-(2´´-*E*-*p*-coumaroylglucoside)}-7-*O*-[rhamnosyl-(1→3)-rhamnosyl-(1→3)-(4´´-*E*-*p*-coumaroylrhamnoside)]
Planchonia grandis (lv, Lecythidaceae) (Crublet et al., 2003)
3-*O*-[(2´´-*E*-sinapoylglucosyl)-(1→2)-glucoside]-7-*O*-glucoside
Brassica oleracea (cabbage), *B. napus* (lv, Brassicaceae) (Nielsen et al., 1993; Olsson et al., 1998)
3-*O*-[(2´´-*E*-sinapoylglucosyl)-(1→2)-glucoside]-7-*O*-[glucosyl-(1→4)-glucoside]
Brassica oleracea (cabbage) (lv, Brassicaceae) (Nielsen et al., 1998)
3-*O*-[xylosyl-(1→2)-rhamnoside]-7-*O*-(4´´-acetylrhamnoside)
Kalanchoe streptantha (lv, Crassulaceae) (Costa et al., 1996)
3-*O*-(6´´-*E*-feruloylglucoside)-7-*O*-cellobioside-4´-*O*-glucoside
Heliophila coronopifolia (fl, Brassicaceae) (Saito et al., 2011)

Sulphated glycosides
3-sulphate-7-*O*-arabinoside
Atriplex hortensis (lv, Chenopodiaceae) (Bylka et al., 2001)
8-sulphate
Phyllanthus virgatus (wp, Euphorbiaceae) (Huang et al., 1998)

ap = aerial parts, bd = buds, bl = bulbs, cc = callus cultures, fl = flowers, fr = fronds, ft = fruits, lv = leaves, rt = roots, rz = rhizomes, sd = seeds, sg = stigma, st = stems, wp = whole plants.

Recently, it was shown that an acylated kaempferol tetraglycoside, 3-*O*-(6´´-*E*-feruloylglucoside)-7-*O*-cellobioside-4´-*O*-glucoside acts as copigment substance to acylated delphinidin 3-*O*-sambubioside-5-*O*-glucosides in the blue flowers of *Heliophila coronopifolia* (Brassicaceae) (Saito et al., 2011). The pale purple flowers of *Glycine max* cultivars and their wild species are due to anthocyanins. Kaempferol 3-*O*-gentiobioside is contained in the flowers as a major flavonoid, and may act as copigment and pollinator attractant (Iwashina et al., 2007). Thus, kaempferol glycosides may work on the pollinator attractant in the flowers,

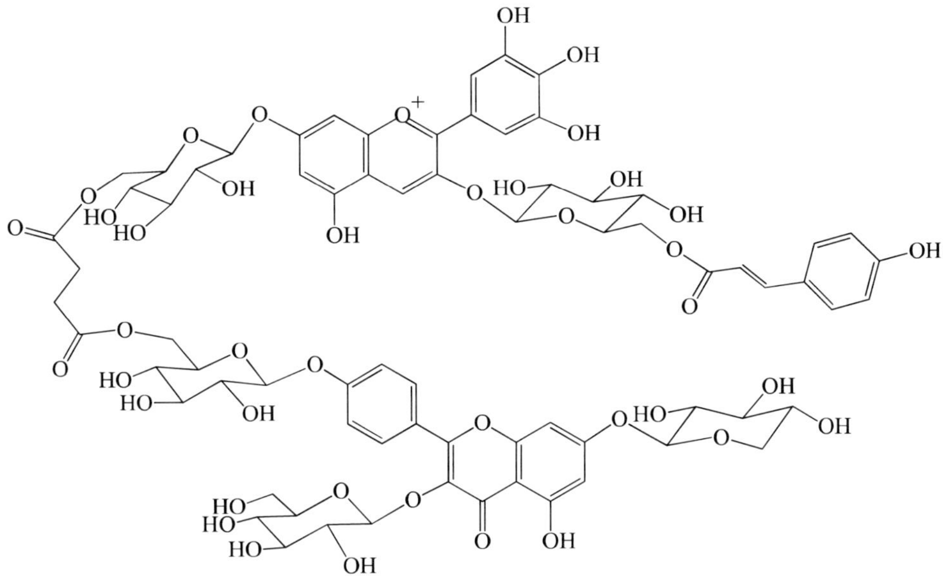

Figure 1. Covalently linked kaempferol glycoside and anthocyanin from *Agapanthus*.

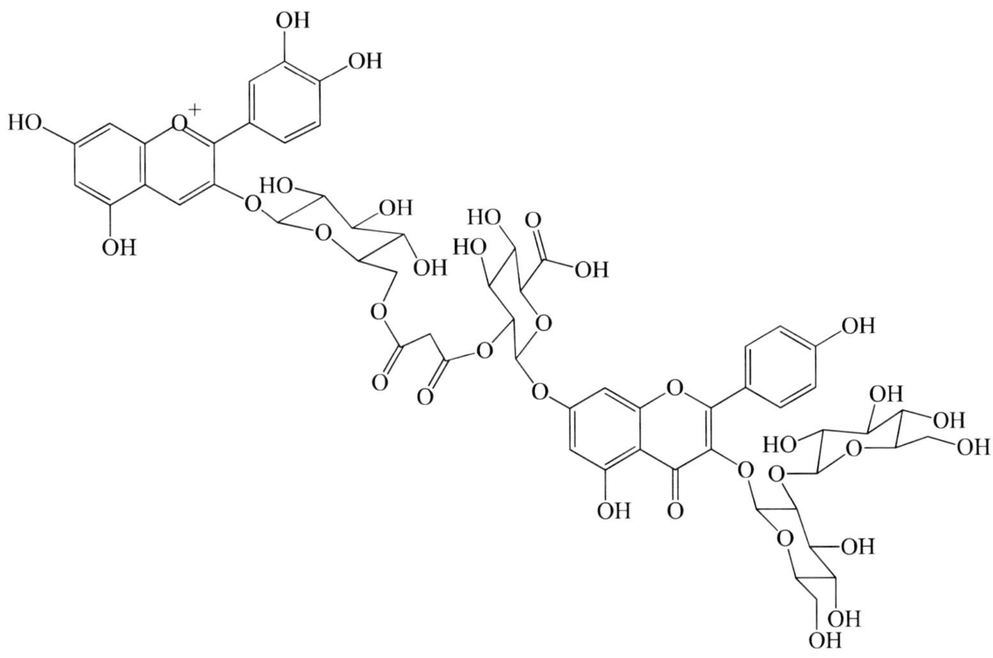

Figure 2. Covalently linked kaempferol glycoside and anthocyanin from *Allium schoenoprasum*.

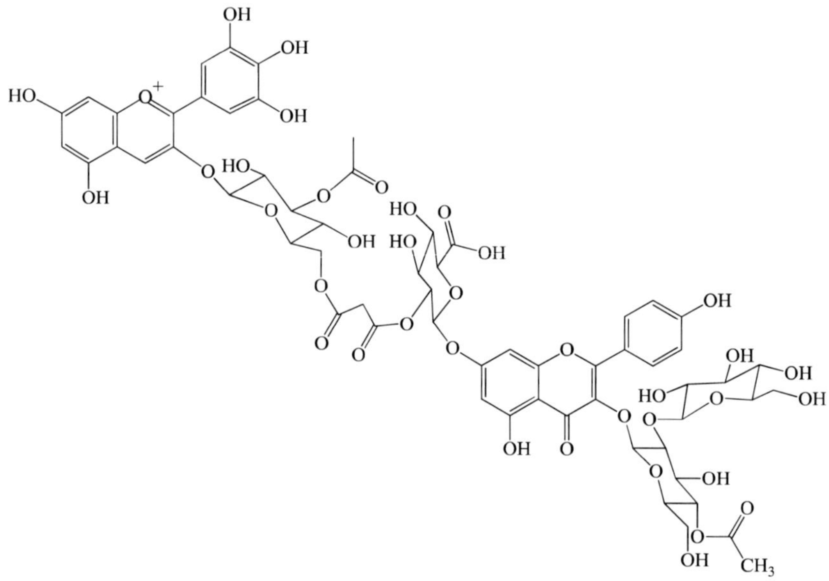

Figure 3. Covalently linked kaempferol glycoside and anthocyanin from *Allium* 'Blue Perfume'.

In recent years, three covalently linked kaempferol glycosides to anthocyanins were reported. They were isolated from the flowers of *Agapanthus praecox* subsp. *orientalis* (Bloor and Falshaw, 2000), *Allium schoenoprasum* (Fossen et al., 2000) and *Allium* cultivar 'Blue Perfume' (Saito et al., 2012), and identified as {6′′-*O*-[delphinidin 3-*O*-(6′′-*p*-coumaroylglucoside)-7-*O*-glucosyl]}-{6′′′′-*O*-[kaempferol 3-*O*-glucoside-7-*O*-xyloside-4′-*O*-glucosyl]}succinate (Figure 1) and {6′′-*O*-[delphinidin 3-*O*-(6′′-*p*-coumaroylglucoside)-7-*O*-glucosyl]}-{6′′′′-*O*-[kaempferol 3,7-di-*O*-glucoside-4′-*O*-glucosyl]}succinate (*Agapanthus*), [6′′-*O*-(cyanidin 3-*O*-glucosyl)]-[2′′′-*O*-(kaempferol 3-*O*-sophoroside-7-*O*-glucuronyl)] malonate (Figure 2) and three analogues (*Allium schoenoprasum*), and {6′′-*O*-[delphinidin 3-*O*-(2′′-acetylglucosyl)]}-[2′′′-*O*-(kaempferol 3-*O*-sophoroside-7-*O*-glucuronyl)]malonate (Figure 3) and two analogues (*Allium* 'BluePerfume'). It was shown that their kaempferol glycosides function as intra-copigment substances.

On the other hand, though kaempferol glycosides such as 3-*O*-neohesperidoside, 3-*O*-sophoroside and 3-*O*-glucosyl-(1→2)-[rhamnosyl-(1→6)-glucoside] contribute to the creamy tone of white flowers in carnation (*Dianthus caryophyllus*) (Iwashina et al., 2010), they did not act as yellow pigments, even if they were abundantly accumulated in the cases of yellow *Clematis* cultivars (Hashimoto et al., 2008). Major pigments of yellow *Clematis* cultivars are quercetin glycosides such as 3-*O*-glucoside, 3-*O*-galactoside and 3-*O*-rutinoside (Hashimoto et al., 2008).

UV Abosorbing Substances

As described above, kaempferol glycosides are important components as copigment and pollinator attractants in flowers. However, almost kaempferol glycosides are contained in leaves. The flavonoids are inducible by numerous and disparate stressors (Gould and Lister, 2006). The best known of these is exposure to UV radiation. In many cases, quercetin glycosides, which have *ortho*-dihydroxyl groups in B-ring, are effective to UV radiation and the leaves of plants are increased quercetin contents than kaempferol by UV-B radiation (Ryan et al., 1998, 2001, 2002).

Practically, kaempferol glycosides are minor compounds as UV shields in plants, e.g. translucent bracts of the Himalayan alpine plant, *Rheum nobile* (Polygonaceae) (Iwashina et al., 2004) and Japanese alpine plants, *Geum calthifolium* var. *nipponicum* and *Sieversia pentapetala* (Rosaceae) (Murai and Iwashina, 2010).

The translucent bracts of *Davidia involucrata* (Davidiaceae) are acts as UV absorbing organs (Takemura et al., 2009). However, their ultraviolet absorbing substances are various quercetin glycosides and kaempferol glycosides are minor compounds (Takemura et al., 2009). On the other hand, it was shown that the UV-B protective flavonoid in epidermal cells of *Arabidopsis thaliana* was shown to be kaempferol 3-*O*-gentiobioside-7-*O*-rhamnoside and 3,7-di-*O*-rhamnoside (Ormrod et al., 1995). Kaempferol 3-*O*-(3´´,4´´-di-*p*-coumaroylglucoside), and 3-*O*-[(2´´-*E*-sinapoylglucosyl)-(1→2)-glucoside]-7-*O*-glucoside and their quercetin analogues were shown to be UV-shield in the epidermal cells of *Pinus sylvestris* (Pinaceae) (Jungblut et al., 1995; Schnitzler et al., 1996), and *Brassica napus* (Olsson et al., 1998).

From the leaf hairs of *Quercus ilex*, two acylated kaempferol glycosides, 3-*O*-(6´´-*E*-*p*-coumaroylglucoside) and 3-*O*-(2´´,6´´-di-*O*-*E*-*p*-coumaroylglucoside) were reported as UV-absorbing substances (Skaltsa et al., 1994). Kaempferol 3-*O*-glucoside was also shown to be UV-shield in the leaves of *Campanula punctata* (Campanulaceae) by qualitative and quantitative HPLC survey between coastal and inland populations (Hashiba et al., 2006).

Other Kaempferol Function in Plants

Though most flavonoids have antioxidant activities, it is shown that a few kaempferol glycosides, e.g., 3-*O*-(6´´-galloylgalactoside) and 3-*O*-(6´´-galloylgucoside) from *Pemphis acidula* (Lythraceae) (Masuda et al., 2001) and 3-*O*-(6´´-*E*-feruloylglucosyl)-(1→2)-galactoside from *Hedyotis diffusa* (Rubiaceae) (Lu et al., 2000) were certainly considered to antioxidant substances.

Kaempferol 3-*O*-arabinoside-7-*O*-rhamnoside from *Andropogon scoparius* (Poaceae) was shown to be the inhibitor of nitrification (Rice and Pancholy, 1974). Picman et al. (1995) proved kaempferol itself as to be a phytoalexin agaist a crop pathogen, *Verticillium albo-atrum*. It is also shown by El-Gammal and Mansour (1986) that kaempferol itself has antimicrobial activity against different bacterial and fungal strains as test organisms. In contrast, kaempferol itself is also shown to be an inducer of nitrogen fixing bacteria (Bassam et al., 1988).

Moreover, antimicrobial, antibacterial and antiviral activities of kaempferol glycosides were proved by some authors, e.g., kaempferol 3-*O*-(2´´,4´´-di-*O*-*p*-coumaroylrhamnoside) from *Pentachondra pumila* (Epacridaceae) (Bloor, 1995), kaempferol 3-*O*-(6´´-*E*-*p*-coumaroylglucoside) from the buds of *Platanus orientalis* (Mitrokotsa et al., 1993), kaempferol 3-*O*-rhamnoside from *Persea americana* (Lauraceae) (de Almeida et al., 1998), and four acylated kaempferol glycosides, 3-*O*-(3´´,6´´-di-*O*-*Z*-*p*-coumaroylglucoside), 3-*O*-(3´´-*Z*-*p*-coumaryl-6´´-*E*-feruloylglucoside), 3-*O*-(3´´-*Z*-*p*-coumaroyl-6´´-*E*-*p*-coumaroylglucoside) and 3-*O*-(3′′-*E*-*p*-coumaroyl-6′′-*Z*-*p*-coumaroylglucoside) from *Stenochlaena palustris* (Pteridaceae) (Liu et al., 1999).

As another function of kaempferol glycosides in plants, feeding attractants are mentioned. It was shown that kaempferol 3-*O*-xylosyl-(1→2)-galactoside with glucosinolates acts as a feeding stimulant for the flea beetle *Phyllotreta armorasia* in horseradish (*Armoracia rusticana*, Brassicaceae) (Nielsen et al., 1979; Larsen et al.,1982). Furthermore, kaempferol and its glycosides have some effect on seed production (Vogt et al., 1994) and plant development (Ringli et al., 2008).

As described above, ca. 350 kaempferol glycosides have been reported, but their functions in plants are hardly known. In the future, their functions may be proved by some reseachers.

References

Abe, F., Iwase, Y., Yamauchi, T., Yahara, S., Nohara, T. 1995.Flavonol sinapoyl glycosides from leaves of *Thevetia peruviana*. *Phytochemistry* 40, 577-581.

Akabori, Y., Hasegawa, M. 1969. Flavonoid pattern in the Pteridaceae, II. Flavonoid constituents in the fronds of *Adiantum capillus-veneris* and *A. cuneatum*. *Bot. Mag. Tokyo* 82, 294-297.

Alaniya, M.D., Chkadua, N.F. 2000. Flavonoids of *Astragalus tana*. *Chem. Nat. Compds* 36, 537.

Allais, D.P., Simon, A., Bennini, B., Chulia, A.J., Kaouadji, M., Delage, C. 1991. Flavone and flavonol glycosides from *Calluna vulgaris*. *Phytochemistry* 30, 3099-3101.

Asakawa, M., Iwashina, T., Kitajima, J. 2010. Flavonol glycosides from *Callianthemum hondoense* and *Callianthemum alatavicum* in Japan and Tien Shan Mountains. *Biochem. Syst. Ecol.* 38, 250-252.

Bassam, B.J., Djordjevic, M.A., Redmond, J.W., Batley, M., Rolfe, B.G. 1988. Identification of a *nodD*-dependent locus in the *Rhizobium* strain NGR234 activated by phenolic factors secreted by soybeans and other legumes. *Mol. Plant-Microbe Interac.* 1, 161-168.

Berhow, M.A., Bennett, R.D., Poling, S.M., Vannier, S., Hidaka, T., Omura, M. 1994. Acyalted flavonoids in callus cultures of *Citrus aurantifolia*. *Phytochemistry* 36, 1225-1227.

Bloom, M., Vickery, R.K. 1973. Pattern partitioning of floral flavonoids in the *Mimulus luteus* complex. *Phytochemistry* 12, 165-167.

Bloor, S.J. 1995. An antimicrobial kaempferol-diacyl-rhamnoside from *Pentachondra pumila*. *Phytochemistry* 38, 1033-1035.

Bloor, S.J., Bradley, J.M., Lewis, D.H., Davies, K.M. 1998. Identification of flavonol and anthocyanin metabolites in leaves of *Petunia* 'Mitchell' and its *LC* transgenic. *Phytochemistry* 49, 1427-1430.

Bloor, S.J., Falshaw, R. 2000. Covalently linked anthocyanin-flavonol pigments from blue *Agapanthus* flowers. *Phytochemistry* 53, 575-579.

Brunner, G., Burger, U., Castioni, P., Kapetanidis, I., Christen, P. 2000. A novel acyalted flavonol glycoside isolated from *Brunfelsia grandiflora* ssp. *garndiflora*. Structure elucidation by gradient accelerated NMR spectroscopy at 14T. *Phytochem. Anal.* 11, 29-33.

Buttery, B.R., Buzzell, R.I. 1975. Soybean flavonol glycosides: identification and biochemical genetics. *Can. J. Bot.* 53, 219-224.

Bylka, W., Stobiecki, M., Frański, R. 2001. Sulphated flavonoid glycosides from leaves of *Atriplex hortensis*. *Acta Physiol. Plant.* 23, 285-290.

Cao, S.-G., Tanaka, T., Mizuno, M., Inoue, K. 2001. Flavonol glycosides from *Elaeagnus bockii* (Elaeagnaceae). *Nat. Prod. Lett.* 15, 1-8.

Carotenuto, A., De Feo, V., Fattorusso, E., Lanzotti, V., Magno, S., Cicala, C. 1996. The flavonoids of *Allium ursinum*. *Phytochemistry* 41, 531-536.

Carotenuto, A., Fattorusso, E., Lanzotti, V., Magno, S., De Feo, V., Cicala, C. 1997. The flavonoids of *Allium neapolitanum*. *Phytochemistry* 44, 949-957.

Catalano, S., Bilia, A.R., Martinozzi, M., Morelli, I. 1994. Kaempferol 3-*O*-β-D-(6′′-feruloylglucoside) from *Polylepis incana*. *Phytochemistry* 37, 1777-1778.

Champavier, Y., Allais, D.P., Chulia, A.J., Kaouadji, M. 2000. Acetylated and non-acetylated flavonol triglycosides from *Galega officinalis*. *Chem. Pharm. Bull.* 48, 281-282.

Chauhan, D., Rai, R., Chauhan, J.S. 2002. Two flavonoid triglycosides from *Cassia marginata*. *Indian J. Chem.* 41B, 446-448.

Cooper-Driver, G., Swain, T. 1977. Phenolic chemotaxonomy and phytogeography of *Adiantum*. *Bot. J. Linn. Soc.* 74, 1-21.

Costa, S.S., Jossang, A., Bodo, B. 1996. 4′′′′-Acetylsagittatin A, a kaempferol triglycoside from *Kalanchoe streptantha*. *J. Nat. Prod.* 59, 327-329.

Crublet, M.-L., Long, C., Sévenet, T., Hadi, H.A., Lavaud, C. 2003. Acylated flavonol glycosides from leaves of *Planchonia grandis*. *Phytochemistry* 64, 589-594.

de Almeida, A.P., Miranda, M.M.F.S., Simoni, I.C., Wigg, M.D., Lagrota, M.H.C., Costa, S.S. 1998. Flavonol monoglycosides isolated from the antiviral fractions of *Persea americana* (Lauraceae) leaf infusion. *Phytother. Res.* 12, 562-567.

Du, M., Xie, J. 1995. Flavonol glycosides from *Rhodiola crenulata*. *Phytochemistry* 38, 809-810.

El-Desoky, S.K., El-Ansari, M.A., El-Negoumy, S.I. 2001. Flavonol glycosides from *Mentha lavandulacea*. *Fitoterapia* 72, 532-537.

El-Gammal, A.A., Mansour, R.M.A. 1986. Antimicrobial activities of some flavonoid compounds. *Zentralbl. Mikrobiol.* 141, 561-565.

El-Mousallamy, A.M.D. 1998. Leaf flavonoids of *Albizia lebbeck*. *Phytochemistry* 48, 759-761.

El-Mousallamy, A.M.D., Afifi, M.S., Hussein, S.A.M. 2002. Acylated flavonol diglucosides from *Lotus polyphyllos*. *Phytochemistry* 60, 807-811.

El-Sayed, N.H., Awaad, A.S., Hifnawy, M.S., Mabry, T.J. 1999. A flavonol triglycoside from *Chenopodium murale*. *Phytochemistry* 51, 591-593.

El-Sayed, N.H., Omara, N.M., Yousef, A.K., Farag, A.M., Mabry, T.J. 2001. Kaempferol triosides from *Reseda muricata*. *Phytochemistry* 57, 575-578.

El-Sayed, N.H., Wojcińska, M., Drost-Karbowska, K., Matławska, I., Williams, J., Mabry, T.J. 2002. Kaempferol triosides from *Silphium perfoliatum*. *Phytochemistry* 60, 835-838.

Fattorusso, E., Lanzotti, V., Taglialatela-Scafati, O., Cicala, C. 2001. The flavonoids of leek, *Allium porrum*. *Phytochemistry* 57, 565-569.

Fico, G., Braca, A., De Tommasi, N., Tomè, F., Morelli, I. 2001a. Flavonoids from *Aconitum napellus* subsp. *neomontanum*. *Phytochemistry* 57, 543-546.

Fico, G., Braca, A., Bilia, A.R., Tomè, F., Morelli, I. 2001b. New flavonol glycosides from the flowers of *Aconitum napellus* ssp. *tauricum*. *Planta Med*. 67, 287-290.

Fiorini, C., David, B., Fourasté, I., Vercauteren, J. 1998. Acylated kaempferol glycosides from *Laurus nobilis* leaves. *Phytochemistry* 47, 821-824.

Fossen, T., Slimestad, R., Øvstedal, D.O., Andersen, Ø.M. 2000. Covalent anthocyanin-flavonol complexes from flowers of chive, *Allium schoenoprasum*. *Phytochemistry* 54, 317-323.

Francis, F.J., Harborne, J.B. 1960. Anthocyanins and flavonol glycosides of *Magnolia* flowers. *Proc. Amer. Soc. Hort. Sci.* 89, 657-665.

Fuchino, H., Nakamura, H., Wada, H., Hakamatsuka, T., Tanaka, N. 1997. 5-*O*-Glucosylated kaempferols from the fern *Dryopteris dickinsii*. *Nat. Medic*. 51, 537-538.

Fukumoto, H., Ishiguro, K., Murashima, T., Yamaki, M., Isoi, K. 1994. Structure determination of a kaempferol 3-rhamnosyldiglucoside from *Impatiens balsamina*. *Phytochemistry* 37, 1486-1488.

Gripenberg, J. 1962. Flavones. *In* Geissman, T.A. (ed.), *The Chemistry of Flavonoid Compounds*. Pergamon Press, Oxford. pp. 406-440.

Gohar, A.A., Maatooq, G.T., Niwa, M. 2000. Two flavonoid glycosides from *Chenopodium murale*. *Phytochemistry* 53, 299-303.

Gould, K.S., Lister, C. 2006. Flavonoid functions in plants. *In* Andersen, Ø.M., Markham, K.R. (eds.), *Flavonoids. Chemistry, Biochemistry and Applications*. CRC Press, Boca Raton. pp. 397-441.

Guangyao, G., Sibao, C., Junshan, Y., Peigen, X. 2000. A new flavonoid from the aerial part of *Thalictrum atriplex*. *Fitoterapia* 71, 627-629.

Han, Y., Nishibe, S., Noguchi, Y., Jin, Z. 2001. Flavonol glycosides from the stems of *Trigonella foenum-graecum*. *Phytochemistry* 58, 577-580.

Harborne, J.B. 1967. Comparative biochemistry of flavonoids – V. Luteolin 5-glycoside and its occurrence in the Umbelliferae. *Phytochemistry* 6, 1569-1573.

Harborne, J.B., Williams, C.A. 1975. Flavone and flavonol glycosides. *In* Harborne, J.B., Mabry, T.J. and Mabry, H. (eds.), *The Flavonoids*. Chapman and Hall, London. pp. 376-441.

Harborne, J.B., Boardley, M. 1983. Trisubstituted flavonol glycosides in *Coronilla emerus* flowers. *Phytochemistry* 22, 622-625.

Hasan, A., Farman, M., Ahmed, I. 1994. Flavonoid glycosides from *Indigofera hebepetala*. *Phytochemistry* 35, 275-276.

Hasan, A., Ahmed, I., Khan, M.A., Chudhary, M.I. 1996. Two flavonol triglycosides from flowers of *Indigofera hebepetala*. *Phytochemistry* 43, 1115-1118.

Hasegawa, M., Akabori, Y. 1968. Flavonoid pattern in Pteridaceae, I. Flavonoid glycosides obtained from the fronds of *Adiantum aethiopicum* and *A. monochlamys*. *Bot. Mag. Tokyo* 81, 469-472.

Hashiba, K., Iwashina, T., Matsumoto, S. 2006. Variation in the quality and quantity of flavonoids in the leaves of coastal and inland *Campanula punctata*. *Biochem. Syst. Ecol.* 34, 854-861.

Hashimoto, M., Iwashina, T., Kitajima, J., Matsumoto, S. 2008. Flavonol glycosides from *Clematis* cultivars and taxa, and their contribution to yellow and white flower colors. *Bull. Natl. Mus. Nature Sci. Ser. B* 34, 127-134.

Hasler, A., Gross, G.-A., Meier, B., Sticher, O. 1992. Complex flavonol glycosides from the leaves of *Ginkgo biloba*. *Phytochemistry* 31, 1391-1394.

Hayashi, Y., Ohara, S., Takahashi, T. 1987. Isolation of new 3-hydroxylated flavonoid *C*-glucosides from the heartwoods of *Zelkova serrata* Makino and other *Zelkova* species. *Mokuzai Gakkaishi* 33, 511-520.

Hohmann, J., Páll, Z., Günther, G., Máthé, I. 1997. Flavonol acyl glycosides of the aerial parts of *Eryngium campestre*. *Planta Med.* 63, 96.

Huang, Y.-L., Chen, C.-C., Hsu, F.-L., Chen, C.-F. 1998. Tannins, flavonol sulfonates, and a norlignan from *Phyllanthus virgatus*. *J. Nat. Prod.* 61, 1194-1197.

Hukuti, G. 1939. On the constituents of Herba Calystegiae. *Yakugaku Zasshi* 59, 258-260.

Imperato, F. 1992. Kaempferol 3-(2′′-*p*-coumaroylrhamnoside)-7-rhamnoside from *Cheilanthes fragrans*. *Phytochemistry* 31, 3291-3292.

Imperato, F. 1996. Kaempferol 3-*O*-(5′′-feruloylapioside) from *Pteridium aquilinum*. *Phytochemistry* 43, 1421-1423.

Imperato, F., Minutiello, P. 1997. Kaempferol 3-*O*-(6′′-caffeoylglucoside) from *Pteridium aquilinum*. *Phytochemistry* 45, 199-200.

Imperato, F. 1998. Kaempferol 7-*O*-rhamnoside-4′-*O*-glucoside from *Pteridium aquilinum*. *Phytochemistry* 47, 911-913.

Iwashina, T., Ootani, S. 1990. Three flavonol allosides from *Glaucidium palmatum*. *Phytochemistry* 29, 3639-3641.

Iwashina, T., Matsumoto, S., Ozawa, K., Akuzawa, K. 1990. Flavone glycosides from *Asplenium normale*. *Phytochemistry* 29, 3543-3546.

Iwashina, T., Matsumoto, S. 1994. Flavonoid variation and evolution in *Asplenium normale* and related species (Aspleniaceae). *J. Plant Res.* 107, 275-282.

Iwashina, T., Matsumoto, S., Nakaike, T. 1995a. Flavonoid characters of five *Adiantum* species in Pakistan. *In* Watanabe, M., Hagiwara, H. (eds.), *Cryptogams of the Himalayas, Vol. 3. Nepal and Pakistan*. National Science Museum, Tsukuba. pp. 179-191.

Iwashina, T., Matsumoto, S., Nishida, M., Nakaike, T. 1995b. New and rare flavonol glycosides from *Asplenium trichomanes-ramosum* as stable chemotaxonomic markers. *Biochem. Syst. Ecol.* 23, 283-290.

Iwashina, T., Ootani, S. 1998. Flavonoids of the genus *Iris*; structure, distribution and function (review). *Ann. Tsukuba Bot. Gard.* 17, 147-183.

Iwashina, T., Kitajima, J. 2000. Chalcone and flavonol glycosides from *Asarum canadense* (Aristolochiaceaea). *Phytochemistry* 55, 971-974.

Iwashina, T., López-Sáez, J.A., Herrero, A., Kitajima, J., Matsumoto, S. 2000. Flavonol glycosides from *Asplenium foreziense* and its related taxa and *A. incisum*. *Biochem. Syst. Ecol.* 28, 665-671.

Iwashina, T., Konta, F., Kitajima, J. 2001. Anthocyanins and flavonols of *Chimonanthus praecox* (Calycanthaceae) as flower pigments. *J. Jap. Bot.* 76, 166-172.

Iwashina, T., Omori, Y., Kitajima, J., Akiyama, S., Suzuki, T., Ohba, H. 2004. Flavonoids in translucent bracts of the Himalayan *Rheum nobile* (Polygonaceae) as ultraviolet shields. *J. Plant Res.* 117, 101-107.

Iwashina, T., Kitajima, J., Kato, T., Tobe, H. 2005a. An analysis of flavonoid compounds in leaves of *Japonolirion* (Petrosaviaceae). *J. Plant Res.* 118, 31-36.

Iwashina, T., Kitajima, J., Shiuchi, T., Itou, Y. 2005b. Chalcones and other flavonoids from *Asarum* sensu lato (Aristolochiaceae). *Biochem. Syst. Ecol.* 33, 571-584.

Iwashina, T., Githiri, S.M., Benitez, E.R., Takemura, T., Kitajima, J., Takahashi, R. 2007. Analysis of flavonoids in flower petals of soybean near-isogenic lines for flower and pubescence color genes. *J. Heredity* 98, 250-257.

Iwashina, T., Yamaguchi, M., Nakayama, M., Onozaki, T., Yoshida, H., Kawanobu, S., Ono, H., Okamura, M. 2010. Kaempferol glycosides in the flowers of carnation and their contribution to the creamy white flower color. *Nat. Prod. Commun.* 5, 1903-1906.

Iwashina, T., Matsumoto, S. 2011. Flavonoid properties of six *Asplenium* species in Vanuatu and New Caledonia, and distribution of flavonoid and related compounds in *Asplenium*. *Bull. Natl. Mus. Nature Sci. Ser. B* 37, 133-145.

Iwashina, T., Tamura, M.N., Murai, Y., Kitajima, J. 2013. New flavonol glycosides from the leaves of *Triantha japonica* and *Tofieldia nuda*. *Nat. Prod. Commun.* In press.

Jungblut, T.P., Schnitzler, J.-P., Heller, W., Hertkorn, N., Mezger, J.W., Szymczak, W., Sandermann, H. 1995. Structures of UV-B induced sunscreen pigments of Scotpine (*Pinus sylvestris* L.). *Angew.Chem. Int. Ed. Engl.* 34, 312-314.

Kaouadji, M., Morand, J.M., Garcia, J. 1993. Further acylated kaempferol rhamnosides from *Platanus acerifolia* buds. *J. Nat. Prod.* 56, 1618-1621.

Kim, J.E., Jung, M.J., Jung, H.A., Woo, J.J., Cheigh, H.S., Chung, H.Y., Choi, J.S. 2002. A new kaempferol 7-*O*-triglycoside from the leaves of *Brassica juncea* L. *Arch. Pharm. Res.* 25, 621-624.

Kazuma, K., Noda, N., Suzuki, M. 2003. Malonylated flavonol glycosides from the petals of *Clitoria ternatea*. *Phytochemistry* 62, 229-237.

Kubo, I., Yokokawa, Y. 1992. Two tyrosinase inhibiting flavonol glycosides from *Buddleia coriacea*. *Phytochemistry* 31, 1075-1077.

Lakenbrink, C., Myloclam, T., Engelhardt, U.H., Wray, V. 2000. New flavonol triglycosides from tea (*Camellia sinensis*). *Nat. Prod. Lett.* 14, 233-238.

Larsen, L.M., Nielsen, J.K., Sørensen, H. 1982. Identification of 3-*O*-[2-*O*-(β-D-xylopyranosyl)- β-D-galactopyranosyl] flavonoids in horse radish leaves acting as feeding stimulants for a flea beetle. *Phytochemistry* 21, 1029-1033.

Li, T.-M., Li, W.-K., Yu, J.-G. 1997. Flavonoids from *Artabotrys hexapetalus*. *Phytochemistry* 45, 831-833.

Lin, Y.-L., Kuo, Y.-H. 1995. Roripanoside, a new kaempferol rhamnoside from *Rorripa indica* (L.) Hiern. *J. Chin. Chem. Soc. (Taipei)* 42, 973-976.

Liu, Q., Liu, M., Mabry, T.J., Dixon, R.A. 1994. Flavonol glycosides from *Cephalocereus senilis*. *Phytochemistry* 36, 229-231.

Liu, Y., Wu, Y., Yuan, K., Ji, C., Wang, L. 1997a. New galloylated flavonoidal glycoside and gallotannins from the leaves of *Loropetalum chinense* Oliv. *Nat. Prod. Res. Develop.* 9, 12-18.

Liu, Y., Wu, Y., Yuan, K., Ji, C., Hou, A., Yoshida, T., Okuda, T. 1997b. Astragalin 2′′,6′′-di-*O*-gallate from *Loropetalum chinense*. *Phytochemistry* 46, 389-391.

Liu, H., Orjala, J., Sticher, O., Rali, T. 1999. Acylated flavonol glycosides from leaves of *Stenochlaena palustris*. *J. Nat. Prod.* 62, 70-75.

Lu, C.-M., Yang, J.-J., Wang, P.-Y., Lin, C.-C. 2000. A new acylated flavonol glycoside and antioxidant effects of *Hedyotis diffusa*. *Planta Med.* 66, 374-377.

Manguro, L.O.A., Ugi, I., Lemmen, P., Hermann, R. 2003. Flavonol glycosides of *Warburgia ugandensis* leaves. *Phytochemistry* 64, 891-896.

Markham, K.R., Porter, L.J., Campbell, E.O., Chopin, J., Bouillant, M.-L. 1976. Phytochemical support for the existence of two species in the genus *Hymenophyton*. *Phytochemistry* 15, 1517-1521.

Markham, K.R., Porter, L.J. 1979. Flavonoids of the primitive liverwort *Takakia* and their taxonomic and phylogenetic significance. *Phytochemistry* 18, 611-615.

Markham, K.R., Geiger, H., Jaggy, H. 1992. Kaempferol-3-*O*-glucosyl(1→2)rhamnoside from *Ginkgo biloba* and a reappraisal of other gluco(1→2, 1→3 and 1→4)rhamnoside structure. *Phytochemistry* 31, 1009-1011.

Masuda, T., Iritani, K., Yonemori, S., Oyama, Y., Takeda, Y. 2001. Isolation and antioxidant activity of galloyl flavonol glycosides from the seashore plant, *Pemphis acidula*. *Biosci. Biotechnol. Biochem.* 65, 1302-1309.

Merfort, I., Wray, V., Barakat, H.H., Hussein, S.A.M., Nawwar, M.A.M., Willuhn, G. 1997. Flavonol triglycosides from seeds of *Nigella sativa*. *Phytochemistry* 46, 359-363.

Min, B.-S., Tomiyama, M., Ma, C.-M., Nakamura, N., Hattori, M. 2001. Kaempferol acetylrhamnosides from the rhizome of *Dryopteris crassirhizoma* and their inhibitory effects on three different activities of human immunodeficiency virus-1 reverse transcriptase. *Chem. Pharm. Bull.* 49, 546-550.

Mitrokotsa, D., Mitaku, S., Demetzos, C., Harvala, C., Mentis, A., Perez, S., Kokkinopoulos, D. 1993. Bioactive compounds from the buds of *Platanus orientalis* and isolation of a new kaempferol glycoside. *Planta Med.* 59, 517-520.

Mizuno, M., Kyotani, Y., Iinuma, M., Tanaka, T., Iwatsuki, K. 1990a. Kaempferol 3-rhamnoside-7-[6-feruloylglucosyl(1→3)rhamnoside] from *Asplenium prolongatum*. *Phytochemistry* 29, 2742-2743.

Mizuno, M., Kyotani, Y., Iinuma, M., Tanaka, T., Kojima, H., Iwatsuki, K. 1990b. Kaempferol glycosides in *Asplenium scolopendrium* Newm. *Z. Naturforsch.* 45c, 143-146.

Mulinacci, N., Vincieri, F.F., Baldi, A., Bambagiotti-Alberti, M., Sendl, A., Wagner, H. 1995. Flavonol glycosides from *Sedum telephium* subspecies *maximum* leaves. *Phytochemistry* 38, 531-533.

Murai, Y., Kokubugata, G., Yokota, M., Kitajima, J. 2008. Flavonoids and anthocyanins from six *Cassytha* taxa (Lauraceae) as taxonomic markers. *Biochem. Syst. Ecol.* 36, 745-748.

Murai, Y., Iwashina, T. 2010. Flavonol glucuronides from *Geum calthifolium* var. *nipponicum* and *Sieversia pentapetala* (Rosaceae). *Biochem. Syst. Ecol.* 38, 1081-1082.

Murai, Y., Takahashi, R., Rodas, F.R., Kitajima, J., Iwashina, T. 2013. New flavonol triglycosides from the leaves of *Glycine max* cultivars. *Nat. Prod. Commun.* In press.

Nasr, C., Haag-Berrurier, M., Lobstein-Guth, A., Anton, R. 1986. Kaempferol coumaroyl glucorhamnoside from *Ginkgo biloba*. *Phytochemistry* 25, 770-771.

Nielsen, J.K., Larsen, L.M., Sørensen, H. 1979. Host plant selection of the horseradish flea beetle *Phyllotreta armoraciae* (Coleoptera: Chrysomelidae): identification of two flavonol glycosides stimulating feeding in combination with glucosinolates. *Ent. Exp. Appl.* 26, 40-48.

Nielsen, J.K., Olsen, C.E., Petersen, M.K. 1993. Acylated flavonol glycosides from cabbage leaves. *Phytochemistry* 34, 539-544.

Nielsen, J.K., Nørbæk, R., Olsen, C.E. 1998. Kaempferol tetraglucosides from cabbage leaves. *Phytochemistry* 49, 2171-2176.

Nørbæk, R., Kondo, T. 1999. Flavonol glycosides from flowers of *Crocus speciosus* and *C. antalyensis*. *Phytochemistry* 51, 1113-1119.

Nørbæk, R., Nielsen, J.K., Kondo, T. 1999. Flavonoids from flowers of two *Crocus chrysanthus-biflorus* cultivars: "Eye-catcher" and "Spring Pearl" (Iridaceae). *Phytochemistry* 51, 1139-1146.

Okamura, H., Mimura, A., Niwano, M., Takahara, Y., Yasuda, H., Yoshida, H. 1993. Two acylated flavonol glycosides from *Eucalyptus rostrata*. *Phytochemistry* 33, 512-514.

Olea, R.S.G., Roque, N.F., Bolzani, V. da S. 1997. Acylated flavonol glycosides and terpenoids from the leaves of *Alibertia sessilis*. *J. Braz. Chem. Soc.* 8, 257-259.

Olsson, L.C., Veit, M., Weissenböck, G., Bornman, J.F. 1998. Differential flavonoid response to enhanced UV-B radiation in *Brassica napus*. *Phytochemistry* 49, 1021-1028.

Olszewska, M., Wolbiś, M. 2001. Flavonoids from the flowers of *Prunus spinosa* L. *Acta Pol. Pharm.* 58, 367-372.

Olszewska, M., Wolbiś, M. 2002. Flavonoids from the leaves of *Prunus spinosa* L. *Polish J. Chem.* 76, 967-974.

Ormrod, D.P., Landry, L.G., Conklin, P.L. 1995. Short-term UV-B radiation and ozone exposure effects on aromatic secondary metabolite accumulation and shoot growth of flavonoid deficient *Arabidopsis* mutant. *Physiol. Plant.* 93, 602-610.

Özden, S., Dürüst, N., Toki, K., Saito, N., Honda, T. 1998. Acyalted kaempferol glycosides from the flowers of *Delphinium formosum*. *Phytochemistry* 49, 241-245.

Park, S.-H., Oh, S.R., Jing, K.Y., Lee, I.S., Ahn, K.S., Kim, J.H., Kim, Y.S., Lee, J.J., Lee, H.-K. 1999. Acylated flavonol glycosides with anti-complement activity from *Persicaria lapathifolia*. *Chem. Pharm. Bull.* 47, 1484-1486.

Picman, A.K., Schneider, E.F., Picman, J. 1995. Effect of flavonoids on mycelial growth of *Verticillium albo-atrum*. *Biochem. Syst. Ecol.* 23, 683-693.

Rao, K.V., Damu, A.G., Jayaprakasam, B., Gunasekar, D. 1999. Flavonol glycosides from *Cassia hirsuta*. *J. Nat. Prod.* 62, 305-306.

Rice, E.L., Panoholy, S.K. 1974. Inhibition of nitrification by climax ecosystems. III. Inhibitors other than tannins. *Amer. J. Bot.* 61, 1095-1103.

Richardson, P.M., Lorenz-Liburnau, E. 1982. *C*-Glycosylxanthones in the *Asplenium adiantum-nigrum* complex. *Amer. Fern J.* 72, 103-106.

Ringli, C., Bigler, L., Kuhn, B.M., Leiber, R.-M., Diet, A., Santelia, D., Frey, B., Pollmann, S., Klein, M. 2008. The modified flavonol glycosylation profile in the *Arabidopsis rol1* mutants results in alterations in plant growth and cell shape formation. *Plant Cell* 20, 1470-1481.

Romussi, G., Parodi, B., Caviglioli, G. 1992. Flavonoidglycoside aus *Quercus canariensis* Willd. 15. Mitteilung: Über Inhaltsstoffe von Cupuliferae. *Pharmazie* 47, 877.

Ryan, K.G., Markham, K.R., Bloor, S.J., Bradley, J.M., Mitchell, K.A., Jordan, B.R. 1998. UVB radiation induced increase in quercetin:kaempferol ratio in wild-type and transgenic lines of *Petunia*. *Photochem. Photobiol.* 68, 323-330.

Ryan, K.G., Swinny, E.E., Winefield, C., Markham, K.R. 2001. Flavonoids and UV photoprotection in *Arabidopsis* mutants. *Z. Naturforsch.* 56c, 745-754.

Ryan, K.G., Swinny, E.E., Markham, K.R., Winefield, C. 2002. Flavonoid gene expression and UV photoprotection in transgenic and mutant *Petunia* leaves. *Phytochemistry* 59, 23-32.

Saito, N., Tatsuzawa, F., Toki, K., Shinoda, K., Shigihara, A., Honda, T. 2011. The blue anthocyanin pigments from the blue flowers of *Heliophila coronopifolia* L. (Brassicaceae). *Phytochemistry* 72, 2219-2229.

Saito, N., Nakamura, M., Shinoda, K., Murata, N., Nakazawa, T., Koto, K., Toki, K., Kasai, H., Honda, T., Tatsuzawa, F. 2012. Covalent anthocyanin-flavonol complexes from the violet-blue flowers of *Allium* 'Blue Perfume'. *Phytochemistry* 80, 99-108.

Sakaguchi, K., Kitajima, J., Iwashina, T. 2012. Kaempferol tri- and tetraglycosides from the flowers of *Clematis* cultivar. *Nat. Prod. Commun.* 7, 173-174.

Saleh, N.A.M., Majak, W., Towers, G.H.N. 1972. Flavonoids of *Equisetum* species. *Phytochemistry* 11, 1095-1099.

Saleh, N.A.M. 1975. Glycosidic nature of *Equisetum* flavonoids. *Phytochemistry* 14, 286-287.

Sannomiya, M., Vilegas, W., Rastrelli, L., Pizza, C. 1998. A flavonoid glycoside from *Maytenus aquifolium*. *Phytochemistry* 49, 237-239.

Satake, T., Hori, K., Kamiya, K., Saiki, Y., Fujimoto, Y., Kimura, Y., Makut, C., Mekin, T. 1993. Studies on the constituents of Turkish plants. I. Flavonol triglycosides from the fruits of *Rhamnus thymifolius*. *Chem. Pharm. Bull.* 41, 1743-1745.

Schnitzler, J.-P., Jungblut, T.P., Heller, W., Köfferlein, M., Hutzler, P., Heinzmann, U., Schmelzer, E., Ernst, D., Langebartels, C. 1996. Tissue localization of u.v.-B-screening pigments and of chalcone synthase mRNA in needles of Scots pine seedlings. *New Phytol.* 132, 247-258.

Semmar, N., Fenet, B., Gluchoff-Fiasson, K., Hasan, A., Jay, M. 2002a. Four new flavonol glycosides from the leaves of *Astragalus caprinus*. *J. Nat. Prod.* 65, 576-579.

Semmar, N., Fenet, B., Gluchoff-Fiasson, K., Comte, G., Jay, M. 2002b. New flavonol tetaglycosides from *Astragalus caprinus.Chem. Pharm. Bull.* 50, 981-984.

Sharaf, M., El-Ansari, M.A., Saleh, N.A.M. 1997. New kaempferol tri- and tetraglycosides from *Pseuderucaria clavata*. *Fitoterapia* 68, 62-64.

Si, J.Y., Chen, D.H., Chang, Q., Shen, L.G. 1994. Isolation and structure determination of flavonol glycosides from the fresh fruits of *Siraitia grosvenori*. *Acta Pharm. Sinica* 29, 158-160.

Simon, A., Chulia, A.J., Kaouadji, M., Allais, D.P., Delge, C. 1993. Two flavonol 3-[triacetylarabinosyl(1→6)glucosides] from *Calluna vulgaris*. *Phytochemistry* 33, 1237-1240.

Singab, A.N.B. 1998. Acetylated flavonol triglycosides from *Ammi majus* L. *Phytochemistry* 49, 2177-2180.

Skaltsa, H., Verykokidou, E., Harvala, C., Karabourniotis, G., Nanetas, Y. 1994. UV-B protective potential and flavonoid content of leaf hairs of *Quercus ilex*. *Phytochemistry* 37, 987-990.

Slimestad, R., Andersen, Ø.M., Francis, G.W., Marston, A., Hostettmann, K. 1995. Syringetin 3-*O*-(6′′-acetyl)-β-glucopyranoside and other flavonols from needles of Norway Spruce, *Picea abies*. *Phytochemistry* 40, 1537-1542.

Smith, D.M., Harborne, J.B. 1971. Xanthones in the Apparachian *Asplenium* complex. *Phytochemistry* 10, 2117-2119.

Snook, M.E., Chortyk, O.T., Sisson, V.A., Costello, C.E. 1992. The flower flavonols of *Nicotiana* species. *Phytochemistry* 31, 1639-1647.

Straubinger, M., Jezussek, M., Waibel, R., Winterhalter, P. 1997. Two kaempferol sophorosides from *Crocus sativus*. *Nat. Prod. Lett.* 10, 213-216.

Takeda, K., Yamaguchi, S., Iwata, K., Tsujino, Y., Fujimori, T., Fusain, S.Z. 1996. A malonylated anthocyanin and flavonols in the blue flowers of *Meconopsis*. *Phytochemistry* 42, 863-865.

Takemura, T., Takatsu, Y., Kasumi, M., Marubashi, W., Iwashina, T. 2005. Flavonoids and their distribution patterns in the flowers of *Gladiolus* cultivars. *Acta Hort.* 673, 487-493.

Takemura, T. 2008. Flower colors and their pigments in *Gladiolus* cultivars. Degree of Doctor of Philosophy, Thesis. Graduate School of Tokyo University of Agriculture and Technology, Japan.

Takemura, S., Kitajima, J., Iwashina, T. 2009. Ultraviolet-absorbing substances in the translucent bracts of *Davidia involucrata* (Davidiaceae). *Bull. Natl. Mus. Nature Sci. Ser. B* 35, 1-9.

Tanaka, M., Fujimori, T., Uchida, I., Yamaguchi, S., Takeda, K. 2001. A malonylated anthocyanin and flavonols in blue *Meconopsis* flowers. *Phytochemistry* 56, 373-376.

Tang, Y., Lou, F., Wang, J., Li, Y., Zhuang, S. 2001. Coumaroyl flavonol glycosides from the leaves of *Ginkgo biloba*. *Phytochemistry* 58, 1251-1256.

Tanikawa, N., Yoshida, K., Kondo, T., Mizuno, T., Iwashina, T., Nakayama, M. 2011. Urakunoside, a new tetraglycosyl kampferol from petals of the Wabisuke Camellia cv. Tarokaja. *Biosci. Biotechnol. Biochem.* 75, 2046-2048.

Tatsuzawa, F., Aiba, Y., Morino, T., Saito, N., Shinoda, K., Kato, K., Toki, K., Honda, T. 2012. Copigmentation with acylated anthocyanin and kaempferol glycosides in violet and purple flower cultivars of *Aubrieta* ×*cultorum* (Brassicaceae). *J. Japan. Soc. Hort. Sci.* 81, 275-284.

Umikalsom, Y., Grayer-Barkmeijer, R.J., Harborne, J.B. 1994. A comparison of the flavonoids in Athyriaceae and Aspleniaceae. *Biochem. Syst. Ecol.* 22, 587-594.

Urushibara, S., Kitayama, Y., Watanabe, T., Okuno, T., Watarai, A., Matsumoto, T. 1992. New flavonol glycosides, major determinants inducing the green fluorescence in the guard cells of *Allium cepa*. *Tetrahedron Lett.* 33, 1213-1216.

Veit, M., Czygan., F.-C., Witte, L., Markham, K.R., Geiger, H. 1993. New kaempferol glycosides from *Equisetum* species. *Z. Naturforsch.* 48b, 1398-1400.

Vogt, T., Pollak, P., Tarlyn, N., Taylor, L.P. 1994. Pollination- or wound-induced kaempferol accumulation in *Petunia* stigmas enhances seed production. *Plant Cell* 6, 11-23.

Wang, J.-H., Lou, F.-C., Wang, Y.-L., Tang, Y.-P. 2003. A flavonol tetraglycoside from *Sophora japonica* seeds. *Phytochemistry* 63, 463-465.

Williams, C.A., Greenham, J., Harborne, J.B. 2001. The role of lipophilic and polar flavonoids in the classification of temperate members of the Anthemidae. *Biochem. Syst. Ecol.* 29, 929-945.

Williams, C.A. 2006. Flavone and flavonol *O*-glycosides. *In* Andersen, Ø.M., Markham, K.R. (eds.), *Flavonoids. Chemistry, Biochemistry, and Applications*. CRC Press, Boca Raton. pp. 749-856.

Wu, T., Kong., D.Y., Li, H.T. 2002. Chemical components of *Helicia nilagirica* Beed. I. Structure of three new flavonol glycosides. *Chin. Chem. Lett.* 13, 1071-1074.

Wu, Y.Q., Li, S., Li, Y., Li, Y. 2003. A new acylated flavonoid from *Anaphalis aureo punctata*. *Chin. Chem. Lett.* 14, 66-67.

Xie, C., Xu, L.-Z., Luo, X.-Z., Zhong, Z., Yang, S.-L. 2002. Flavonol glycosides from *Lysimachia capillipes*. *J. Asian Nat. Prod. Res.* 4, 17-23.

Yadava, R.N., Singh, A. 1993. A novel flavone glycoside from the seeds of *Crotalaria laburnifolia* Linn. *J. Indian Chem. Soc.* 70, 273.

Yahara, S., Kohjyouma, M., Hohoda, H. 2000. Flavonoid glycosides and saponins from *Astragalus shikokianus*. *Phytochemistry* 53, 469-471.

Yoshida, K., Kitahara, S., Ito, D., Kondo, T. 2006. Ferric ions involved in the flower color development of the Himalayan blue poppy, *Meconopsis grandis*. *Phytochemistry* 67, 992-998.

Yoshitama, K. 1997. Flavonoids of *Ginkgo biloba*. *In* Hori, T., Ridge, R.W., Tulecke, W., Tredici, D.P., Tremouillaux-Guiller, J., Tobe, H. (eds.), *Ginkgo biloba–A global Treasure. From Biology to Medicine*. Springer, Tokyo. pp. 287-299.

Zahid, M., Ali, A., Ishurd, O., Ahmed, A., Ali, Z., Ahmed, V.U., Pan, Y. 2002. New cycloartane and flavonol glycosides from *Corchorus depressus*. *Helv. Chim. Acta* 85, 689-697.

Zallocchi, E.M., Pomilio, A.B. 1994. Evolution of flavonoids in the Phaseolinae. *Phytochemistry* 37, 449-453.

Zhou, Y.-J., Xu, S.-X., Che, Q.-M., Sun, Q.-S. 2001. Flavonoids from the leaves of *Quercus dentata*. *Indian J. Chem.* 40B, 394-398.

In: Kaempferol
Editors: Garion Villers and Yves Fougere
ISBN: 978-1-62618-515-9

Chapter 4

Analytical Approaches for Kaempferol Determination

Ondrej Biba, [1,*] ***Miroslav Strnad*** [1,2] ***and Jiri Gruz*** [1,2,†]
[1]Centre of the Region Hana for Biotechnological and Agricultural Research, Faculty of Science, Palacky University, Olomouc, Czech Republic
[2]Laboratory of Growth Regulators, Institute of Experimental Botany AS CR, Olomouc, Czech Republic

Abstract

The rapid development of technology and analytical instruments in the last 20 years allowed scientists to detect and quantify thousands of compounds which were difficult to analyse before. Kaempferol, a common biologically active phytochemical, occurs mostly in complex matrices together with other plant metabolites that complicate the analytical procedure. Various extraction and purification steps, including solid phase extraction, were therefore developed to facilitate the determination of kaempferol in complex biological samples. This chapter is an overview of currently used analytical instruments and their basic settings used to detect and quantify kaempferol in various plants, plant-derived sources and other biological samples.

Introduction

Kaempferol is a common flavonol which is present in most species of land plants. It is intensively studied due to its biological effects on human health. In plants, kaempferol is

* Ondrej Biba: Centre of the Region Hana for Biotechnological and Agricultural Research, Faculty of Science, Palacky University, Slechtitelu 11, CZ 783 71 Olomouc, Czech Republic.
† Corresponding author: Jiri Gruz. Centre of the Region Hana for Biotechnological and Agricultural Research, Faculty of Science, Palacky University, Slechtitelu 11, CZ 783 71 Olomouc, Czech Republic. Laboratory of Growth Regulators, Institute of Experimental Botany AS CR, Slechtitelu 11, CZ 783 71 Olomouc, Czech Republic. E-mail: jiri.gruz@ gmail.com.

biosynthesized by the phenylpropanoid pathway which is the major source of plant secondary metabolites. Although kaempferol and its derivatives are present at high concentrations in many species, they are rarely the major secondary metabolites in plants. In most cases, other structurally related and potentially interfering phenolic compounds, such as flavonoids, phenolic acids and lignans, are present at the same time. The matrix is further complicated by the presence of various kaempferol derivatives, such as glycosides and esters. Thus the detection and determination of kaempferol mostly require sophisticated analytical systems consisting of a separation module online coupled with a selective detector. This is a review of natural occurrence and current analytical approaches for kaempferol determination in complex plant samples.

Natural Occurrence

Kaempferol is a typical flavonol, a class of flavonoids characterized by 3-hydroxyflavone backbone. Flavonoids are biosynthesized by the phenylpropanoid pathway which also produces other secondary metabolites, such as phenolic acids, coumarins, and lignans. The key enzyme in the phenylpropanoid pathway is phenylalanine ammonia-lyase (PAL) which converts L-phenylalanine into *trans*-cinnamic acid, the basic precursor of all phenylpropanoids. Cinnamate is hydroxylated by cinnamate-4-hydroxylase (monophenolmonooxygenase) to 4-coumaric acid which is conjugated with malonyl-CoA and transformed into naringenin, a basal flavonoid, by chalcone synthase and chalcone isomerase activity. Kaempferol is then formed from its precursor aromadendrin which is biosynthesized by hydroxylation of naringenin. While flavonoid biosynthetic enzymes are generally located in the cytosol, flavonoids themselves are actively transported to the vacuole and other subcellular or extracellular locations.

The phenylpropanoid pathway is widely distributed among land plants but is not generally present in other organisms. From the evolutionary point of view, phenylpropanoid pathway development was triggered by PAL acquisition by early land plants. Such acquisition was likely mediated by horizontal gene transfer during plant symbioses with soil bacteria and fungi (Emiliani et al., 2009). Subsequently, the ability to produce various phenolic compounds became highly important and the current knowledge suggests that PAL activity is essential for common flowering plants, such as *Arabidopsis thaliana* (Huang et al., 2010). Kaempferol may be found in plant species containing not only PAL but also further biosynthetic enzymes which are typical for plants, such as cinnamte-4-hydroxylase, chalcone synthase, chalcone isomerase, flavanone 3-hydroxylase and flavonol synthase. Therefore, kaempferol was not yet found in PAL-containing fungi and yeasts such as Basidiomycetes and Ascomycetes, which are lacking some of these enzymes and produce only unusual phenylpropanoids and flavonoids if any. According to the current knowledge, PAL is present in all land plants, including mosses, liverworts, and hornworts. The distribution of kaempferol or its derivatives mostly follows PAL and is illustrated by Figure 1 which contains only plant species where kaempferol or its derivatives were detected. Kaempferol is uniformly distributed among main classes and subclasses of land plants (Embryophyta), including angiosperms, gymnosperms, monocots and dicots.

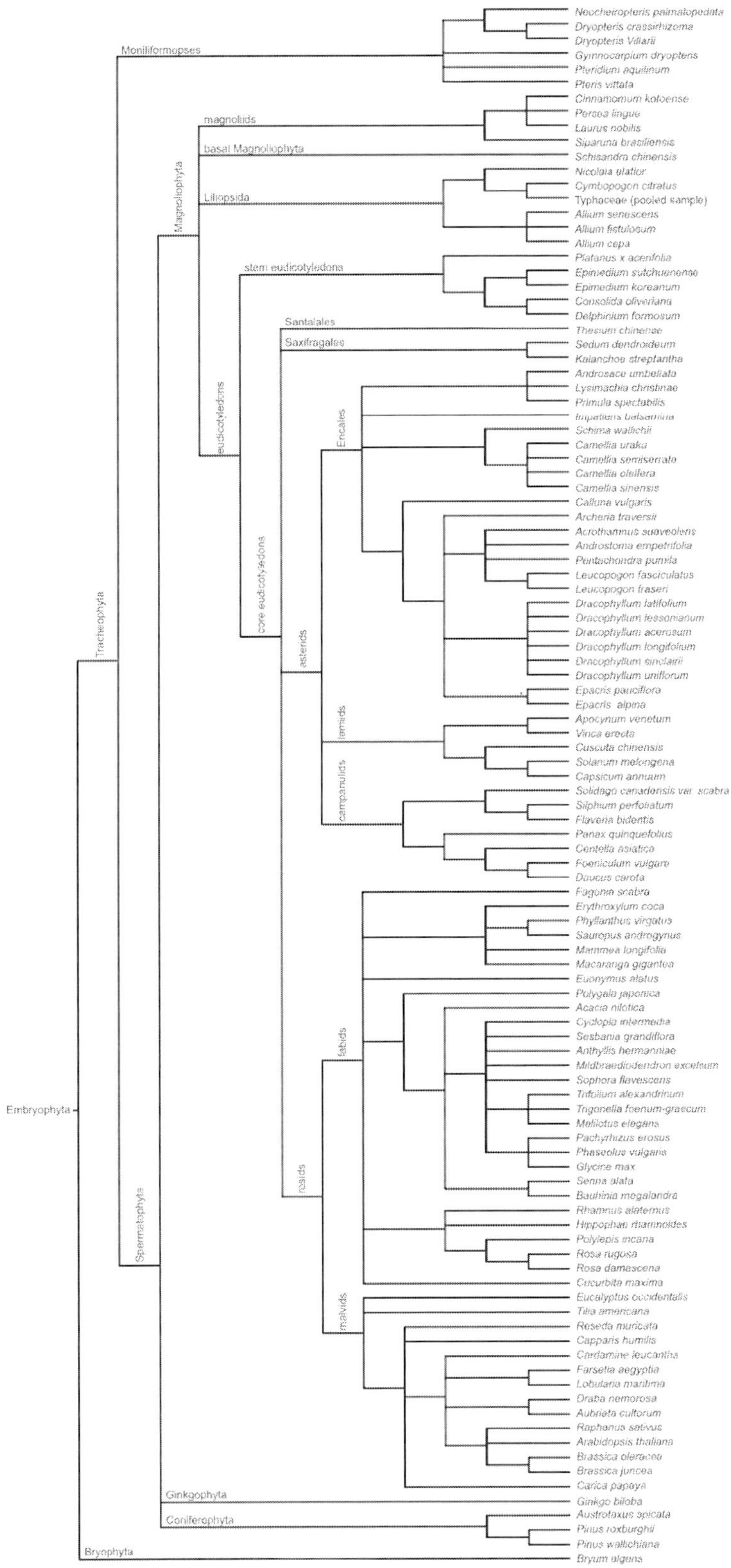

Figure 1. Phylogenetic tree showing plant species where kaempferol and/or its derivatives were identified.

Table 1. Kaempferol derivatives identified in plants

1	3,7,4´´-tri-O-methylkaempferol *Siparuna apiosyce* De Candolle (Leitao et al., 2000)	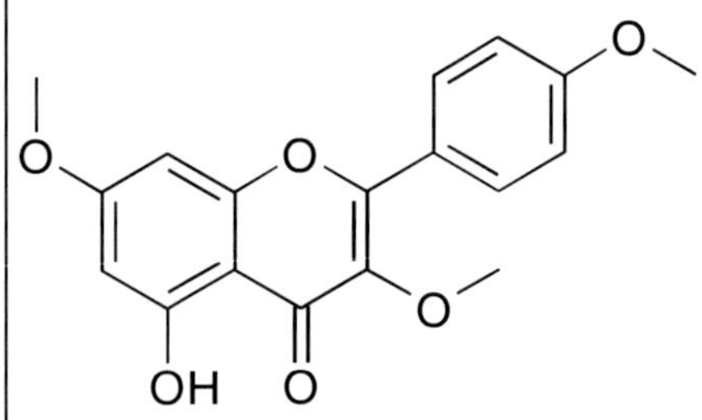
2	4´´´´-acetylsagittatin A *Kalanchoe streptantha* Baker (Costa et al., 1996)	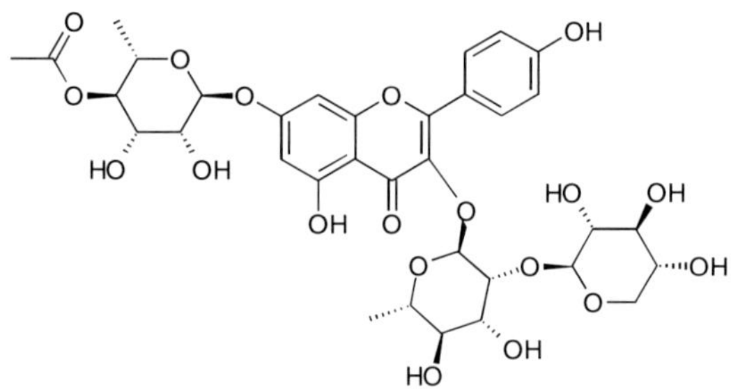
3	5-hydroxy-2-(4-hydroxyphenyl)-4-oxo-7-(α-L-rhamnopyranosyloxy)-4H-chromen-3-yl 2-O-acetyl-3-O-ß-D-glucopyranosyl-α-L-rhamnopyranoside *Androsace umbellata (Lourr.) Merr.* (Lei et al., 2009)	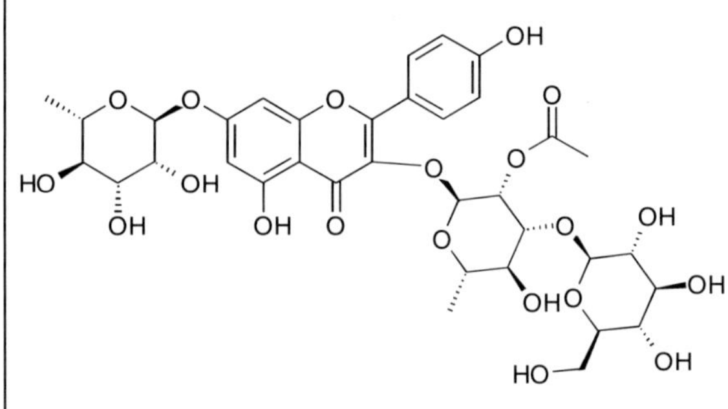
4	5-hydroxy-2-(4-hydroxyphenyl)-4-oxo-7-(α-L-rhamnopyranosyloxy)-4H-chromen-3-yl ß-D-glucopyranosyl-(1›2)-6-O-[(2E)-3-(4-hydroxyphenyl)prop-2-enoyl]-ß-D-glucopyranosyl-(1›2)-ß-D-glucopyranoside *Androsace umbellata (Lourr.) Merr.* (Lei et al., 2009)	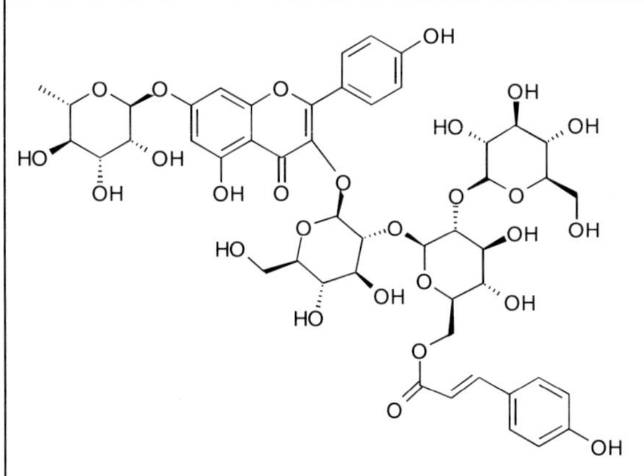

5	6,8-di-C-methylkaempferol 3,4´-dimethyl ether *Eucalyptus occidentalis* (Benyahia et al., 2004)	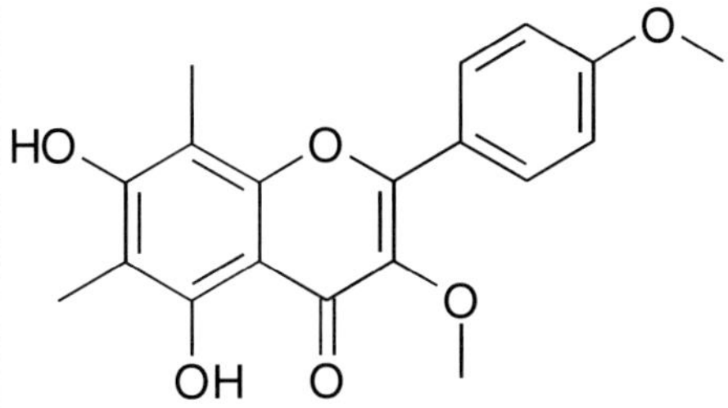
6	6,8-di-C-methylkaempferol 3-methyl ether *Eucalyptus occidentalis* (Benyahia et al., 2004)	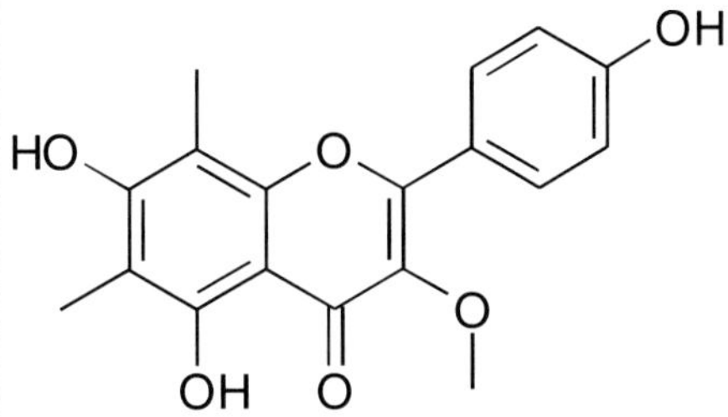
7	8-lavandulylkaempferol *Sophora flawescens* (Shen et al., 2006)	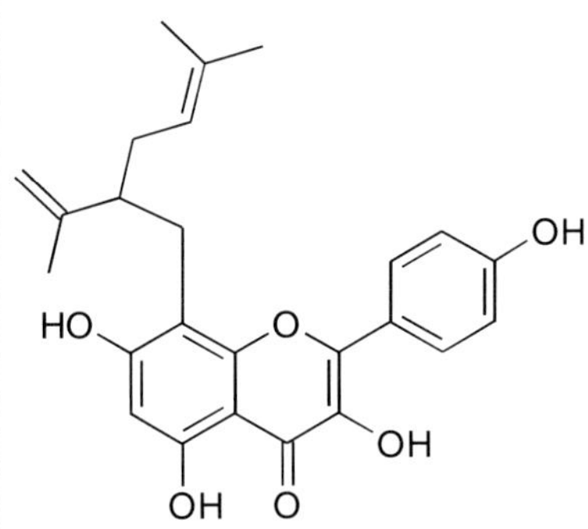
8	afzelin *Foeniculum vulgare, F. dulce* (Soliman et al., 2002)	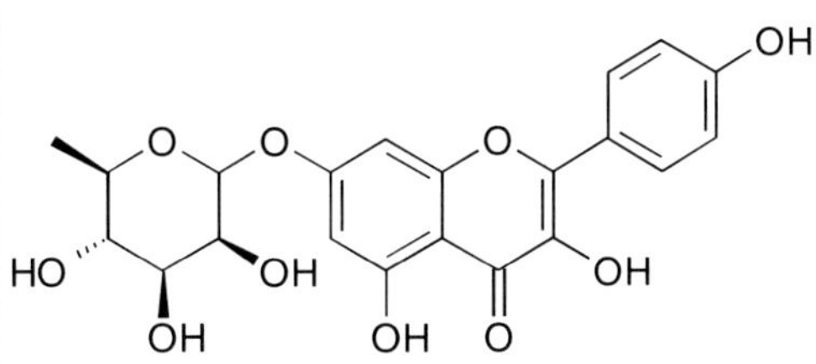
9	α-rhamnoisorobin *Sedum dendroideum* (De Melo et al., 2009)	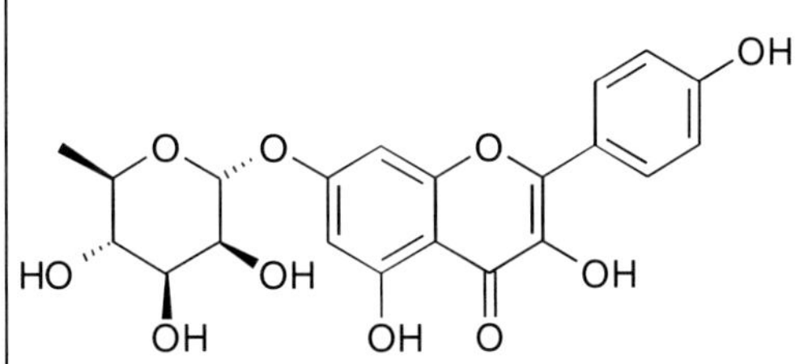

Table 1. (Continued)

10	Astragalin *Pteridium aquilinum (L.) Kuhn subspecies aquilinum* (Imperato, 1995)	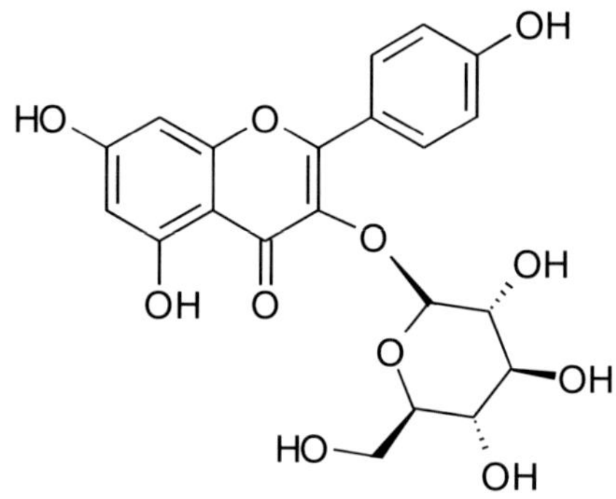
11	Biorobin *Consolida oliveriana* (Diaz et al., 2008)	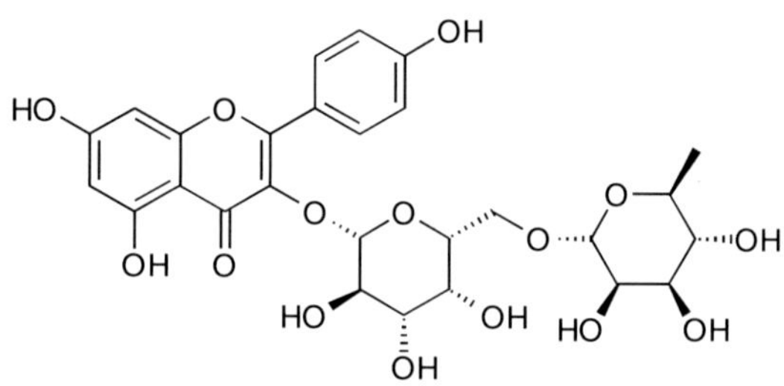
12	crassirhizomosides A *Dryopteris crassirhizoma* (Min et al., 2001)	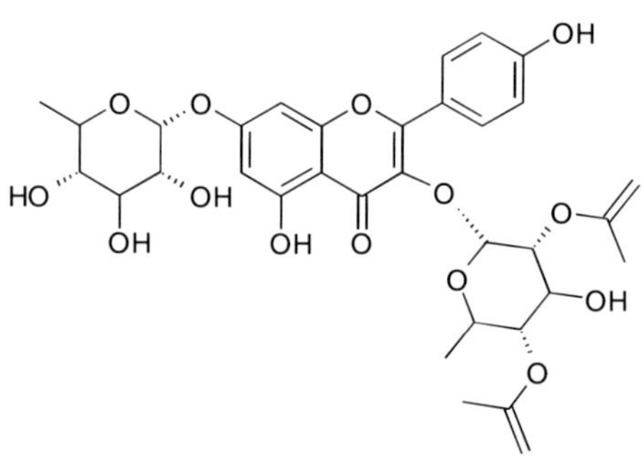
13	crassirhizomosides B *Dryopteris crassirhizoma* (Min et al., 2001)	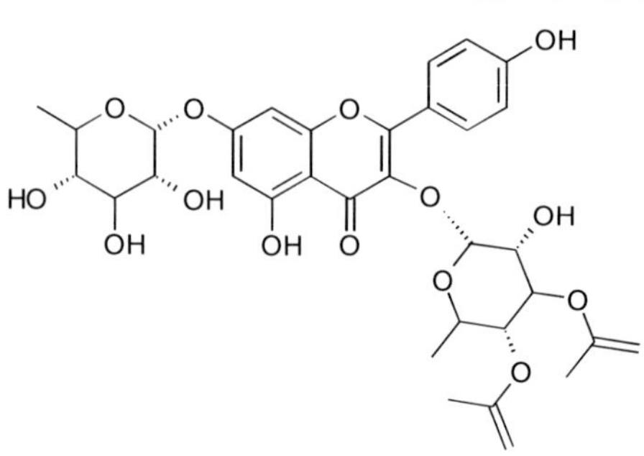

14	crassirhizomosides C *Dryopteris crassirhizoma* (Min et al., 2001)	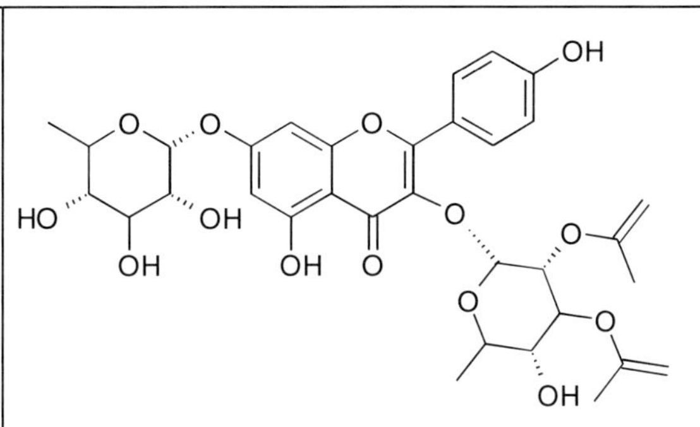
15	drabanemoroside *Draba nemorosa* (Moon et al., 2010)	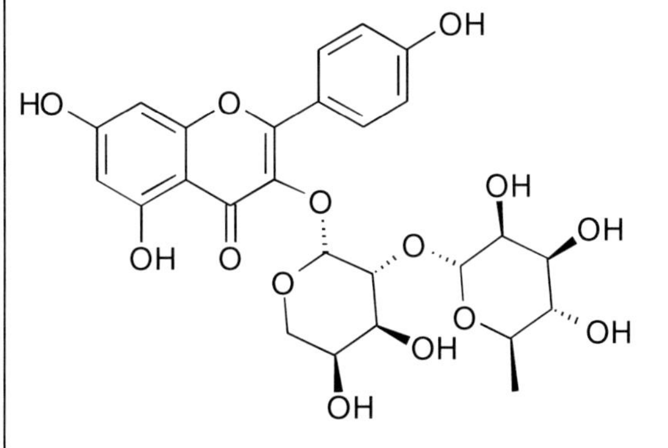
16	epimedin A *Epimedium koreanum* NAKAI (Sun et al., 1998)	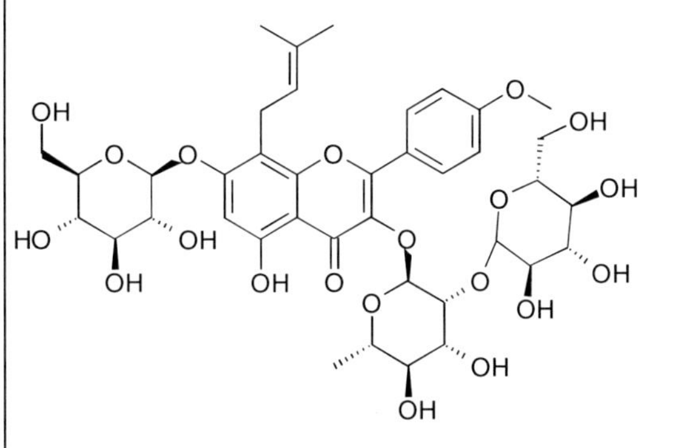
17	epimedin C *Epimedium sutchuenense* (Mizuno et al., 1991)	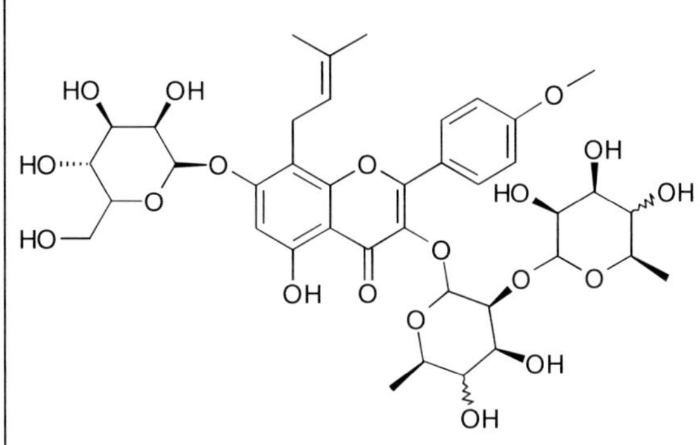

Table 1. (Continued)

18	glyasperin A *Macaranga gigantea* (Tanjung et al., 2009)	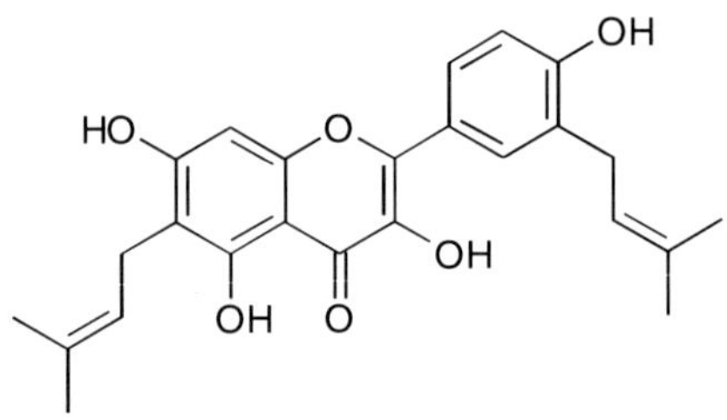
19	Icariin *Epimedium sutchuenense* (Mizuno et al., 1991)	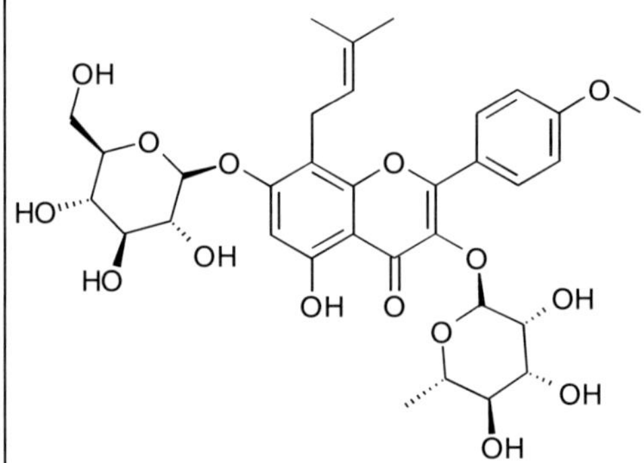
20	juglanin *Rosa rugosa* Thunb (Xiao et al., 2006)	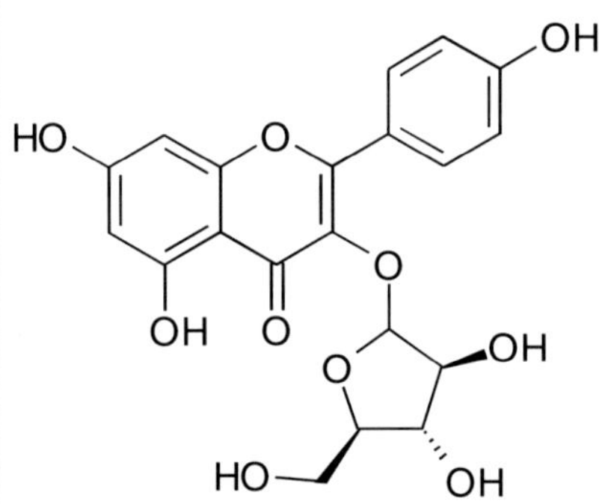
21	kaempferitrin *Tilia americana* L. var. mexicana (Schltdl.) Hardin (Aguirre-Hernandez et al., 2010)	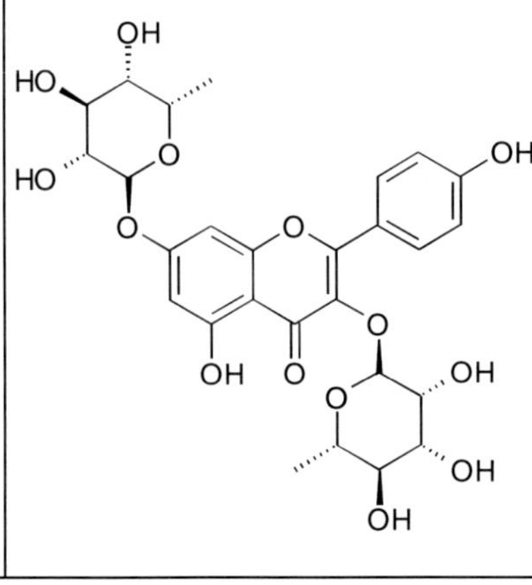

22	kaempferol 3-(4´´, 6´´-diacetylglucoside)7-rhamnoside *Delphinium formosum* Boiss. et Huet. (Ozden et al., 1998)	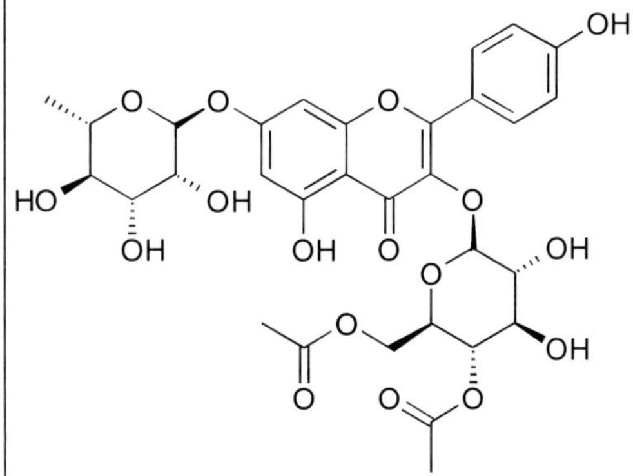
23	kaempferol 3-(6´´-acetylglucoside)7-rhamnoside *Delphinium formosum* Boiss. et Huet. (Ozden et al., 1998)	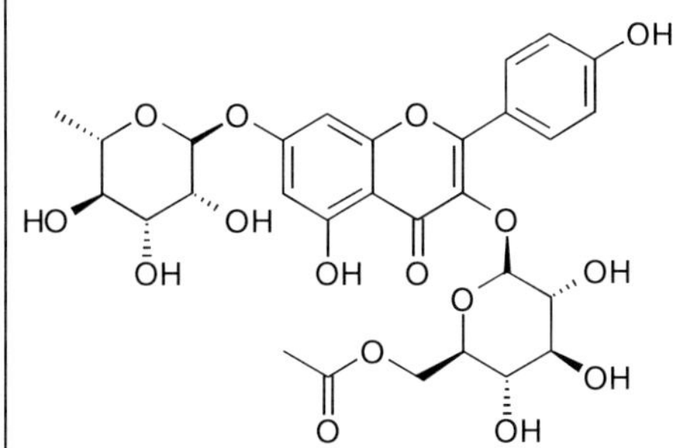
24	kaempferol 3-(6”-acetyl) glucoside *Polygala japonica* Houtt. (Jung et al., 1993)	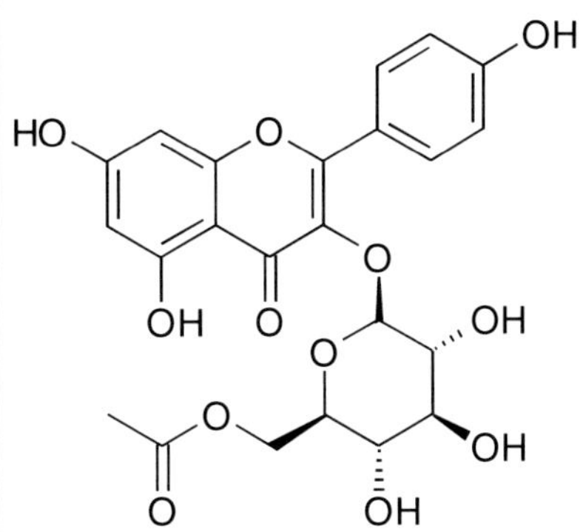
25	kaempferol 3-O-(2,6-di-O-α-rhamnopyranosyl)-ß-galactopyranoside *Glycine max* L. Merr. (Ho et al., 2002)	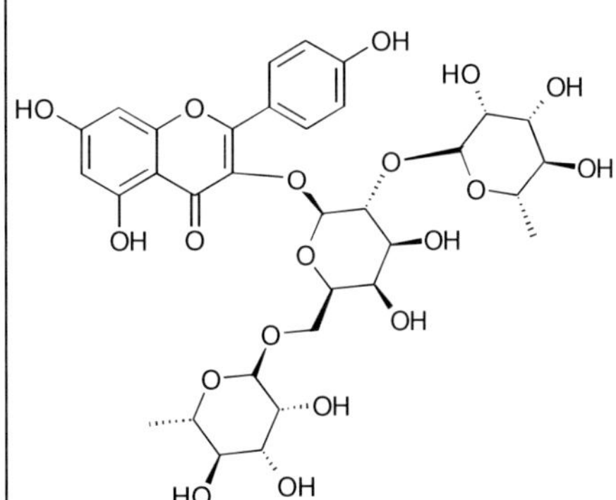

Table 1. (Continued)

26	kaempferol 3-O-(2´´,3´´-di-O-p-coumaroyl)-glucoside *Pteris vittata* L. (Imperato, 2003)	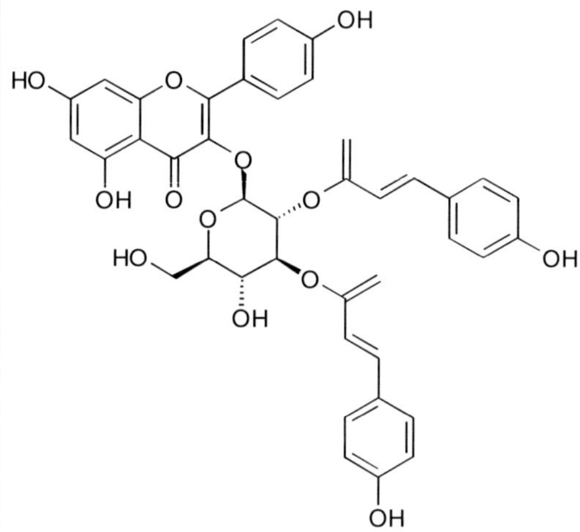
27	kaempferol 3-O-(2´´-O-(6´´-O-(p-coumaroyl)-ß-D-glucosyl)-α-L-rhamnosyl) *Ginkgo biloba* (van Beek, 2002)	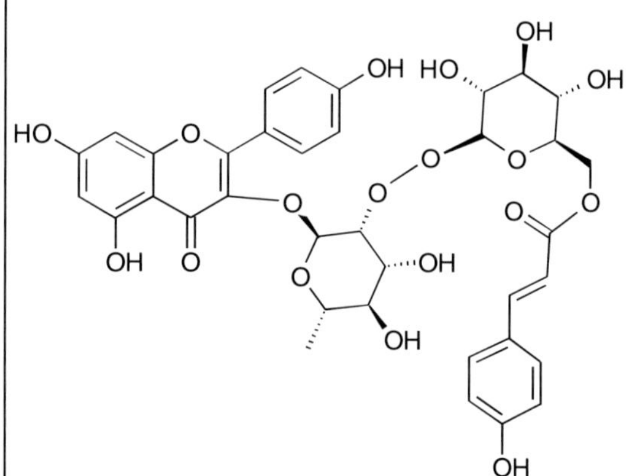
28	kaempferol 3-O-(2´´-O,6´´-O-bis(α-L-rhamnosyl)-ß-D-glucosyl) *Ginkgo biloba* (van Beek, 2002)	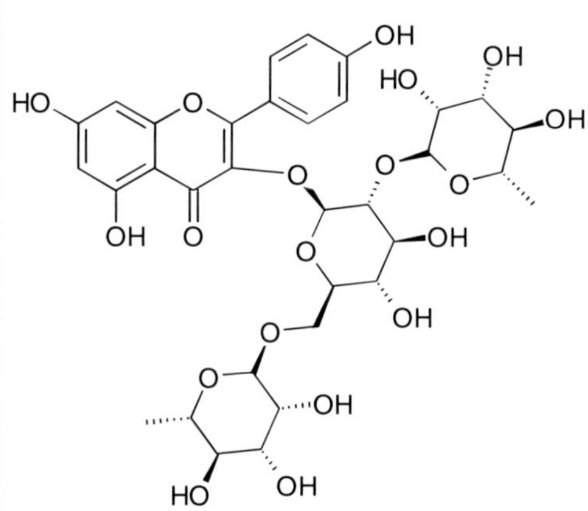

29	kaempferol 3-O-(3´´,4´´-di-E-p-coumaroyl)-α-L-rhamnopyranoside *Mammea longifolia* Planch and Triana syn (Rao et al., 2002)	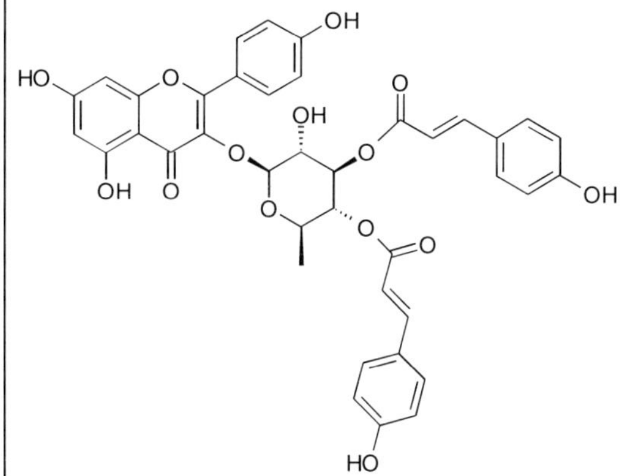
30	kaempferol 3-O-(5´´-feruloylapioside) *Pteridium aquilinum* (L.) Kuhn subspecies aquilinum (Imperato, 1996)	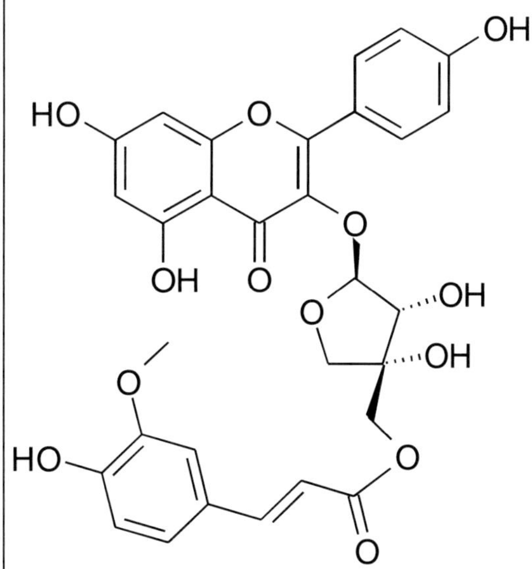
31	kaempferol 3-O-(6´´-O-(α-L-rhamnosyl)-ß-D-glucosyl) *Ginkgo biloba* (van Beek, 2002)	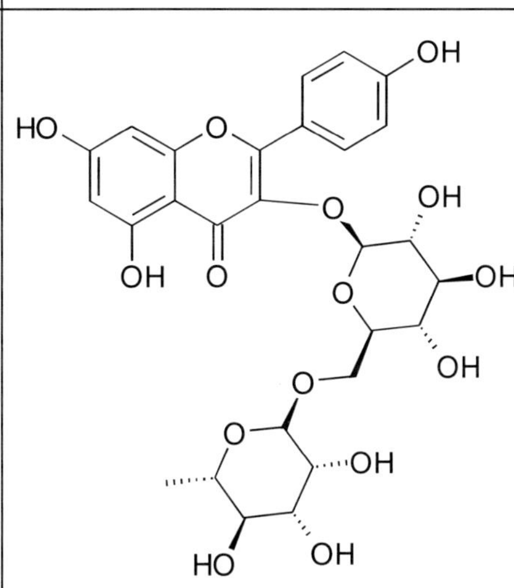

Table 1. (Continued)

32	kaempferol 3-O-(caffeoylrhamnoside) *Dryopteris villarii* (Bellardi) Schinz and Thell (Imperato, 2007)	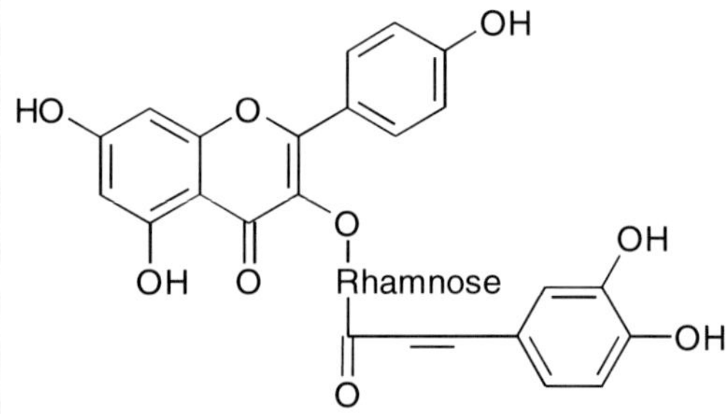
33	kaempferol 3-O-(x´´-O-p-coumaroyl-x´´-O-feruloyl)-glucose *Pteris vittata* L. (Imperato, 2003)	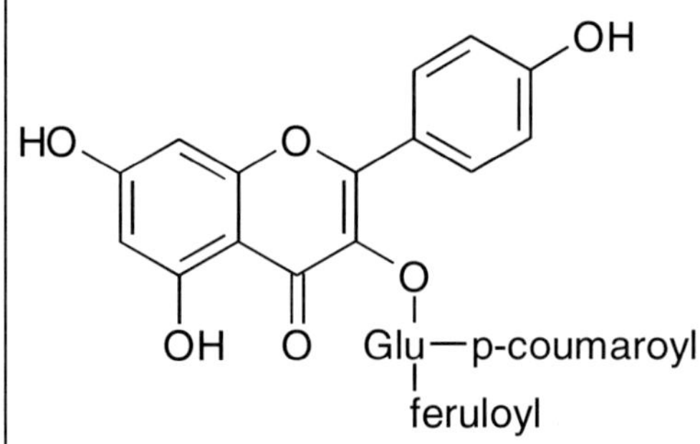
34	kaempferol 3-O-[(2-O-acetyl)-α-L-rhamnopyranosyl(1›3)(4-O-acetyl)-α-L-rhamnopyranosyl(1›6)-β-D-glucopyranoside *Camellia semiserrata* Chi (Tang et al., 2011)	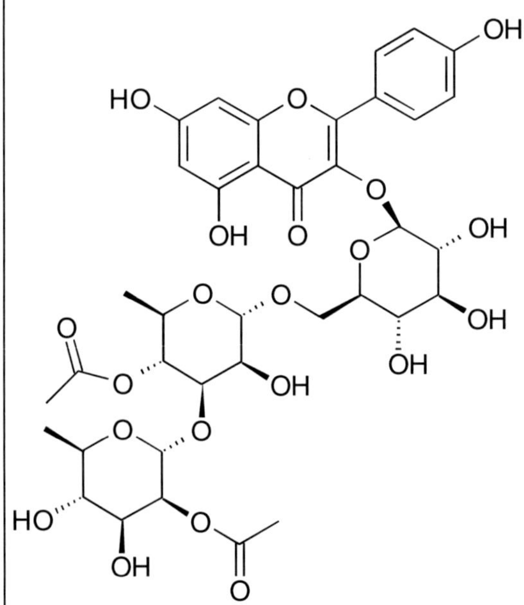

35	kaempferol 3-O-[(3-O-acetyl)-α-L-rhamnopyranosyl(1›3)(4-O-acetyl)-α-L-rhamnopyranosyl (1›6)-β-D-glucopyranoside *Camellia semiserrata* Chi (Tang et al., 2011)	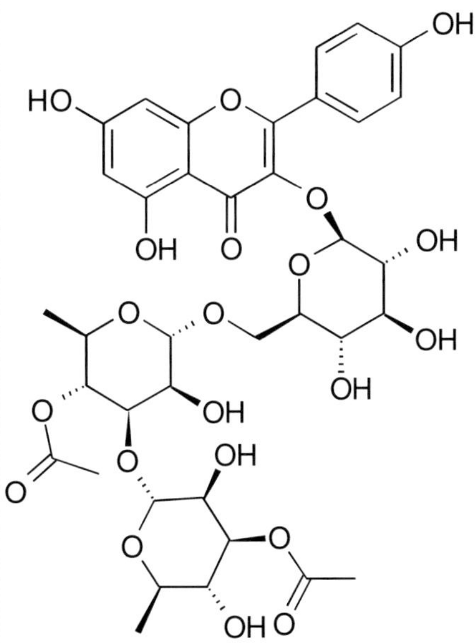
36	kaempferol 3-O-[2-O-(ß-glucopyranosyl)-6-O-(trans-sinapoyl)-ß-glucopyranoside] *Aubrieta* × *cultorum* 'Royal Red' (Tatsuzawa et al., 2012)	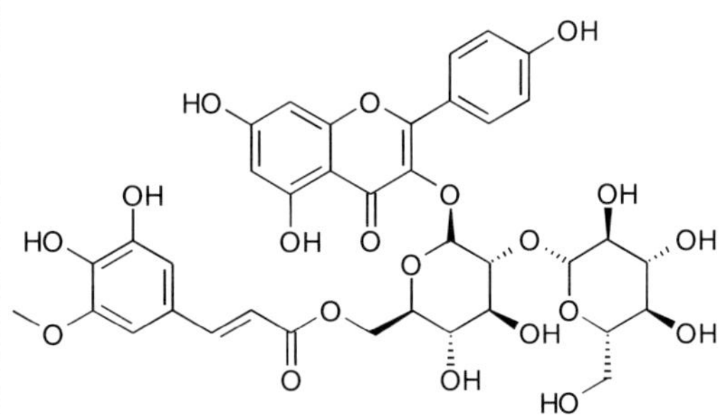
37	kaempferol 3-O-[2-O-(ß-glucopyranosyl)-6-O-(trans-sinapoyl)-ß-glucopyranoside]-7-O-[4-O-(ß-glucopyranosyl)-6-O-(trans-sinapoyl)-ß-glucopyranoside] *Aubrieta* × *cultorum* 'Royal Violet' (Tatsuzawa et al., 2012)	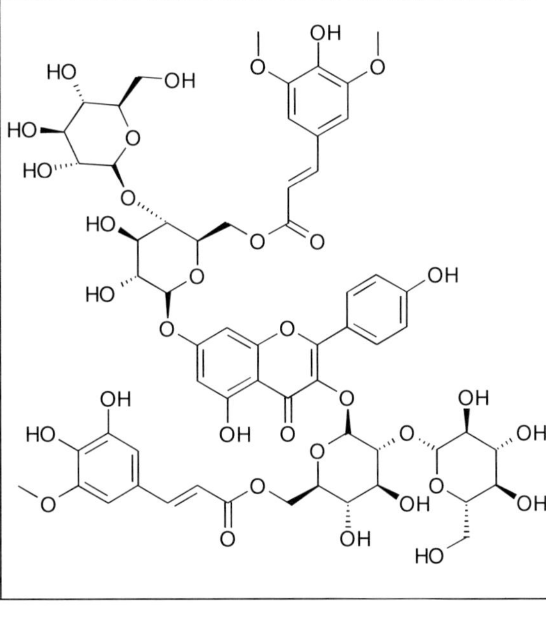

Table 1. (Continued)

38	kaempferol 3-O-[6”-O-(3-hydroxy-3-methylglutaroyl)-glucoside *Polygala japonica* Houtt. (Jung et al., 1993)	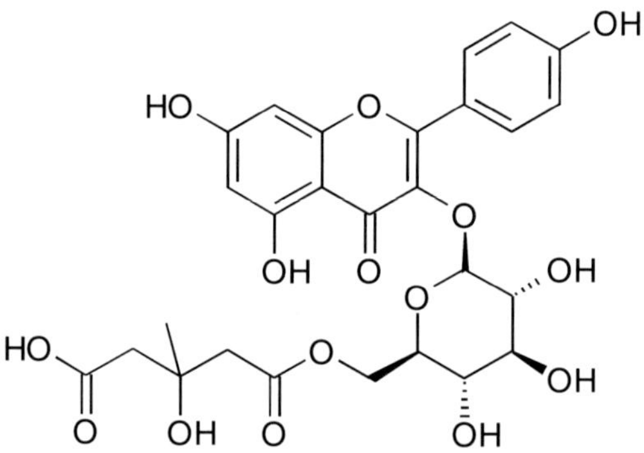
39	kaempferol 3-O-[6-O-(3-O-ß-glucopyranosyl-α-rhamnopyranosyl)-ß-glucopyranosyl] *Camellia uraku* Tarokaja (Tanikawa et al., 2011)	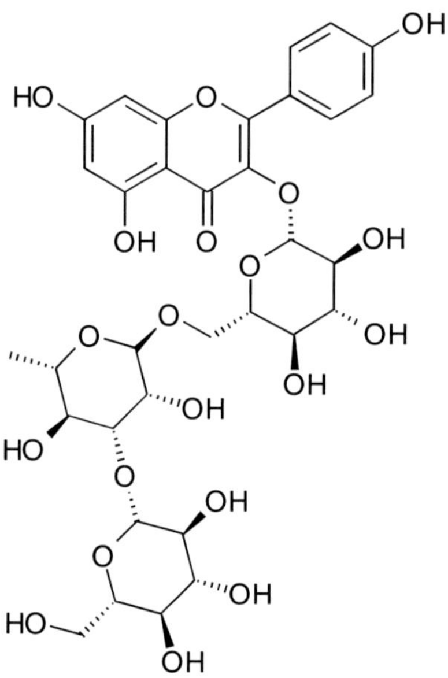
40	kaempferol 3-O-[α-L-rhamnopyranosyl-(1›6)]-ß-D-glucopyranosyl-7-O-α-L-rhamnopyranoside *Androsace umbellata (Lourr.) Merr.* (Lei et al., 2009)	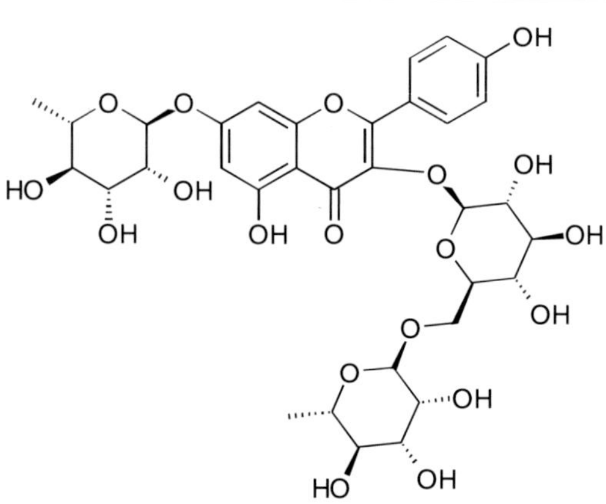

41	kaempferol 3-O-[α-L-rhamnopyranosyl-(1›2)-α-L-arabinopyranoside]-7-O-α-L-rhamnopyranoside *Anthyllis hermanniae* L. (Halabalaki et al., 2011)	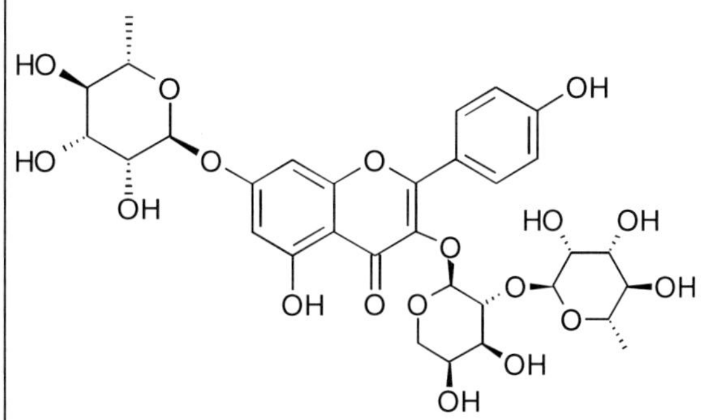
42	kaempferol 3-O-[ß-d-glucopyranosyl-(1›3)]-α-L-rhamnopyranosyl-7-O-α-L-rhamnopyranoside *Androsace umbellata (Lourr.) Merr.* (Lei et al., 2009)	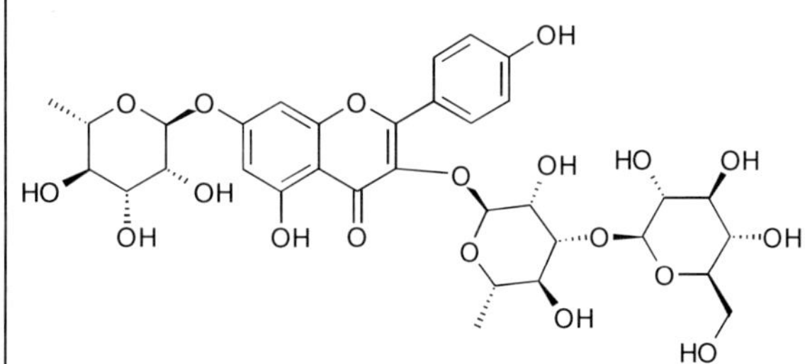
43	kaempferol 3-O-α-(2”-galloyl)rhamnoside *Bauhinia megalandra* (Rodriguez et al., 2010)	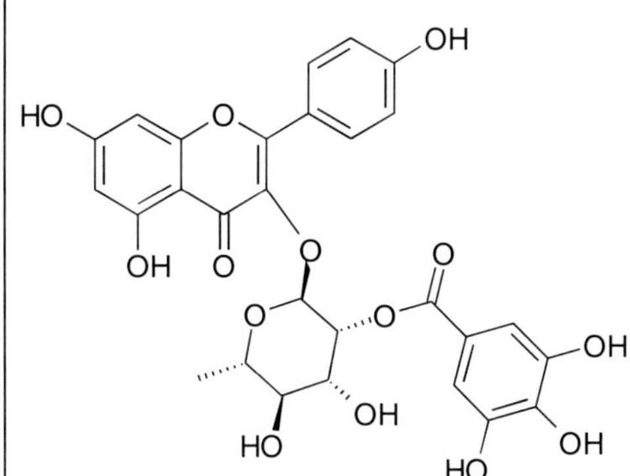
44	kaempferol 3-O-α-L-(2-E-p-coumaroyl) rhamnoside *Platanus acerifilia* Willd. (Kaouadji et al., 1993)	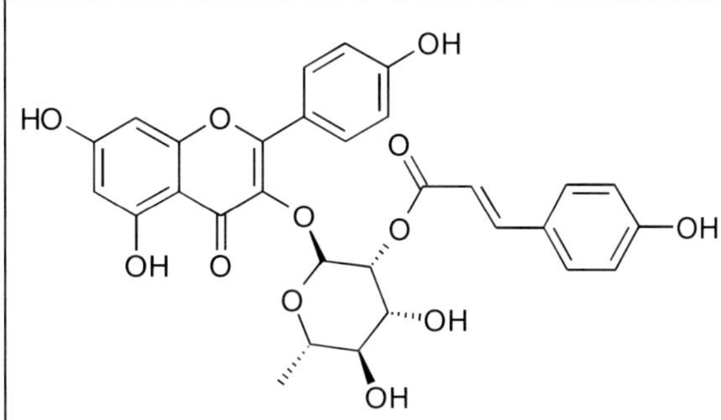

Table 1. (Continued)

45	kaempferol 3-O-α-L-(2-Z-p-coumaroyl) rhamnoside *Platanus acerifilia* Willd. (Kaouadji et al., 1993)	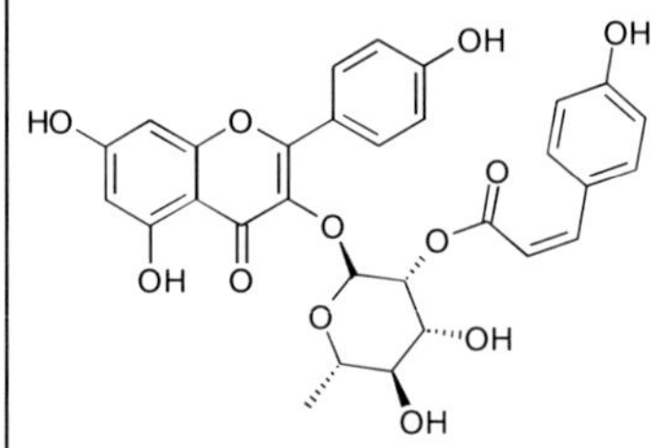
46	kaempferol 3-O-α-L-[2-(Z)-p-coumaroyl-4-(E)-p-coumaroyl] rhamnopyranoside *Cinnamomum kotoense* Kanehira and Sasaki (Kuo et al., 2005)	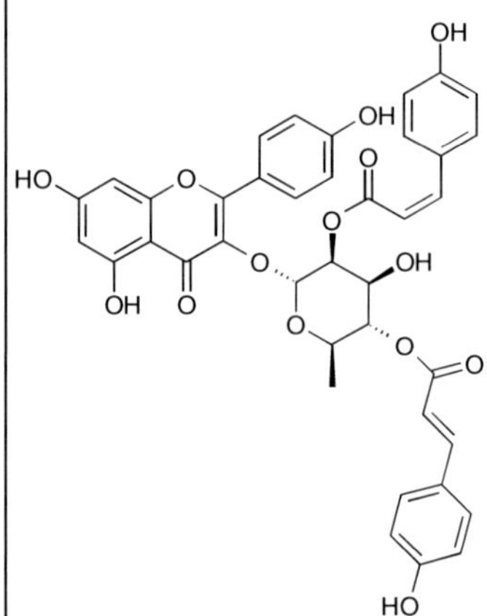
47	kaempferol 3-O-α-L-[2´´,4´´-di-(E)-p-coumaroyl] rhamnopyranoside *Cinnamomum kotoense* Kanehira and Sasaki (Kuo et al., 2005)	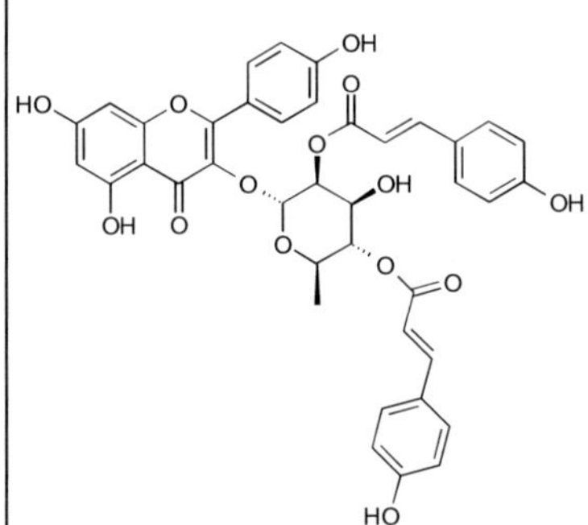
48	kaempferol 3-O-α-L-rhamnopyranosyl-(1›2)-ß-D-glucopyranosyl (1›6)-ß-D-galactopyranoside *Glycine max* L. Merr. (Ho et al., 2002)	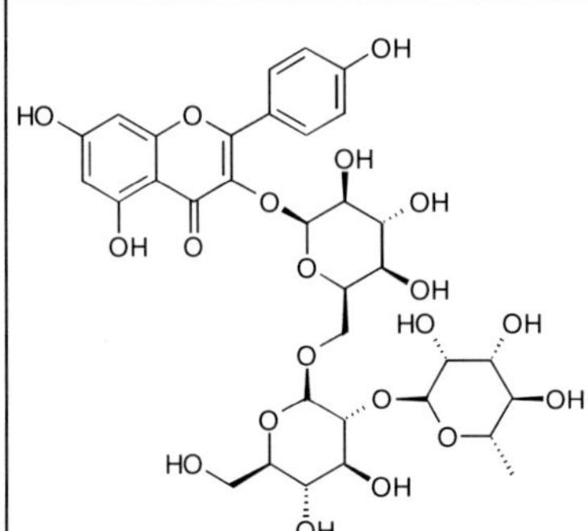

49	kaempferol 3-O-α-L-rhamnopyranosyl-(1›6)-ß-D-galactopyranoside *Glycine max* L. Merr. (Ho et al., 2002)	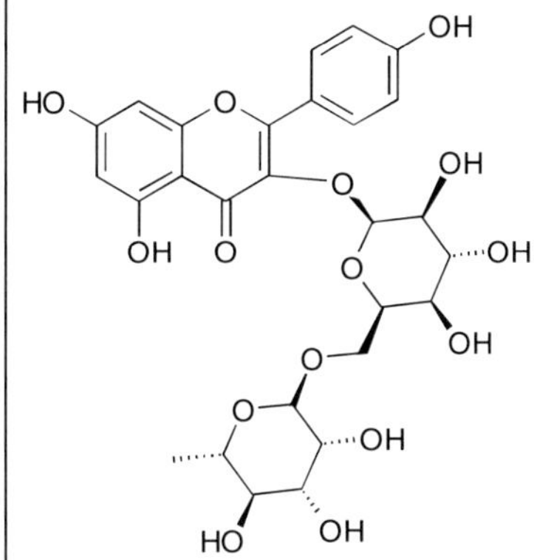
50	kaempferol 3-O-α-rhamnoside *Bauhinia megalandra* (Rodriguez et al., 2010)	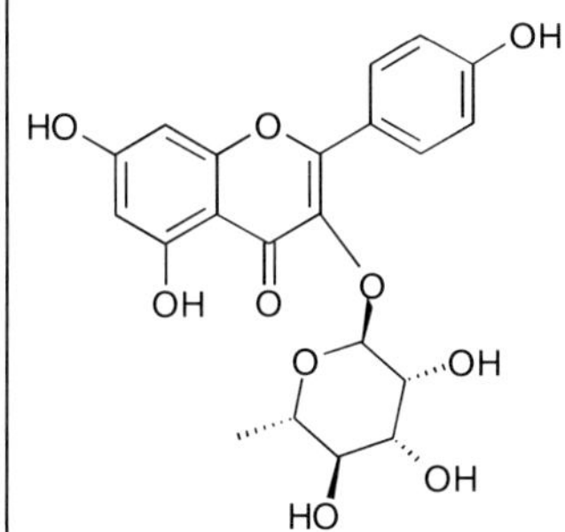
51	kaempferol 3-O-digalactopyranoside *Glycine max* L. Merr. (Ho et al., 2002)	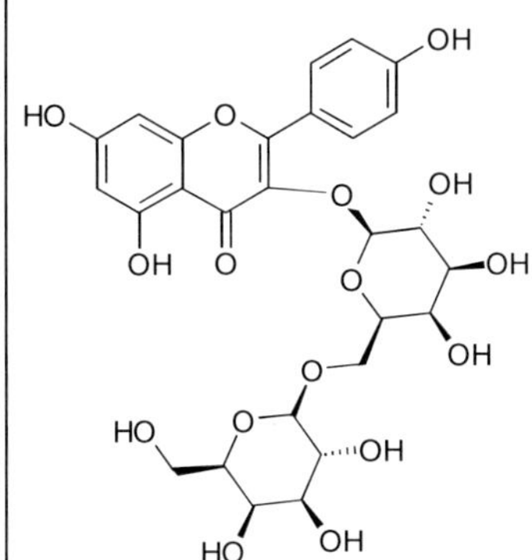
52	kaempferol 3-O-isorhamninoside *Rhamnus alaternus* L. (Bhouri et al., 2011)	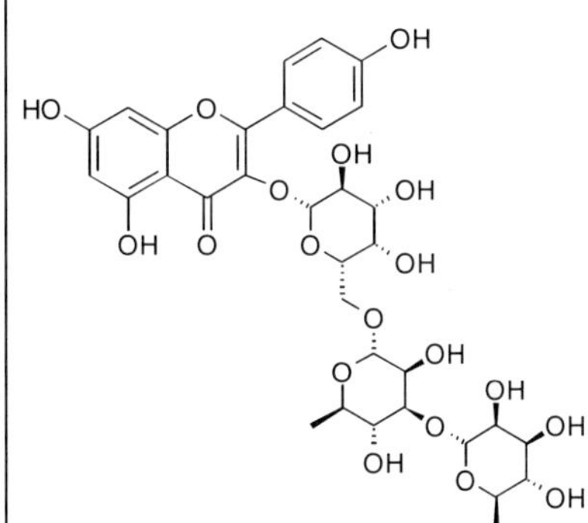

Table 1. (Continued)

53	kaempferol 3-O-neohesperidoside-7-O-α-rhamnopyranoside *Sedum dendroideum* (De Melo et al., 2009)	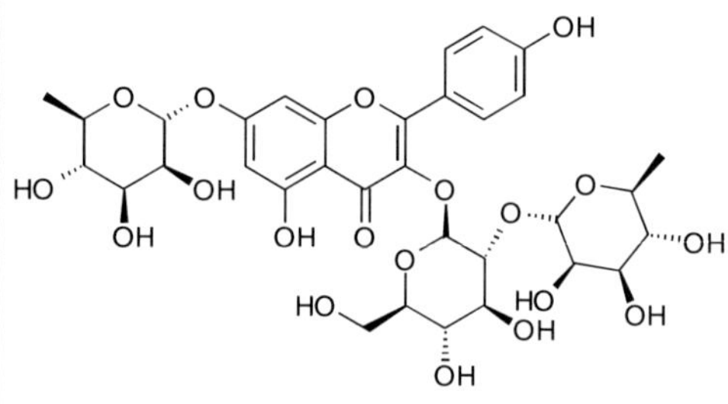
54	kaempferol 3-O-neohesperidoside-7-O-ß-glucopyranoside *Sedum dendroideum* (De Melo et al., 2009)	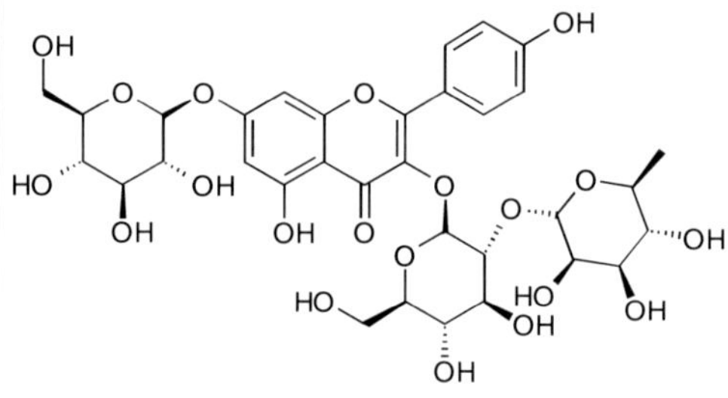
55	kaempferol 3-O-rutenoside *Glycine max* L. Merr. (Ho et al., 2002)	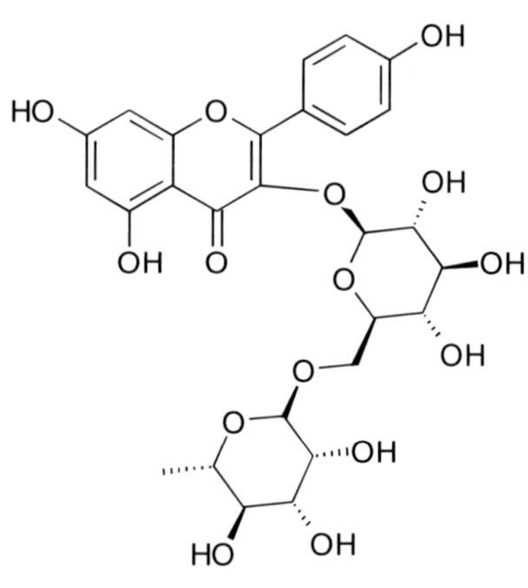
56	kaempferol 3-O-rutinoside 7-O-ß-D-apiofuranoside *Solidago altissima* L. (Jin et al., 2007)	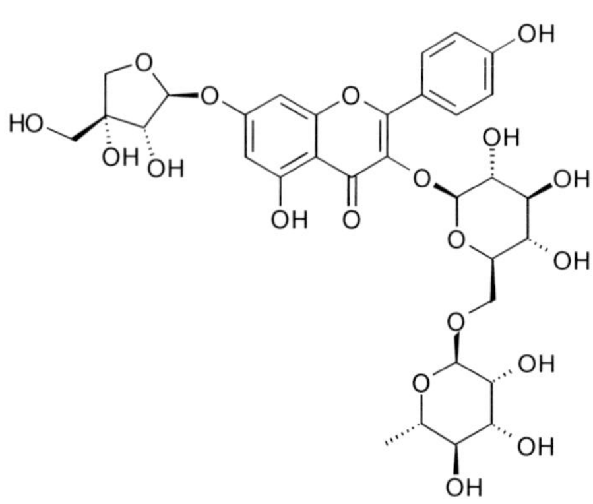

57	kaempferol 3-O-ß- (6´´-caffeoylglucoside) *Pteridium aquilinum (L.) Kuhn subspecies aquilinum* (Imperato and Minutiello, 1997)	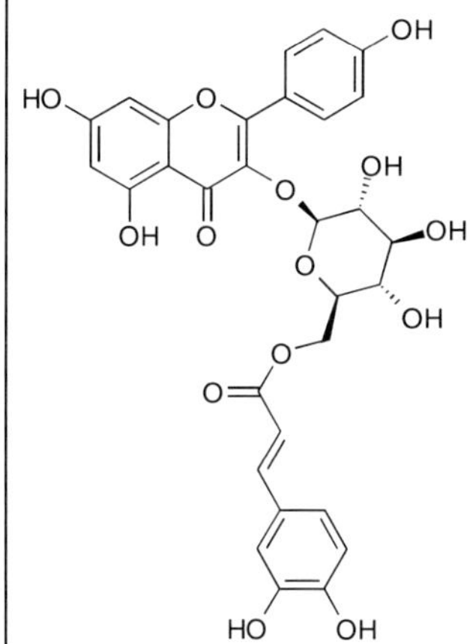
58	kaempferol 3-O-ß- (6´´-feruloylglucoside) *Polylepis incana* H.B.K. (Catalano et al., 1994)	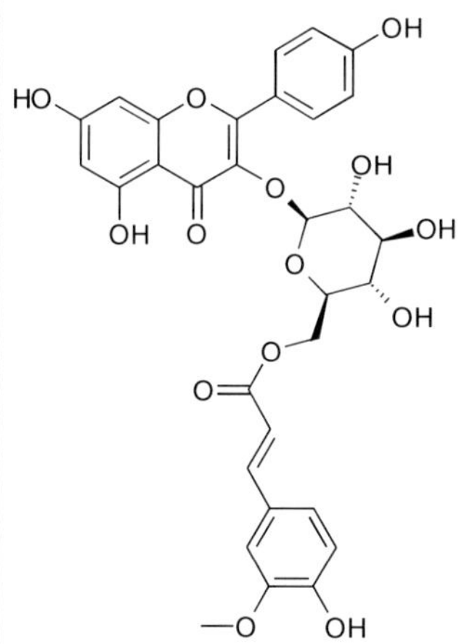
59	kaempferol 3-O-ß-D-(2-O-α-L-rhamnopyranosyl) galactopyranoside *Glycine max* L. Merrill. ´Jindai´ (Zang et al., 2011)	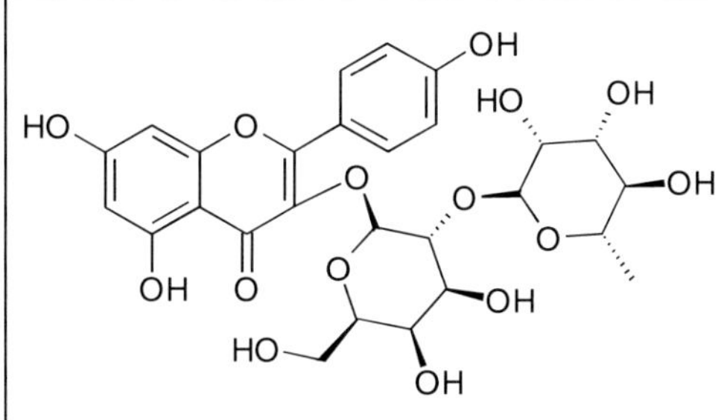
60	kaempferol 3-O-ß-D-(2-O-ß-D-glucopyranosyl) galactopyranoside *Glycine max* L. Merrill. ´Jindai´ (Zang et al., 2011)	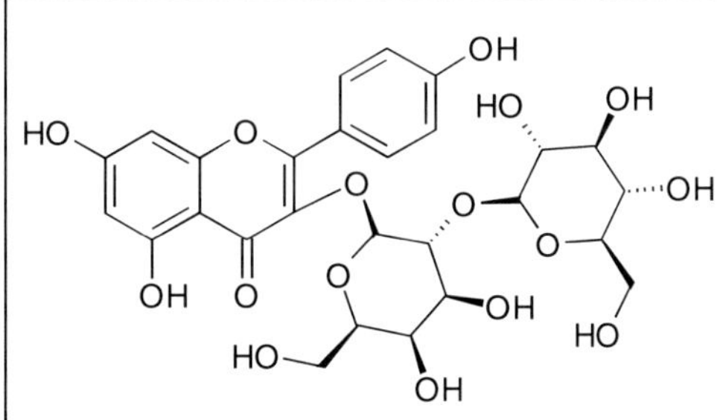

Table 1. (Continued)

61	kaempferol 3-O-ß-D-apiofuranoside 7-O-α-L-rhamnosyl-(1´´´´›6´´´)-O-ß-D (2´´´-O-E-caffeoylgalactopyranoside) *Silphium perfoliatum* L. tribe Heliantheae (El-Sayed et al., 2002)	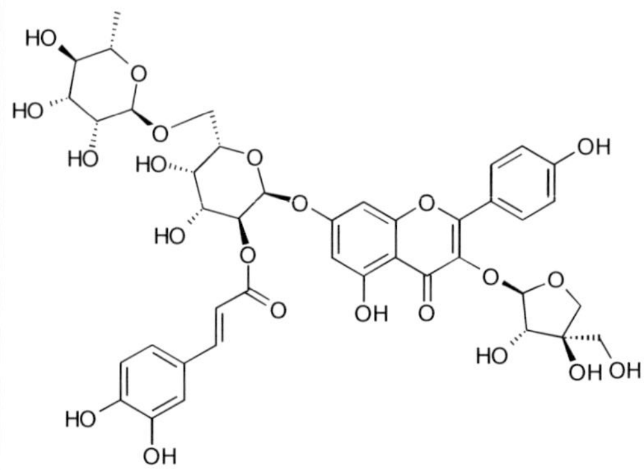
62	kaempferol 3-O-ß-d-apiofuranoside 7-O-α-L-rhamnosyl(1´´´´›6´´´)-O-ß-D-galactopyranoside *Silphium perfoliatum* L. tribe Heliantheae (El-Sayed et al., 2002)	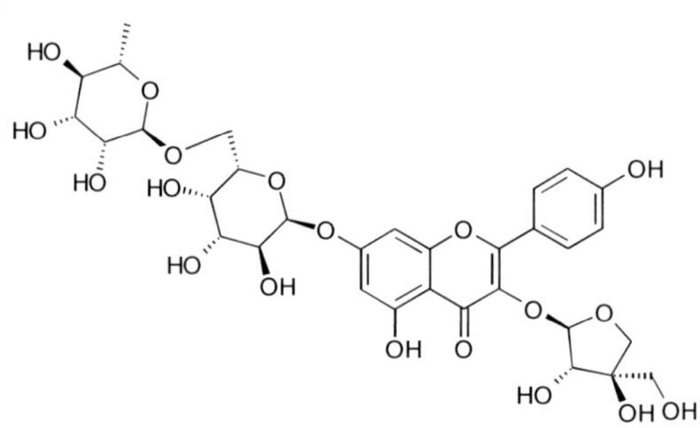
63	kaempferol 3-O-α-L-rhamnopyranosyl-(1›2)-O-ß-d-galactopyranoside-7-α-L-rhamnopyranose *Lobularia maritima* (L.) Desv. (Fiorentino et al., 2009)	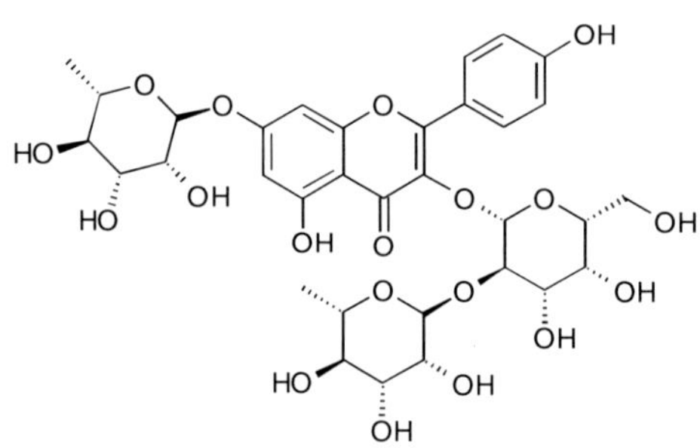
64	kaempferol 3-O-ß-D-glucopyranosyl-(1›2)-ß-D-galactopyranoside-7-α-L-rhamnopyranose *Lobularia maritima* (L.) Desv. (Fiorentino et al., 2009)	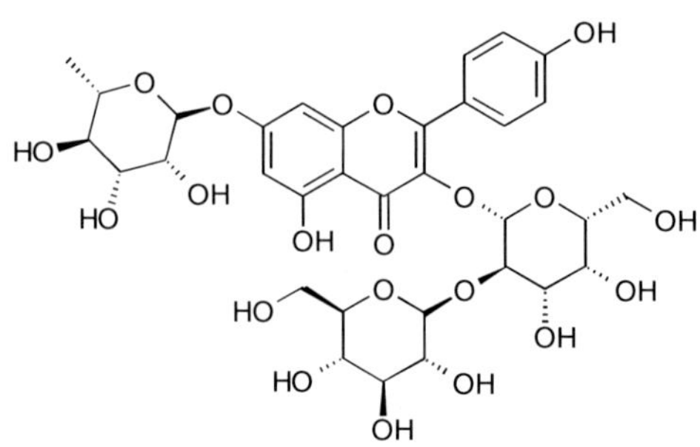

65	kaempferol 3-O-ß-D-glucopyranosyl (1›2)-O-[α-L-rhamnopyranosyl (1›6)]-ß-D-galactopyranoside *Glycine max* L. Merrill. 'Jindai' (Zang et al., 2011)	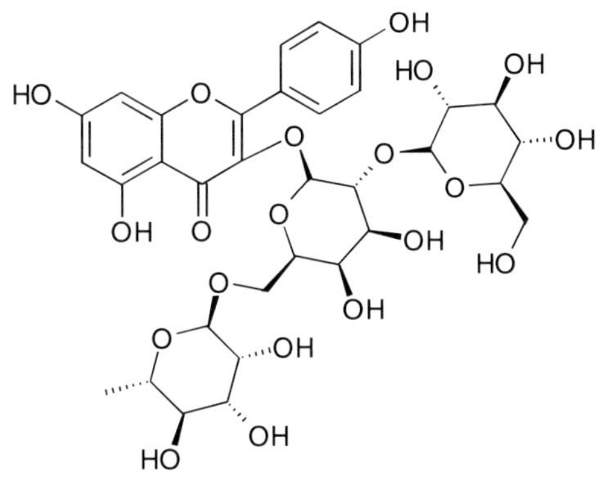
66	kaempferol 3-O-ß-D-glucopyranosyl (1›2)-O-[α-L -rhamnopyranosyl (1›6)]-ß-D-glucopyranoside *Glycine max* L. Merrill. 'Jindai' (Zang et al., 2011)	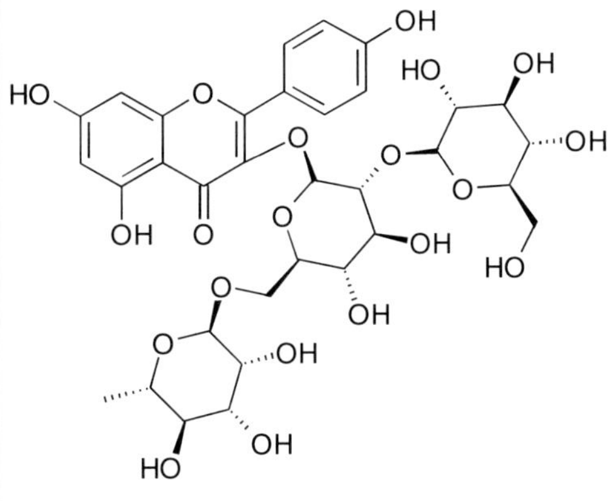
67	kaempferol 3-O-ß-D-glucopyranosyl-(1′′′›2′′)-O-α-L-rhamnopyranoside 7-O-ß-D-(6′′′′-O-E-coumaryl-glucopyranoside) *Reseda muricata* C. Presl. (El-Sayed et al., 2001)	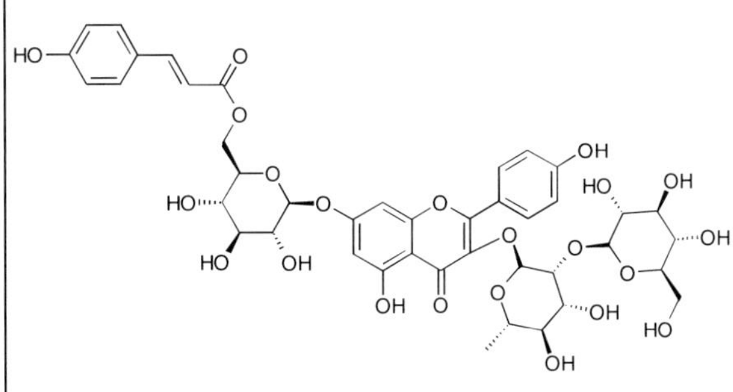
68	kaempferol 3-O-ß-D-glucopyranosyl-(1′′′›2′′)-O-α-L-rhamnopyranoside-7-O-ß-D-glucopyranoside *Reseda muricata* C. Presl. (El-Sayed et al., 2001)	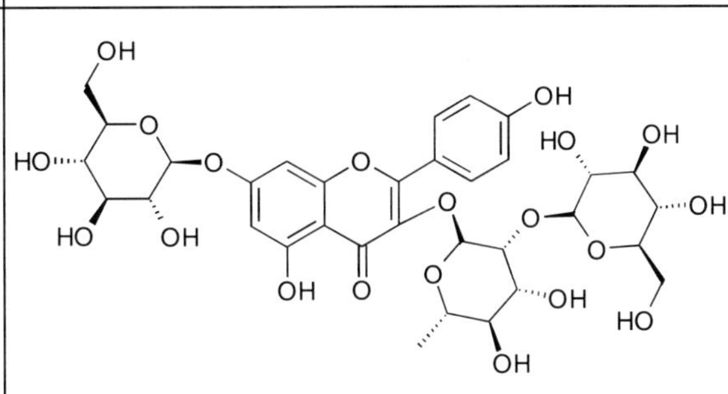

Table 1. (Continued)

69	kaempferol 3-O-α-L-rhamnopyranosyl-(1›2)-O-ß-D-glucopyranoside-7-α-L-rhamnopyranose *Lobularia maritima* (L.) Desv. (Fiorentino et al., 2009)	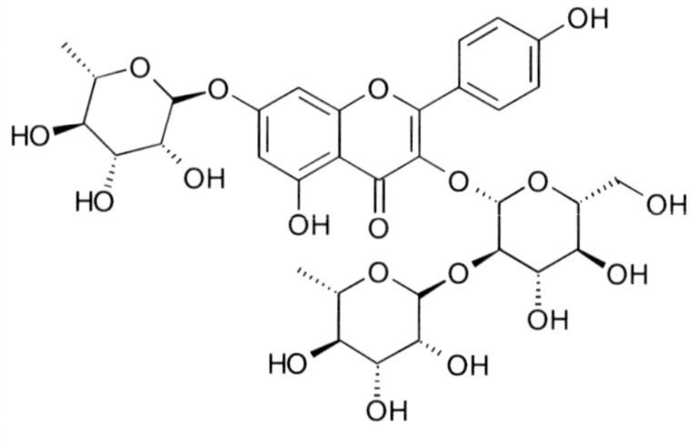
70	kaempferol 3-O-ß-D-glucopyranosyl-(1›2)-O-α-L-xylopyranoside *Lobularia maritima* (L.) Desv. (Fiorentino et al., 2009)	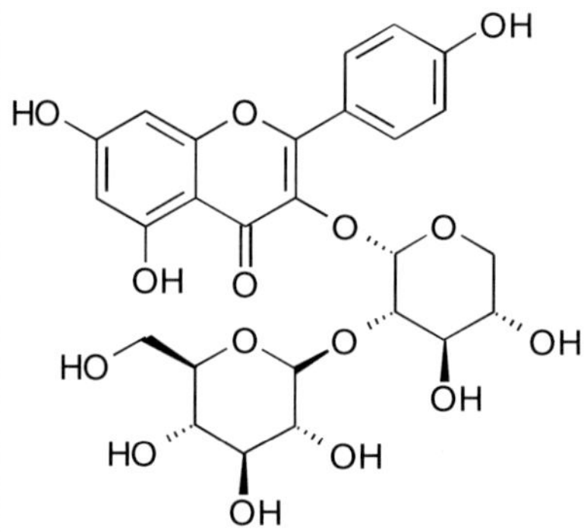
71	kaempferol 3-O-ß-D-glucosyl-(1›6)-ß-D-glucoside *Sauropus androgenus* (Yu et al., 2006)	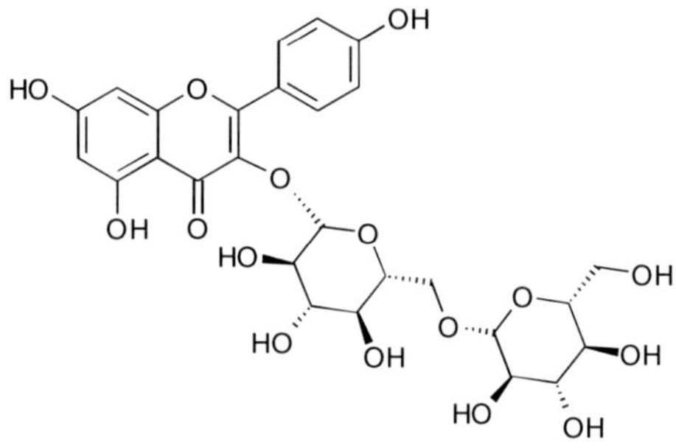
72	kaempferol 3-O-ß-D-glucosyl-7-O-α-L –rhamnoside *Sauropus androgenus* (Yu et al., 2006)	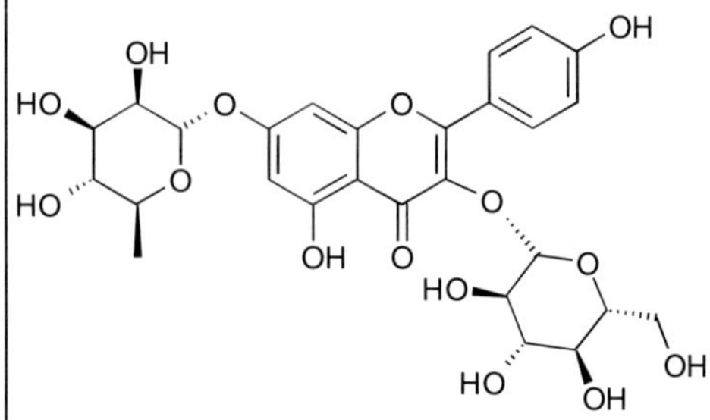

73	kaempferol 3-O-ß-L-arabinopyranosyl-(1›4)-α-L-rhamnopyranoside-7-O-α-L-rhamnopyranoside *Fagonia taeckholmiana* (Ibrahim et al., 2008)	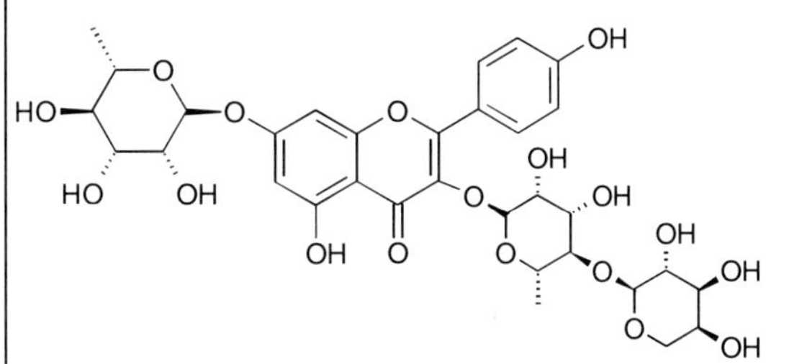
74	kaempferol 4´-O-(rhamnosyl)glucoside *Erythroxylum coca* var. *ipadu* Plowman. (Johnson et al., 2003)	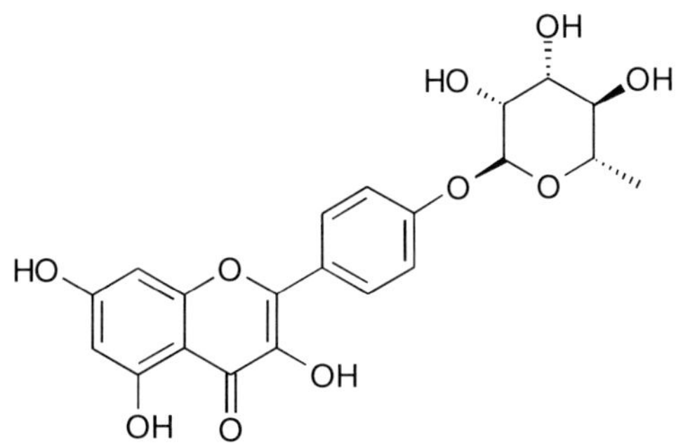
75	kaempferol 7-O-rhamnoside-4'-O-glucoside *Pteridium aquilinum (L.) Kuhn subspecies aquilinum* (Imperato, 1998)	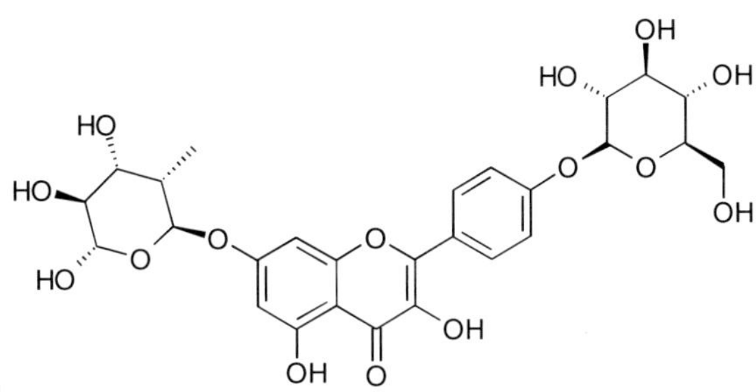
76	kaempferol 7-O-ß-D-glucopyranosyl-(1›3)-[ß-D-glucopyranosyl-(1› 6)]-glucopyranoside *Brassica juncea* L. (Kim et al., 2002)	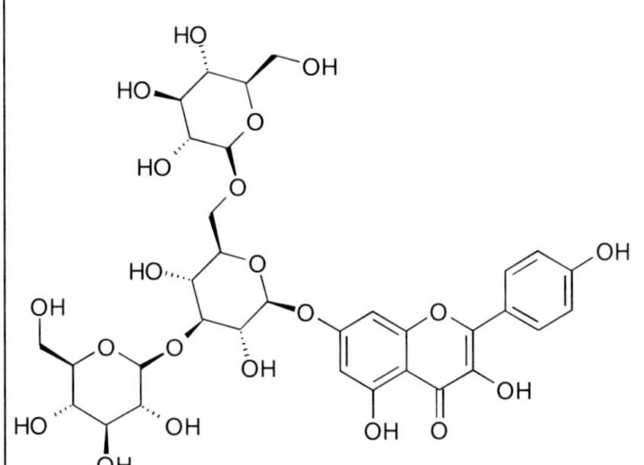

Table 1. (Continued)

77	kaempferol 8-C-glucoside *Primula spectabilis* Tratt. (Vitalini et al., 2011)	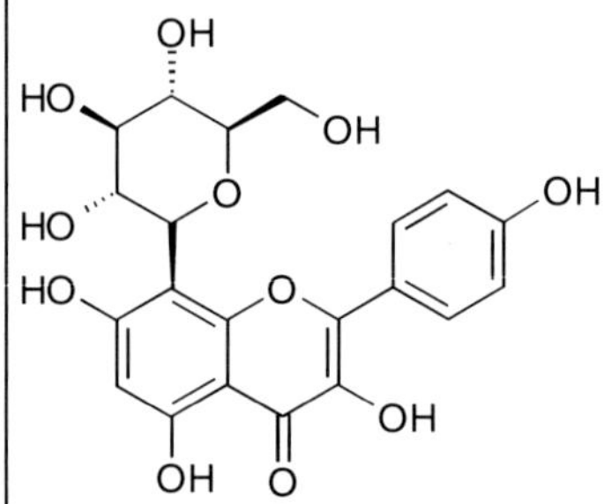
78	kaempferol *Acacia nilotica* Willd. Ex. Del. (Singh et al., 2008)	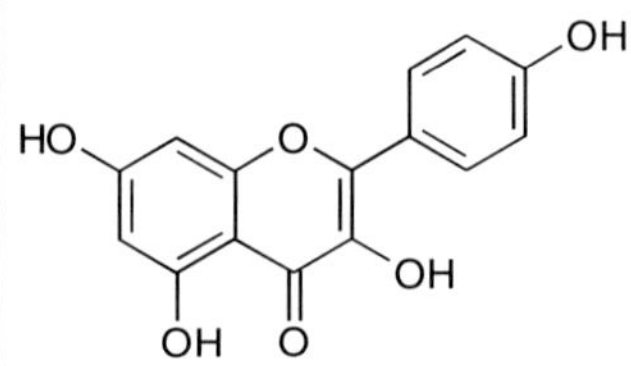
79	kaempferol-3,7-di-O-α-L-rhamnopyranoside *Farsetia aegyptia* Turra. (Marzouk et al., 2009)	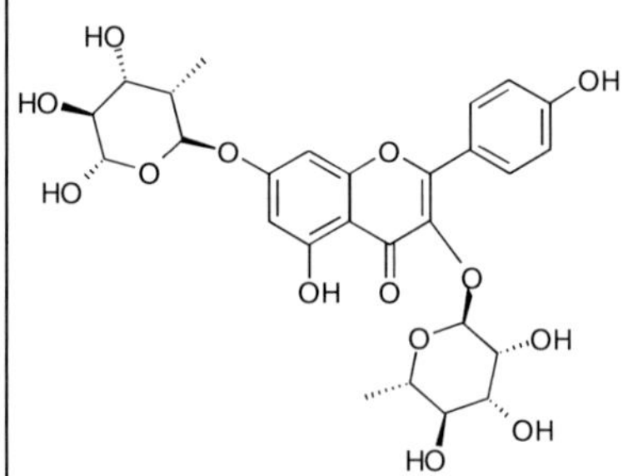
80	kaempferol-3-O-[2”-O-α-L-rhamnopyranosyl-3”-O-ß-D-glucopyranosyl]-ß-D-glucopyranoside *Impatiens balsamina* L. (Fukumoto et al., 1994)	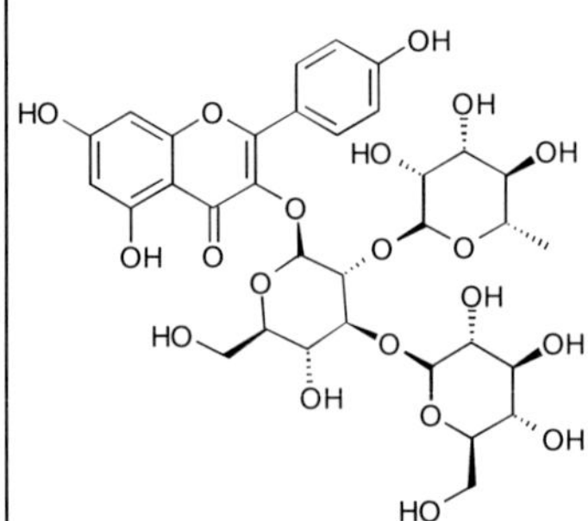
81	kaempferol-3-O-β-glucuronide *Pteris vittata* L. (Imperato, 2003)	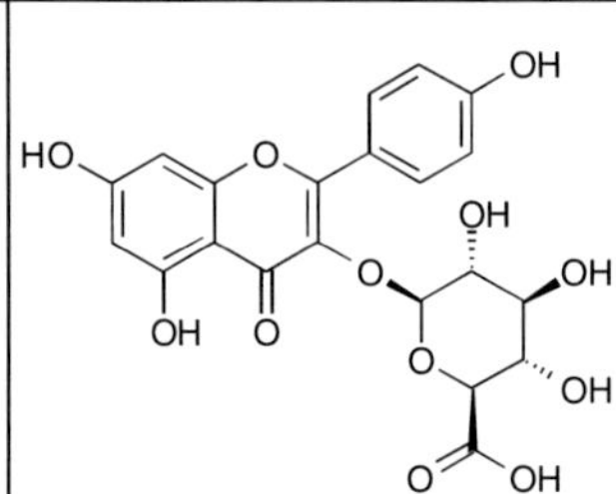

82	kaempferol-3-O-(2′′-O-β-D-glycopyranosyl)-β-D-glucopyranoside *Rosa rugosa* Thunb (Xiao et al., 2006)	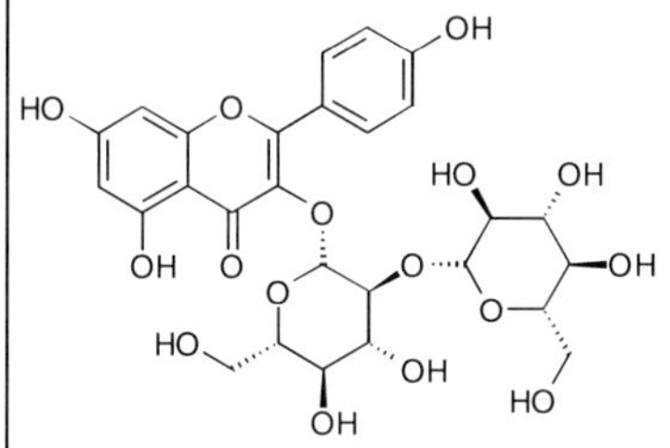
83	kaempferol-3-O-(2"-α-L-arabinopyranosyl)-α-L-rhamnopyranoside-7-O-α-L-rhamnopyranoside *Farsetia aegyptia* Turra. (Marzouk et al., 2009)	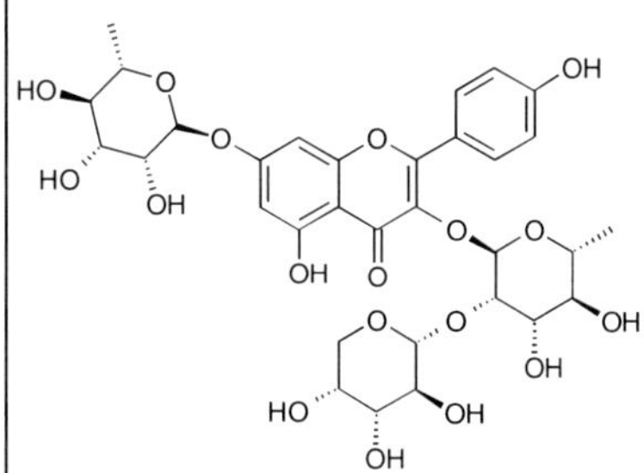
84	kaempferol-3-O-(2-O-feruloyl-ß-D-glucopyranosyl-(1›2)-ß-D-glucopyranoside)-7-O-glucopyranoside) *Brassica juncea* L. (Kim et al., 2002)	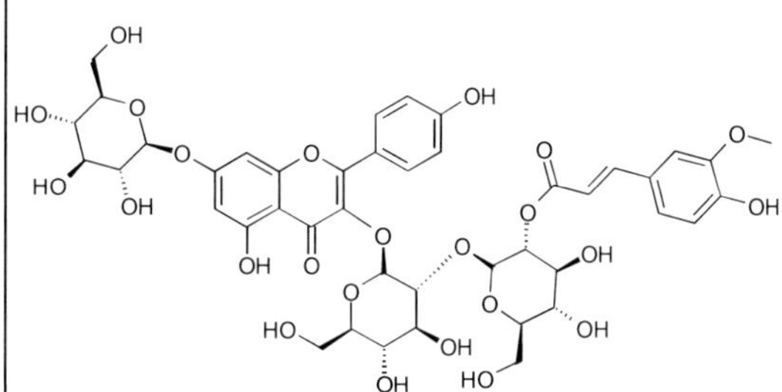
85	kaempferol-3-O-(2-O-sinapoyl)-ß-D-glucopyranosyl-(1›2)-ß-D-glucopyranoside-7-O-ß-D-glucopyranoside *Brassica juncea* (Jung et al., 2009)	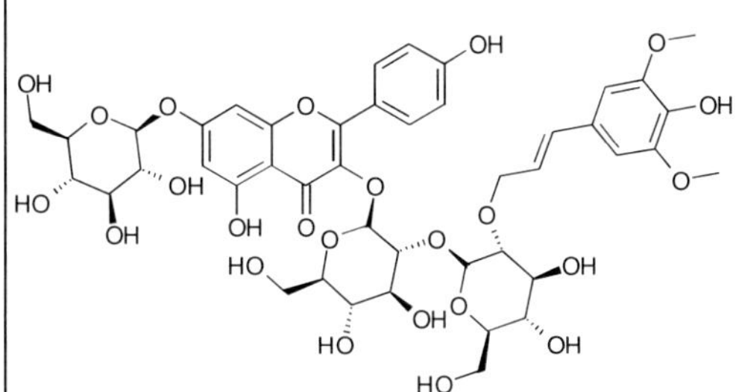
86	kaempferol-3-O-(2-O-sinapoyl)-ß-D-glucopyranosyl-(1›2)-ß-D-glucopyranoside-7-O-ß-D-glucopyranosyl-(1›6)-ß-D-glucopyranoside *Brassica juncea* (Jung et al., 2009)	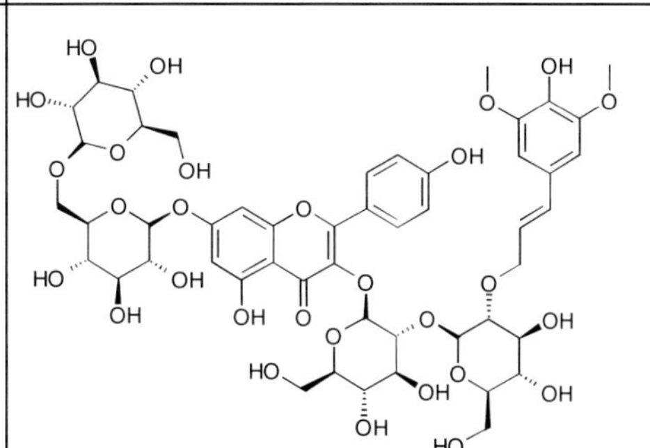

Table 1. (Continued)

87	kaempferol-3-O-(2"-ß-D-glucopyranosyl)-α-L-rhamnopyranoside-7-O-α-L-rhamnopyranoside *Farsetia aegyptia* Turra. (Marzouk et al., 2009)	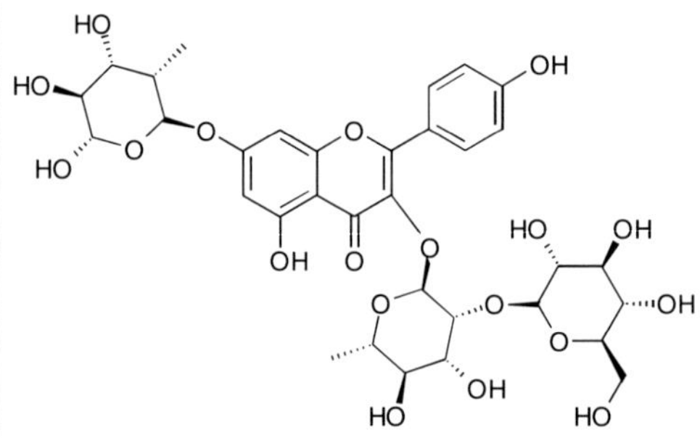
88	kaempferol-3-O-[2-O-ß-D-galactopyranosyl-6-O-α-L-rhamnopyranosyl]-ß-D-glucopyranoside *Camellia oleifera* Abel. (Zhu et al., 2011)	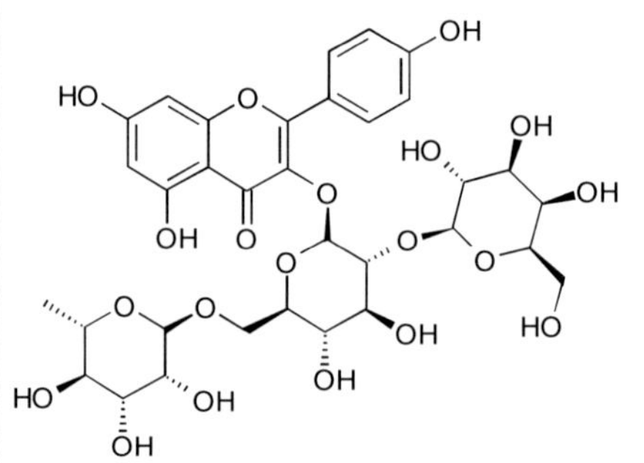
89	kaempferol-3-O-[2-O-ß-D-xylopyranosyl-6-O-α-L-rhamnopyranosyl]-ß-D-glucopyranoside *Camellia oleifera* Abel. (Zhu et al., 2011)	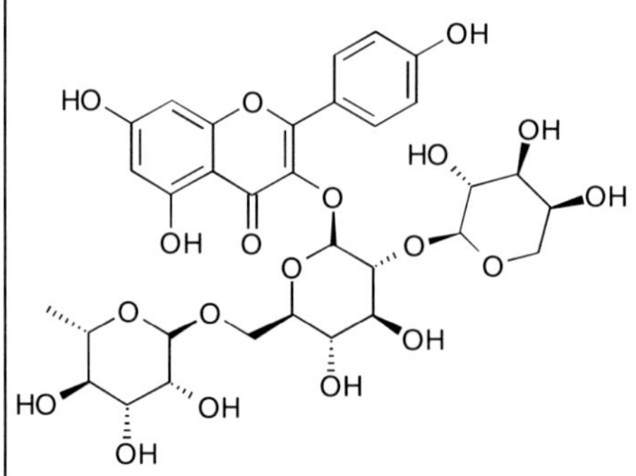
90	kaempferol-3-O-[4′′′′-O-acetyl-α-L-rhamnopyranosyl-(1›6)]-[ß-D-glucopyranosyl-(1›2)]-ß-D-glucopyranoside *Camellia oleifera* Abel. (Zhu et al., 2011)	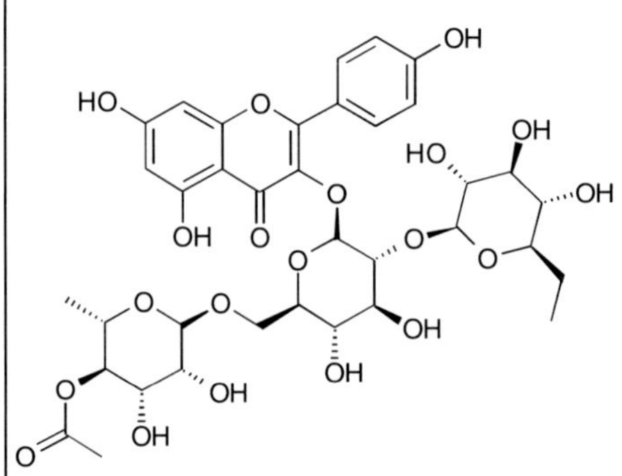

91	kaempferol-3-O-[4´´´´-O-acetyl-α-L-rhamnopyranosyl-(1›6)]-[ß-D-xylopyranosyl-(1›2)]-ß-D-glucopyranoside *Camellia oleifera* Abel. (Zhu et al., 2011)	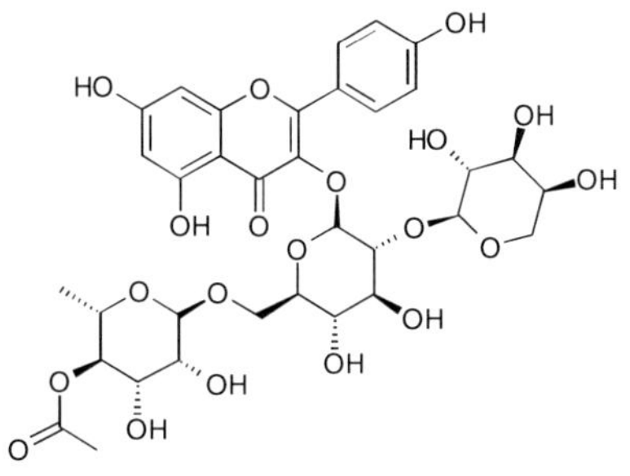
92	kaempferol-3-O-α-L-(2,4-di-E-p-coumaroyl)-rhamnoside *Petachondra pumila* (Bloor, 1995)	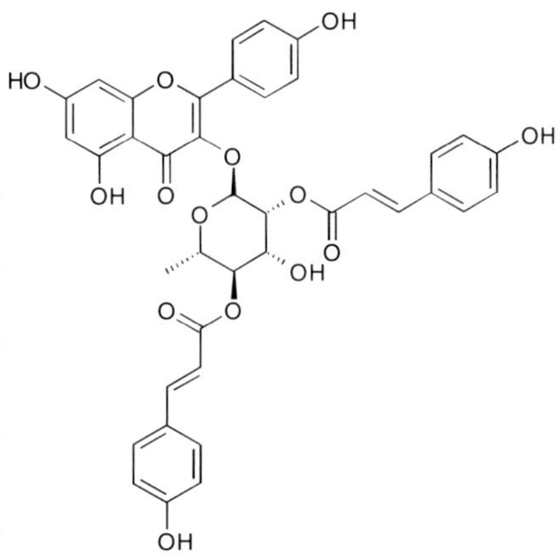
93	kaempferol-3-O-α-L-(2´´,3´´-di-E-p-coumaroyl)-rhamnoside *Foeniculum vulgare, F. dulce* (Soliman et al., 2002)	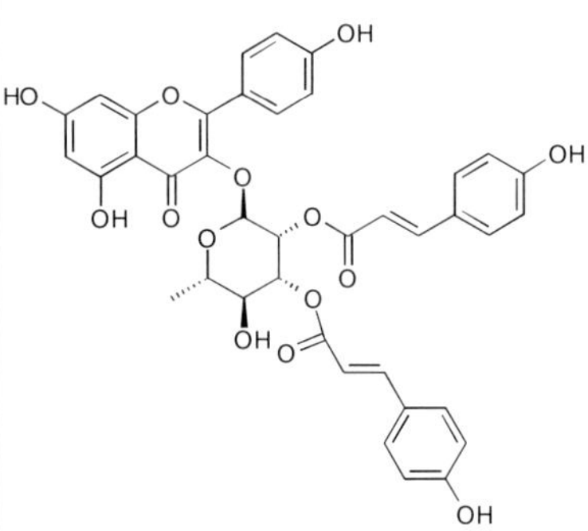
94	kaempferol-3-O-α-L-(2´´,4´´-di-Z-p-coumaroyl)-rhamnoside *Laurus nobilis* L. (Fiorini et al., 1998)	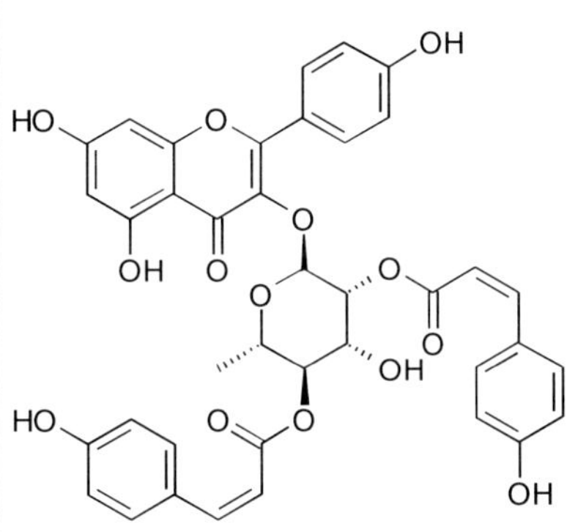

Table 1. (Continued)

95	kaempferol-3-O-α-L-(2´´-E-p-coumaroyl)-rhamnoside *Laurus nobilis* L. (Fiorini et al., 1998)	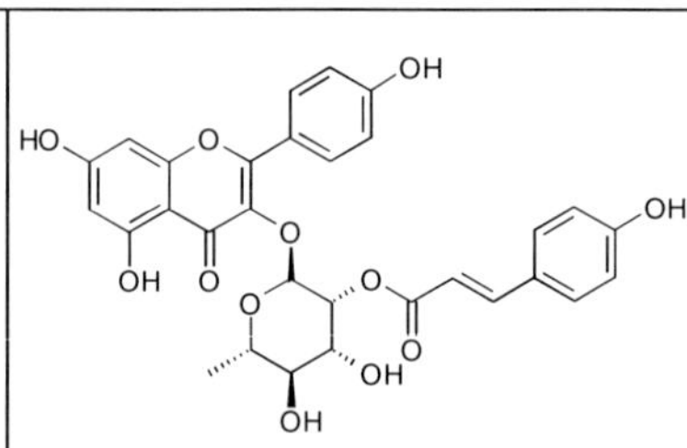
96	kaempferol-3-O-α-L-rhamnosyl (1›2)-ß-D-glucopyranoside *Lysimachia christinae* Hance (Wei et al., 2008)	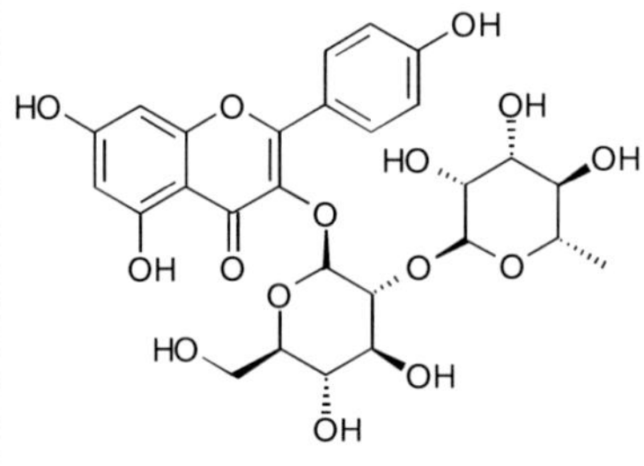
97	kaempferol-7-O-arabinoside *Dryopteris villarii* (Bellardi) Schinz and Thell (Imperato, 2006)	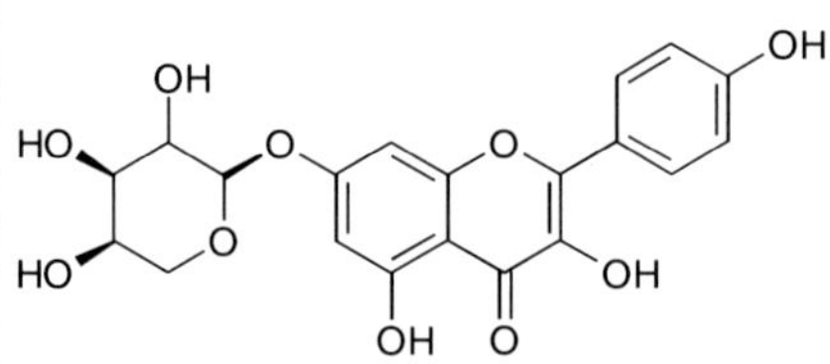
98	kaempferol-3-O-ß-D-[2-E-caffeoyl-ß-D -glucopyranosyl(1›2) glucopyranoside]-7-O-ß-D-[ß-D-glucopyranosyl(1›4) glucopyranoside] *Brassica oleracea* (Nielsen et al., 1998)	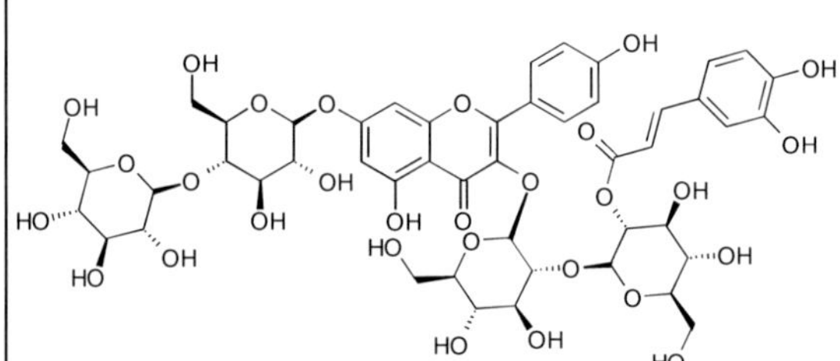
99	kaempferol-3-O-ß-D-[2-E-feruloyl-β-D-glucopyranosyl(1›2) glucopyranoside]-7-O-ß-D-[ß-D-glucopyranosyl(1›4) glucopyranoside] *Brassica oleracea* (Nielsen et al., 1998)	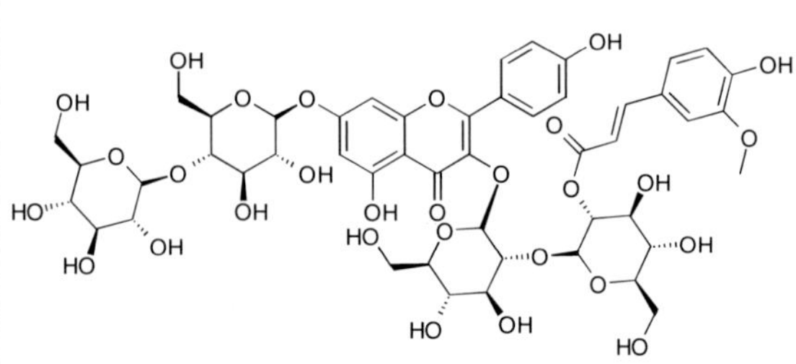

100	kaempferol-3-O-ß-D-[2-E-sinapoyl-ß-D-glucopyranosyl(1›2) glucopyranoside]-7-O-ß-D-[ß-D-glucopyranosyl(1›4) glucopyranoside] *Brassica oleracea* (Nielsen et al., 1998)	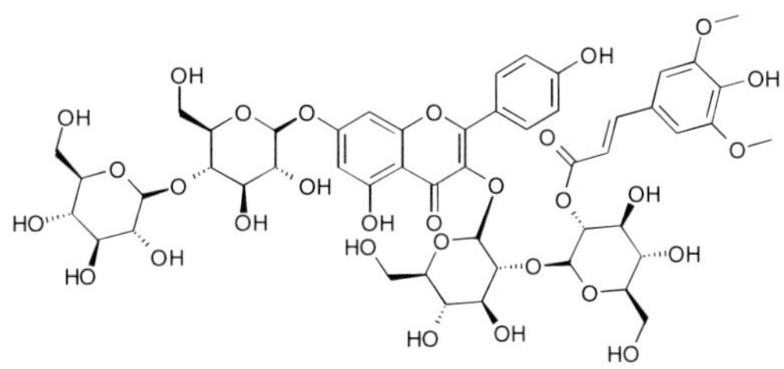
101	kaempferol-3-O-ß-D-[ß-D-glucopyranosyl(1›2) glucopyranoside]-7-O-ß-D-[ß-D-glucopyranosyl(1›4) glucopyranoside] *Brassica oleracea* (Nielsen et al., 1998)	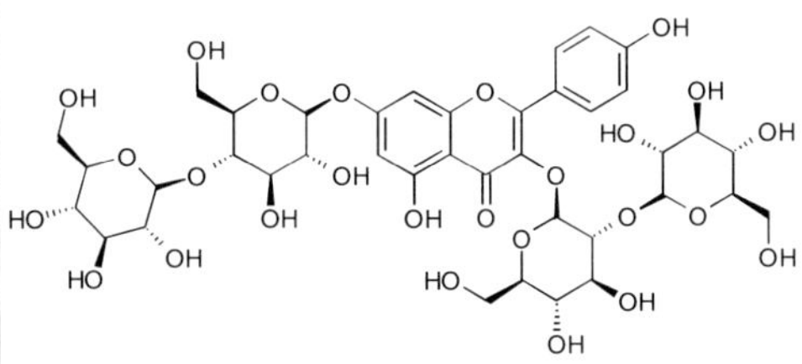
102	kaempferol-3-O-ß-D-glucopyranosyl-(1›2)-ß-D-glucopyranoside-7-O-ß-D-glucopyranoside *Brassica oleracea* (Nielsen et al., 1998)	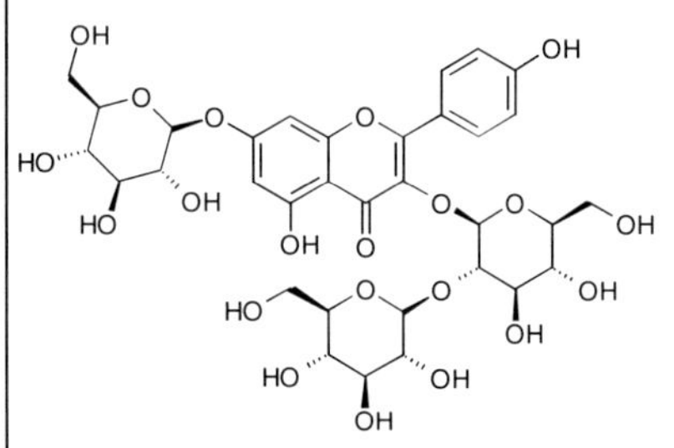
103	kaempferol-3-O-ß-D-glucopyranosyl-(1›2)-ß-D-glucopyranoside-7-O-ß-D-glucopyranosyl-(1›6)-ß-D-glucopyranoside *Brassica juncea* (Jung et al., 2009)	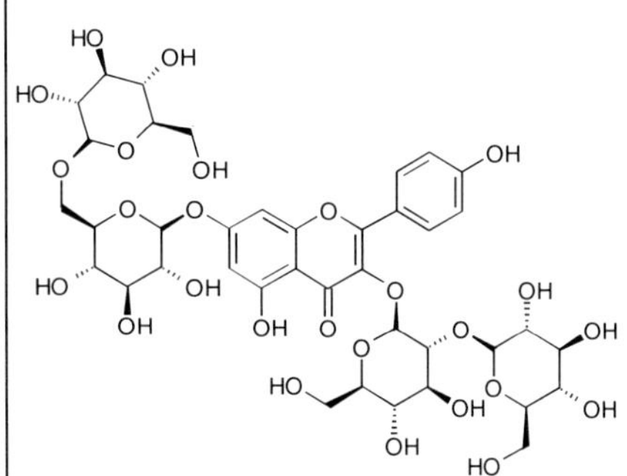
104	kaempferol-3-O-ß-D-glucosyl(1›2)-ß-D-galactoside-7-O-α-L-rhamnoside *Panax quinquefolium* L. (Liu et al., 2009)	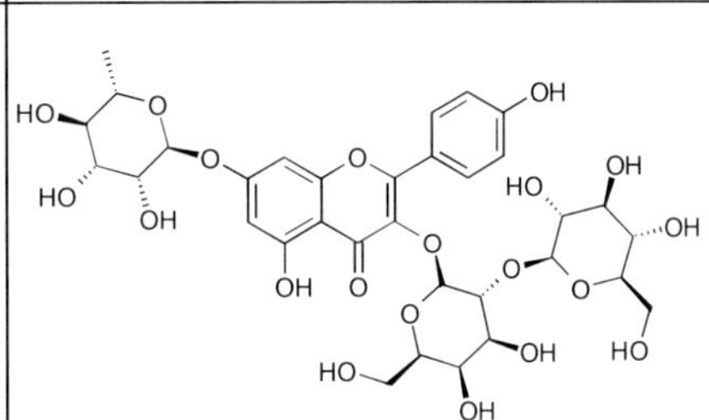

Table 1. (Continued)

105	kaempferol-3-O-ß-neohesperidoside-7-O-[2-O-(trans-feruloyl)]-ß-D–glucopyranoside *Allium senescens* L. s. str. (Selyutina et al., 2008)	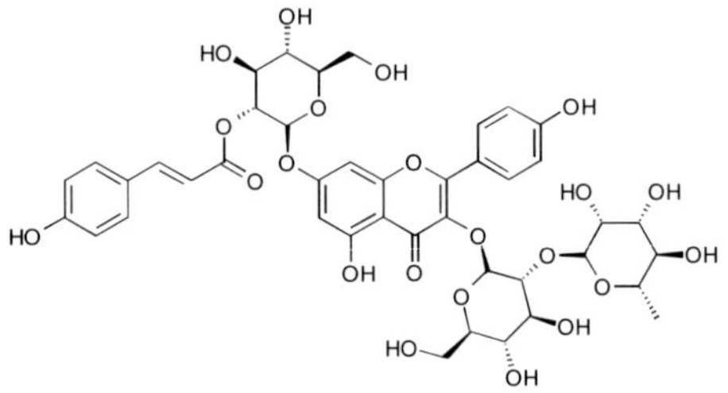
106	kaempferol-8-sulfonate *Phyllanthus virgatus* Forst. f. (Huang et al., 1998)	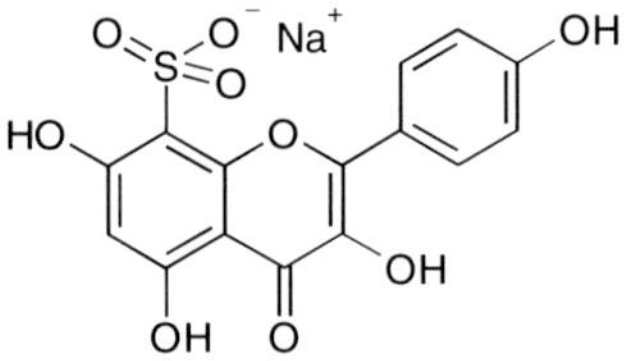
107	lespedin *Raphanus sativus* L. var. *radicula* Pers. (Muminova et al., 2006)	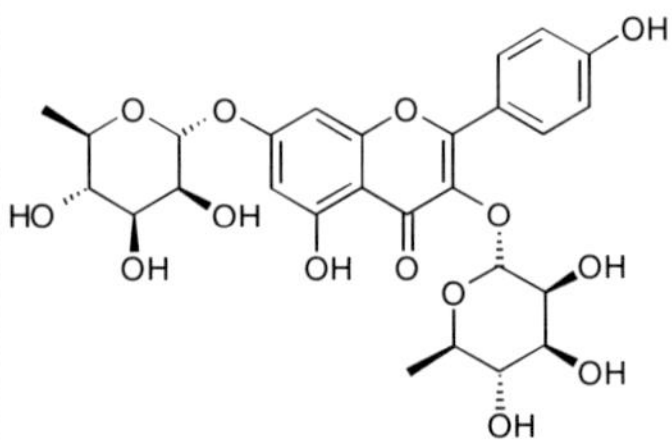
108	macagigantin *Macaranga gigantea* (Tanjung et al., 2009)	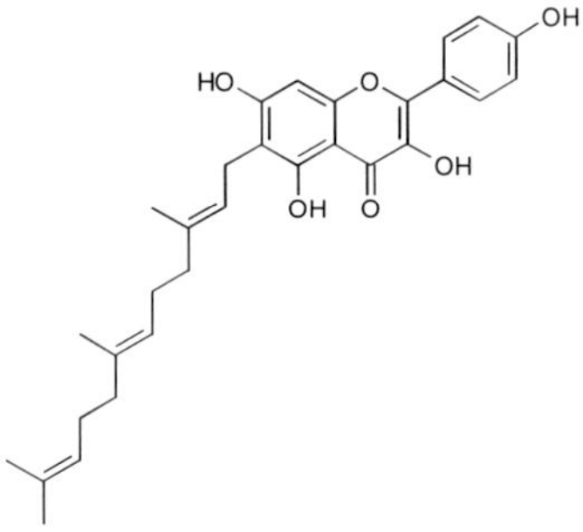
109	Mindbraedin *Mildbraediodendron excelsum* Harms (Veitch et al., 2005)	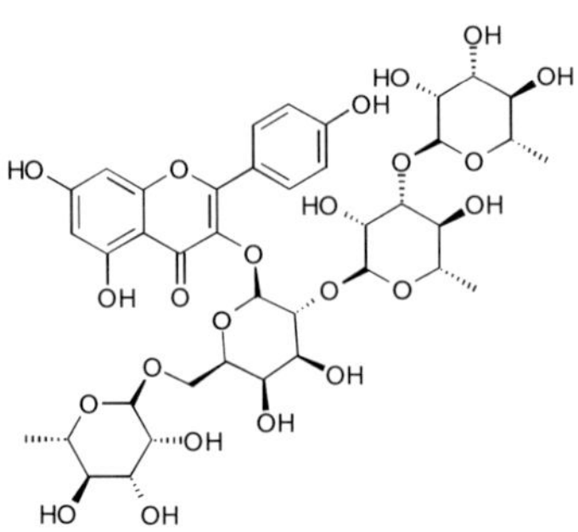

110	multiflorin A *Neocheiropteris palmatopedata* (Baker) Christ (Yang et al., 2010)	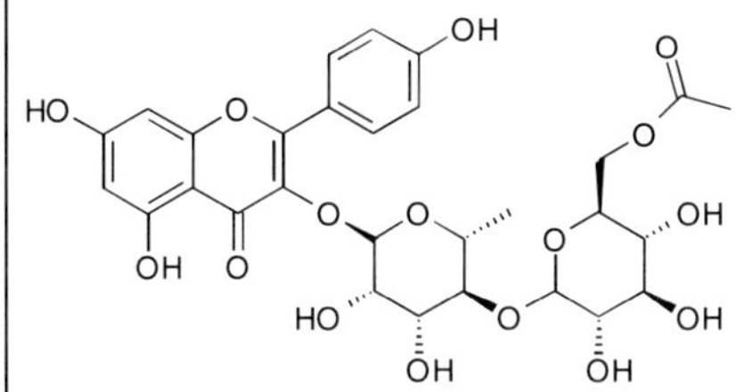
111	multiflorin B *Neocheiropteris palmatopedata* (Baker) Christ (Yang et al., 2010)	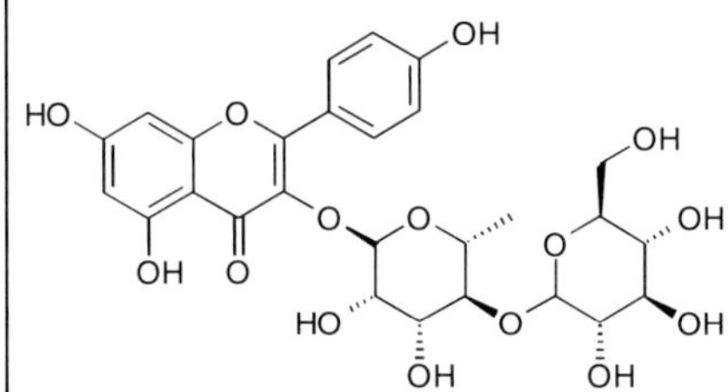
112	palmatoside A *Neocheiropteris palmatopedata* (Baker) Christ (Yang et al., 2010)	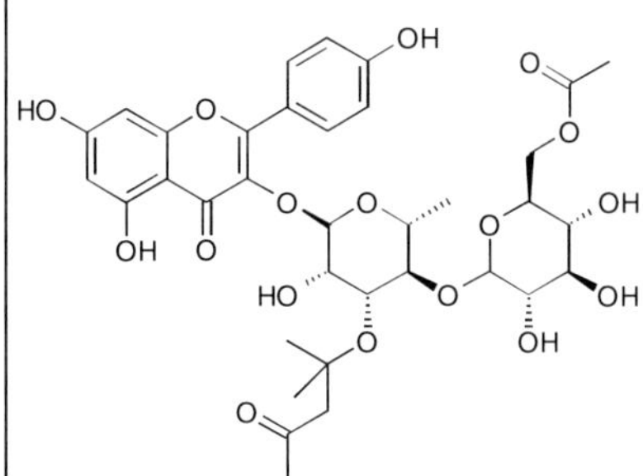
113	palmatoside B *Neocheiropteris palmatopedata* (Baker) Christ (Yang et al., 2010)	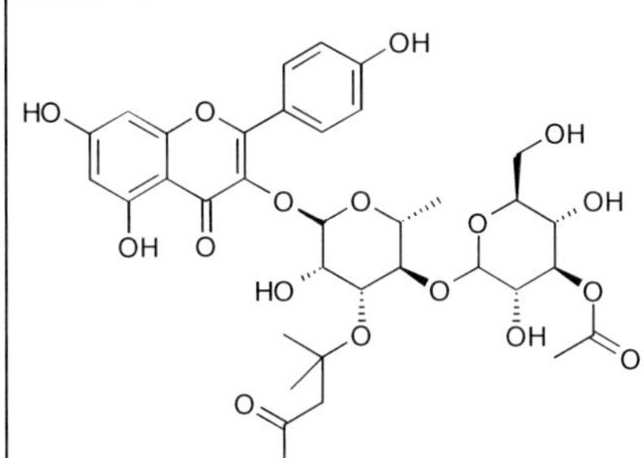
114	palmatoside C *Neocheiropteris palmatopedata* (Baker) Christ (Yang et al., 2010)	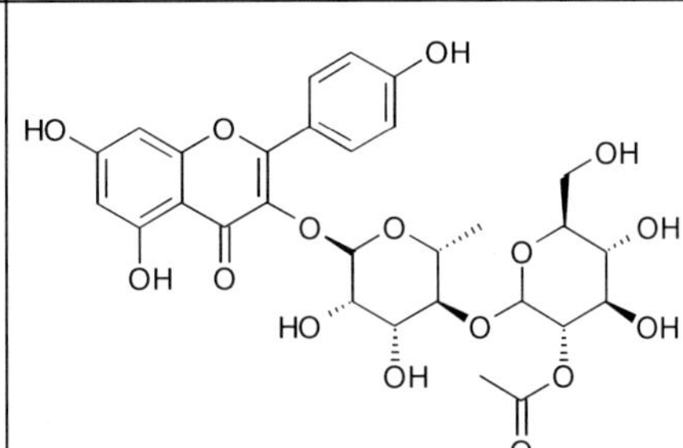

Table 1. (Continued)

115	Panasenoside *Panax quinquefolium* L. (Liu et al., 2009)	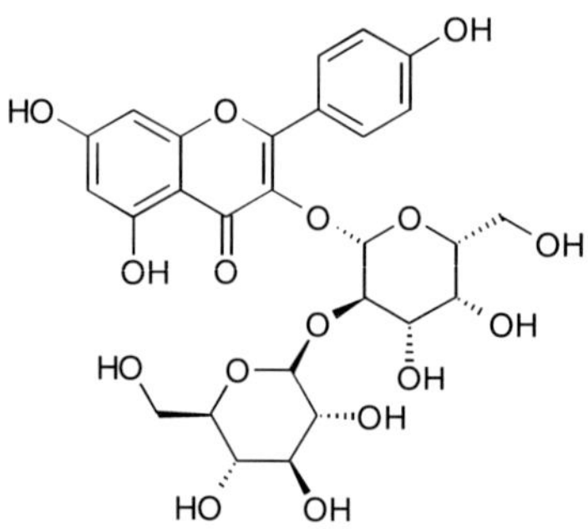
116	platanoside *Platanus acerifilia* Willd. (Kaouadji et al., 1993)	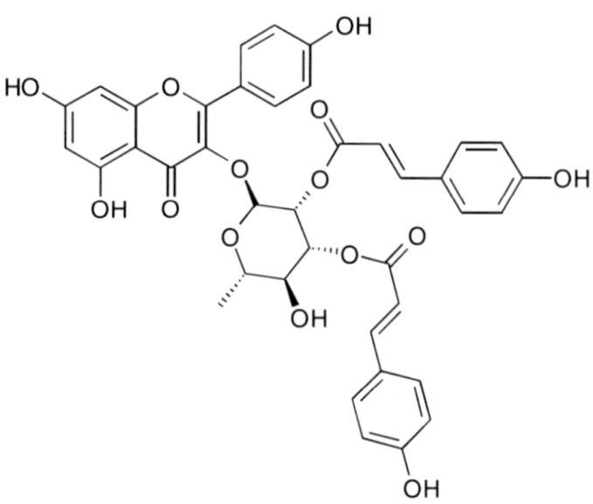
117	Robinin *Consolida oliveriana* (Diaz et al., 2008)	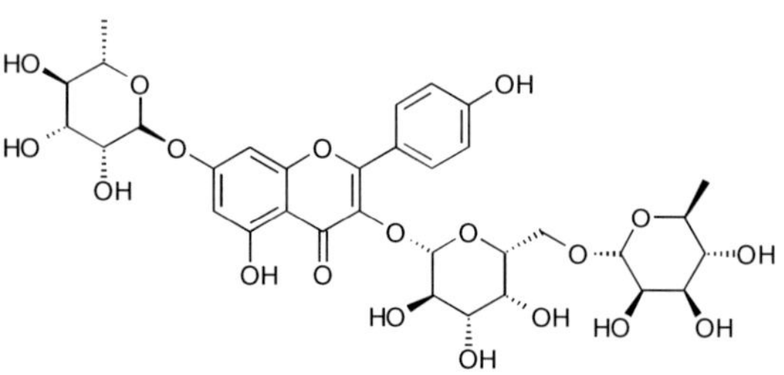
118	sagittatin A *Kalanchoe streptantha* Baker (Costa et al., 1996)	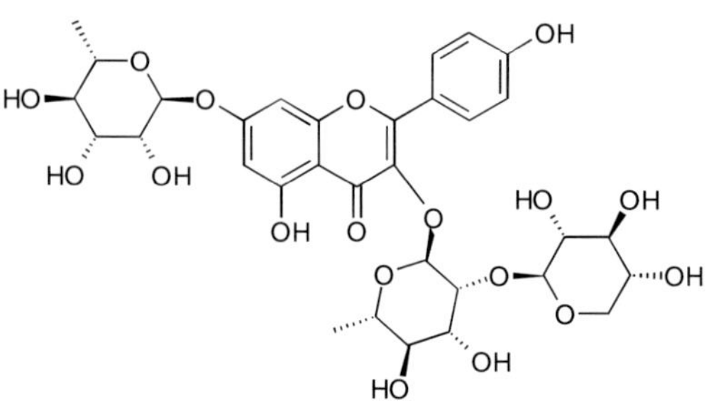

119	siparunoside *Siparuna apiosyce* De Candolle (Leitao et al., 2000)	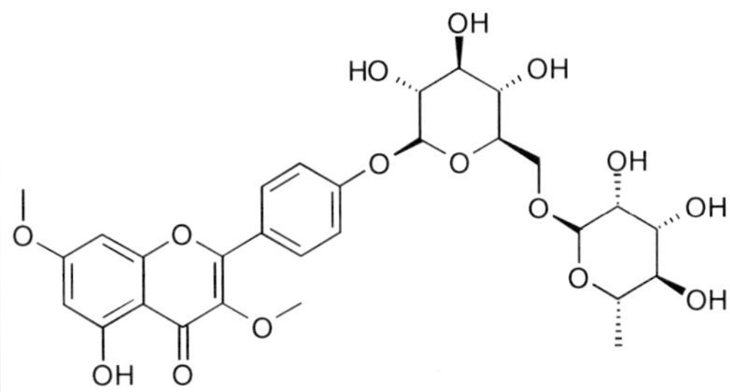
120	sutchuenoside A *Epimedium sutchuenense* (Mizuno et al., 1991)	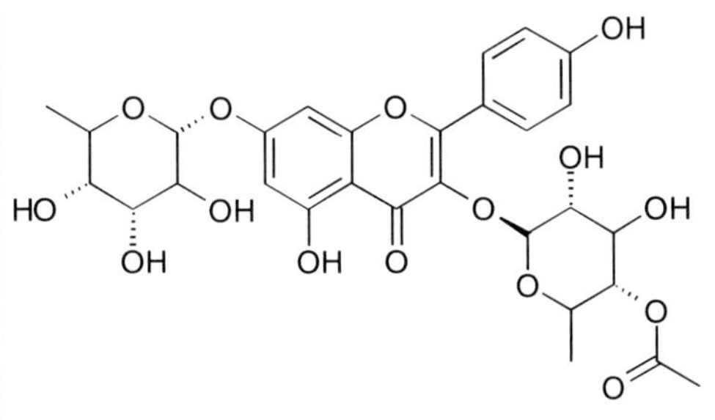
121	tiliroside *Tilia americana* L. var. mexicana (Schltdl.) Hardin (Aguirre-Hernandez et al., 2010)	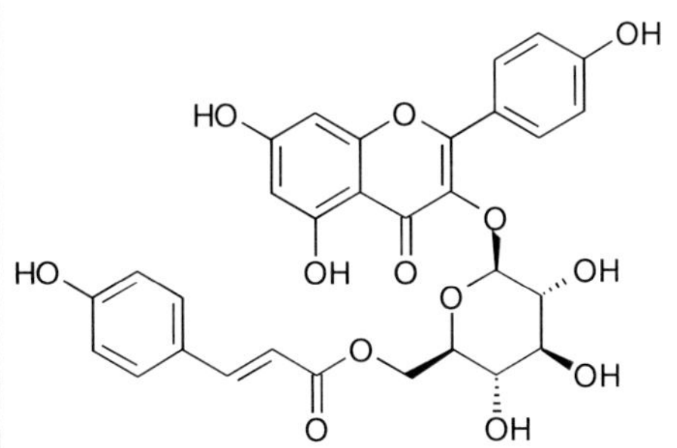
122	trans-tiliroside *Solidago altissima* L. (Jin et al., 2007)	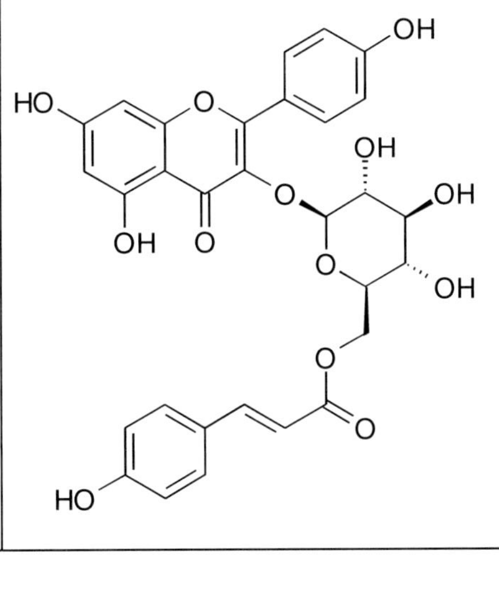

Table 1. (Continued)

123	Trifolin *Pteridium aquilinum (L.) Kuhn subspecies aquilinum* (Imperato, 1995)	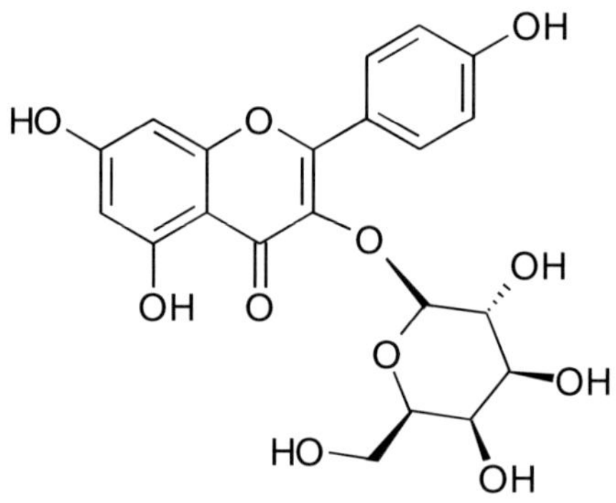
124	Urakunoside *Camellia uraku* Tarokaja (Tanikawa et al., 2011)	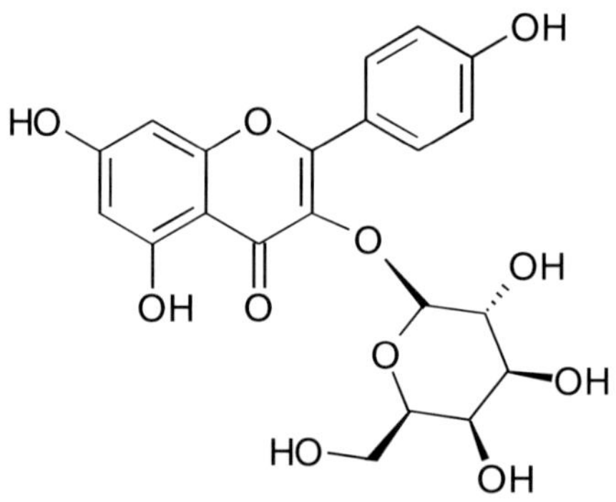

Kaempferol is also produced by some mosses, although flavonoids are rarely detected in Bryophyta. On the other hand, kaempferol was not identified in red or green algae where PAL is also absent.

Structural Diversity

Kaempferol is not only present in its free form but commonly as various glycosides, esters and prenylated derivatives (Table 1). The largest group of structural isomers are glycosylated derivatives whose number is further increased by acylation of sugar moieties by organic acids, such as coumaric, ferulic, sinapic, caffeic, gallic, glutaric and acetic acids, resulting in more than 100 individual kaempferol glycosides identified until now.

Another group of kaempferol derivatives are prenylated compounds, which are usually less polar and soluble than kaempferol itself.Prenylation results in covalent C-C bond joining the flavonoid skeleton and an isoprenoid side chain. The structural diversity greatly complicates identification of individual compounds due to the absence of commercially available standards. Thus, highly specific detection methods, such as NMR and MS^n, in combination with isolation procedures are necessary to identify these compounds in real samples.

Another approach is to simplify the matrix by acid hydrolysis which leads to release of kaempferol aglycone from its respective glycosides. With this method, only glycosidic, but not C-C bonds, for instance, in prenylated derivatives, are cleaved.

Extraction

Extraction, the first step in isolating compounds from plants, affects not only the amount of extracted compound but also instrumental method performance in real samples. Good repeatability and precision of quantification can never be achieved without a proper extraction method. Extraction generally consists of three principal steps, namely homogenisation, primary extraction and preconcentration/purification,which will be discussed separately in this chapter.

1. Plant Material and Homogenisation

The variety of plants and plant parts that are shown in Figure 1 were used for isolation of kaempferol, its derivatives and glycosides. However, different materials require different homogenization methods. Different material not only means different plant parts, such as roots, flowers, leaves, buds, and fruits, but also various storage conditions and eventually applied drying. To avoid analyte degradation, the extraction of fresh plant parts and storage under -80 °C is generally suggested. In practice, dried samples are used more frequently due to their easy handling and storage. When dried samples have to be used, the recommended procedure should be freeze drying which affects unstable compounds in plant material to a lesser degree.

Both dried and fresh plant parts are usually homogenised prior to primary extraction to improve extraction effectivity. Homogenisation can be performed with or without the presence of extraction solvent. Crushing under liquid nitrogen and trituration in a food processor are methods applied almost exclusively to fresh material (De Melo et al., 2009). Both fresh and dried material can be homogenized by grinding samples with or without extraction solvent in oscillation ball mill (Prokudina et al., 2012). Such a system provides better precision for replicated samples due to its ability to homogenise simultaneously 24–48 samples by using a carousel for eppendorf tubes. Dried material can also be crushed in mortar with pestle or in any rotor-stator homogeniser.

2. Primary Extraction

Primary extraction ensures transfer of analysed compounds to the primary extract for further purification. To improve the stability of isolated compounds, inhibitors of enzymes are sometimes added, unless a high percentage ethanol or methanol is present in the extraction solvent. Effectivity of primary extraction depends on particle size, solvent polarity, solvent pH, temperature and mechanical assistance. Kaempferol is usually extracted using a high percentage methanol or ethanol (Table 2).

Table 2. An overview of extraction methods used to extract kaempferol and/or its derivatives

plant species	material	primary extraction	preconcentration	target compound	reference
Aubrieta x cultorum	dried flowers	5% HOAc at RT for 5 h	Diaion HP-20 column (90 × 150 mm) → preparative paper chromatography → HPLC C18 (19 × 150 mm)	kaempferol glycosides	(Tatsuzawa et al., 2012)
Eucalyptus occidentalis	dried leaves	80% EtOH at RT for 24 h	filtration → $CHCl_3$ extraction → silica gel column → Sephadex LH-20 → silica gel CC	kaempferol derivatives	(Benyahia et al., 2004)
Gymnocarpium dryopteris	dried aerial parts	100% EtOH at RT	evaporation → dissolution in 30% EtOH → ethylether extraction → evaporation of aqueous ethanol layer → *n*-BuOH extraction → evaporation → hydrolysis in boiling 6% HCl → EtOAc extraction → evaporation → silica gel column	kaempferol and its glycosides	(Kovganko et al., 2004)
Flaveria bidentis	dried aerial parts	80% EtOH (refluxed) for 75 min	High Performance Counter Current Chromatography	kaempferol and its glycosides	(Wei et al., 2011)
Sedum dendroideum	fresh leaves	leaves were homogenized without solvent addition; total cellular fluid separated	partition with EtOAc and *n*-BuOH → RP-2 column (36.0 cm × 2.2 cm)	kaempferol glycosides	(De Melo et al., 2009)
Trigonella foenum-graecum	dried leaves	80% MeOH and sonicated for 60 min at RT	hydrolysis of glycosides in 37% HCl	kaempferol	(Gikas et al., 2011)
Camellia oleifera	fresh plant	60% EtOH (refluxed), defatted sample for 3 h at 70 °C	none	kaempferol glycosides	(Zhu et al., 2011)
Camellia oleifera	fresh plant	100% H_2O using AGO-2 high-intensity planetary activator	none	kaempferol glycosides	(Zhu et al., 2011)
Macaranga gigantea	dried leaves	100% acetone at RT	evaporation → silica gel → centrifugal planar chromatography	prenylated kaempferol	(Tanjung et al., 2009)

For example, extraction of kaempferol from *Lycium barbarum* was most effective by using 70% ethanol for extraction (Dong et al., 2011). However nonpolar (farnesylated or prenylated derivatives) and highly polar (high glycosylated flavonoids) derivatives, can be extracted by acetone and water, respectively.

The advantage of extraction with pure water might be the decrease in coextracted nonpolar compounds (e.g. fats, chlorophyll). Nonpolar compounds can also be removed by preextraction with hexane, ether or chloroform, as often reported.

Another solvent PEG-200, was also used as an effective extraction solvent in microwave-assisted extraction (Liu et al., 2012). Increased temperature can enhance the extraction process, but may also result in thermal degradation of unstable compounds. The study of ultrasonic assisted extraction revealed that increasing temperature enhances the effectivity until 60 °C but further increasing the temperature conversely decreases the effectivity of extraction (Dong et al., 2011).

The authors also compared microwave-ultrasonic-assisted extraction (M-UAE), ultrasonic-assisted extraction (UAE), ultrasonic-reflux extraction (URE) and simple extraction (SE). M-UAE was considered to be the most effective approach. As an example, UAE needed twice the extraction time and two times more solvent volume than M-UAE to reach the same recovery. SE was even less effective, requiring four times longer extraction time and four times more solvent.

Additionally SE and URE were also compared. URE provides better extraction effectivity with four times shorter extraction time at even lower temperatures than SE.

Further effective and advisable approaches to extract kaempferol are pressurised liquid extraction (Yi et al., 2012) and supercritical fluid extraction (Leal et al., 2008) which are generally more effective than common procedures due to the increased pressure.

3. Preconcentration

The main goal of preconcentration is to either increase the analyte/background ratio in analyzed samples or isolate sufficient amounts of target compound. The selection of preconcentration steps depends on the final analytical instrumentation (HPLC-MS, NMR, etc.) which may require different amounts and purities of analytes.

The crude extract can be preconcentrated by a few methods, including liquid-liquid extraction, column chromatography and solid phase extraction (SPE) columns. Liquid-liquid extraction, such as ether–water or ethylacetate–water, is a simple and cheap method which can be used to remove salts and highly polar compounds from kaempferol prior to LC analysis (Porgali and Buyuktuncel, 2012).

An automated case of liquid-liquid extraction is High Performance Counter Current Chromatography which was used to purify and isolate kaempferol from traditional Chinese herbs (Wei et al., 2011).

Column chromatography, a simple type of LC, is an ideal tool for isolation of higher quantities of analytes. Various sorbents and elution solvents, including neutral aluminium oxide and hexane–ethylacetate (Singh et al., 2008), silica gel and ethylacetate (Kovganko et al., 2004), silica gel and dichlormethane-methanol (Kim et al., 2002), hydroxypropylated dextran and water-methanol (Costa et al., 1996), polyamide and water-methanol (Ibrahim et al., 2008), polyamide and water-ethanol (Selyutina et al., 2008), macropore resin and water-ethanol (Liu et al., 2009), silanized silica gel and water-methanol (De Melo et al., 2009), styrene-divinylbenzene and methanol (Moriyama et al., 2003), as well as C-18 reversed phase silica gel and methanol (Jung et al., 2009), were used to isolate kaempferol and/or its derivatives from plant sources.

While column chromatography is used mainly to isolate higher quantity of kaempferol or its derivatives, SPE columns are almost exclusively used as a purification step preceding instrumental analysis.

Extraction recovery of kaempferol and several other flavonoids on different reversed-phase SPE columns, including four types of C18 columns, an amide column, and two divinylbenzene/N-vinylpyrrolidone columns, was studied in detail previously (Bajer et al., 2007).

The authors found that divinylbenzene/N-vinylpyrrolidone copolymer SPE column showed the best yields for all tested compounds. The suggested elution solvent was 20% methanol with 0.5% ammonium hydroxide. A principally different approach to preconcentrate free kaempferol in samples is the hydrolysis of its glycosides. Acid hydrolysis with HCl under increased temperature results in the release of kaempferol aglycone from its respective glycosides and an increase of free kaempferol in the sample (Olszewska, 2008).

Liquid Chromatography

Liquid Chromatography (LC) is the most common separation technique used to analyse kaempferol and its derivatives. Chromatographic columns are mostly reversed-phase columns, such as C18, C8 and Phenyl-Hexyl. Narrow bore columns instead of conventional HPLC columns became increasingly popular due to low consumption of organic solvents and better compatibility with MS.

The common mobile phases consist of an aqueous phase and methanol or acetonitrile. Mobile phase modifiers, such as formic and acetic acids, are frequently used due to their compatibility with mass spectrometry. However, less common additives, such as tetrahydrofuran, were also used to improve separation (Wang and Huang, 2004).

HPLC, the most common type of LC, was used to separate kaempferol from its derivative robinin by isocratic elution (Tsiklauri et al., 2012). The authors found that a mobile phase consisting of an uncommon mixture of THF:MeOH:*i*-PrOH with phosphate buffer (pH 2) resulted in most effective separation.

Such mobile phase composition can be used with PDA but not MS detectors due to the incompatibility of phosphate buffers with MS.

An evolved form of HPLC is Ultra High Performance Liquid Chromatography (UHPLC or UPLC) which provides improved resolution, shorter retention times and higher sensitivity. UPLC was used for quantification of 26 flavonoids and other phenylpropanoids including kaempferol in mung bean (*Vigna radiata*) (Prokudina et al., 2012) as well as for quantification of kaempferol, quercetin and isorhamnetin in rabbit blood plasma after intragastric administration of pollen Typhae (Li et al., 2009).

In general, chromatographic runs of reported UPLC methods are approximately 2–3 times shorter than those of corresponding HPLC methods. Further examples of HPLC methods are described in Table 3.

Table 3. An overview of instrumental analytical methods used to determine kaempferol and/or its derivatives

separation method	detection	method description	LOD	linear range	reference
none	PDA (200–400 nm)	Derivative spectroscopy, zero–crossing	NA	NA	(Baranowska and Rarog, 2001)
GC	ESI(+)–TOF–MS, ESI(+)–MS, FID	MSq: BPX5 column (5% phenylmethylpolysilphenylenesiloxane (32 m × 0.25 mm; 0.25 μm d_f), FID: BPX5 (30 m × 0.25 mm; 0.25 μm d_f) as ^{1}D, BPX50 phase (50% phenyl) (1.0 m × 0.1 mm; 0.1 μm d_f) as ^{2}D, MS-TOF: BPX50 (30 m × 0.25 mm ID; 0.25 μm d_f) as ^{1}D, BPX5 (1 m × 0.1 mm ID; 0.1 μm d_f) as ^{2}D	3.81 μg mL^{-1}	NA	(Gao et al., 2010)
CZE	PDA (250 nm)	35 mM borate buffer, pH 8.9, 70 cm (effective length 45 cm) × 75 m capillary, and electric field strength of 240 V cm^{-1}	0.08 μg mL^{-1}	0.1–140 μg mL^{-1}	(Wang and Huang, 2004)
none	CV	pyrolitic-graphite and hemoglobin/polysorbate-20 electrodes, saturated calomel electrode as reference electrode and platinum counter electrode, pH 5.0, 0.1 M Britton–Robinson buffer	0.01 mM	0.04–4 mM	(Zheng et al., 2008)
HPLC	PDA (200–450 nm), ESI(+)–MS^n	C18 column (150 × 4.6 mm, 5 μm), gradient of 20–50% aq. MeOH (containing 1% HOAc), 30 min	NA	NA	(Aguirre-Hernandez et al., 2010)
HPTLC	TLC Scanner (254 nm)	silica gel 60 F254 TLC, EtOAc–MeOH–HOAc–H_2O (4:0.5:0.5:1 v/v/v/v)	40 ng $spot^{-1}$	120–280 ng $spot^{-1}$	(Singh et al., 2009)
none	DPV	measuring: C-PVC electrode, carbon fibber electrode, carbon fibber electrode coated with Nafion, reference Ag/AgCl electrode, platinum counter electrode, MeOH–MeCN–$NaClO_4$ (0.1 M) (30:30:40, v/v/v) for RP, dioxane–hexane–MeOH (40:40:20, v/v/v) for NP	0.1 μg mL^{-1}	NA	(Aguilar-Sanchez et al., 2005)
HPLC	PDA (370 nm)	endcapped C18 column (250 x 4.6 mm, 5-μm), MeCN and phosphate buffer (pH 3.4, adjusted with glacial HOAc) 60:40 (v/v), isocratic, 15 min	0.10 μg mL^{-1}	1.0–10 μg mL^{-1}	(Thakur et al., 2009)
TLC	UV lamp (366 nm)	EtOAc-HCOOH-H_2O (65:15:20 v/v/v)	NA	NA	(Males and Medic-Saric, 2001)
UPLC	ESI(-)–MS	C18 column (50 × 2.1 mm, 1.7 μm), MeOH and HCOOH (0.1%), gradient	0.78 ng mL^{-1}	9.1–910 ng mL^{-1}	(Li et al., 2012)
HPLC	CAD (100-800 mV)	C18 column (250 × 2.0 mm , 5 μm), MeCN, MeOH, 70 mM KH_2PO_4 adjusted to pH 2.4 with phosphoric acid, gradient, graphite working electrode with palladium reference electrode	NA	NA	(Aaby et al., 2004)
UPLC	ESI(+)–MS/MS, MRM	C8 column (150 × 2.1 mm, 1.7 μm), MeOH and 10 mM aqueous HCOOH, gradient	4.35 mM	15–500 mM	(Prokudina et al., 2012)
HPLC	PDA (250–360 nm), APCI(+) (-)–MS^n	C18 column (250 × 4.0 mm, 5 μm), MeOH, H_2O, gradient	NA	NA	(Grayer et al., 2000)
HPLC	PDA (250, 270, 290, 350, and 365 nm), CL	C8 column (150 × 4.6 mm, 5 μm), MeOH and aqueous 1.0% HOAc (37:63 v/v), CL: 0.1 mol L^{-1} sulfuric acid, 5.06 μM cerium(IV), 5.06 μM rhodamine G6	3.56 ng mL^{-1}	6–7000 ng mL^{-1}	(Zhang and Cui, 2005)
HPLC (offline)	NMR (2D COSY, NOESY and HMBC)	C18 column (4.6 x 250 mm), 1.5% H_3PO_4, A: 1.5% H_3PO_4 in H_2O, B: 20% HOAc, 25% MeCN in H_2O, gradient	NA	NA	(Tatsuzawa et al., 2012)

Gas Chromatography

Gas Chromatography (GC) is a separation method used for compounds that can be vaporized without decomposition. GC mostly provides better separation effectivity than LC, but only volatile compounds can be analyzed directly without derivatization. Vaporized compounds are partitioned between stationary phase (polymer or liquid firmly bonded to column walls) and mobile phase (helium and nitrogen) based on their affinity to the stationary phase. GC can be hyphenated to a number of detectors such as Flame Ionisation Detector (FID), Thermal Conductivity Detector and MS.

Conventional GC–MS as well as comprehensive 2D GC–FID and 2D GC–TOF–MS were applied to analyse flavonoids including kaempferol in propolis, chrysanthemum and dark chocolate samples (Gao et al., 2010). All samples were derivatized with trimethylsilyl to ensure better volatility and thermal stability. The identification of individual compounds was based on a comparison of measured MS spectra with NIST05 and an in-house MS library. The authors concluded that 2D GC-TOF-MS is an excellent tool for analysis of flavonoids in real samples, providing not only better separation than conventional GC but also better potency for structural elucidation than common GC-FID. Further method details and LOD can be found in Table 3.

Capillary Zone Electrophoresis

Capillary Zone Electrophoresis (CZE) is a separation technique which separates ionic compounds based on their charge, frictional forces and hydrodynamic radius. Its separation effectivity is usually better than that of common HPLC but its application is limited to ionizable compounds. Another disadvantage is its complicated hyphenation with mass spectrometry, which is the detector of choice these days. Therefore CZE is mostly used with photodiode detectors. As an illustration, caffeic acid and several flavonoids including kaempferol were identified and quantified in commercially available wine samples by using CZE coupled to PDA (Wang and Huang, 2004). Borax was chosen as the buffer, the most important part of CZE, due to its ability to buffer at pH>8 at which all analyzed flavonoids were ionized. Different buffer pHs, borax concentrations and electric field strengths were applied during method optimisation. Optimised conditions as well as the LOD and linear range are summarized in Table 3.

Thin Layer Chromatography

Thin Layer Chromatography (TLC) is a separation technique performed on sorbent (stationary phase) firmly bonded to glass/aluminium/plastic. The analyte is eluted by mobile phase which slowly rises up the TLC plate via capillary action. Compared to other separation techniques, TLC is a simple, cheap, and solvent efficient method which can quickly separate various compounds including flavonoids. However, TLC cannot simply be connected to any online detector. Detection can be only achieved offline by UV lamp or densitometry, which can serve for quantification as well.

Conventional TLC was used to analyse a methanolic extract of *Helleborus atrorubens* Waldst. et Kit. leaves (Males and Medic-Saric, 2001). The extract contained 15 compounds, including kaempferol. The system was optimized by comparing 13 different mobile phases that were prepared by combining the following solvents: ethylacetate, formic acid, water, methanol, 1-propanol, 1-butanol and methylethylketone. The authors applied sophisticated mathematical approaches, such as calculation of information content, determination of discriminating power and formation of clusters/dendrograms, to find the best conditions. As a result (DP=0,952, I=3,774), ethylacetate-formic acid-water (65:15:20 v/v/v) was considered to be the best mobile phase for separation of several natural compounds including kaempferol.

More precise but more instrumentally dependent is High-Performance Thin Layer Chromatography (HPTLC), an automated TLC. HPTLC was applied to quantify kaempferol, apigenin and quecetin in herbal raw material as well as in pharmaceutical formulations (Singh et al., 2009). The mixture of toluene-ethylacetate-methanol-formic acid (6:3:0.2:0.4 v/v/v/v) was used as mobile phase to quantify kaempferol. The method was validated with an accuracy of 98.5% and an LOQ of 120 ng spot $^{-1}$.

Photodiode Array

This detector measures the amount of light absorbed by the sample. In principle, a photodiode array detector (PDA) can detect any compound that absorbs in the UV–VIS range. Unlike normal UV–VIS spectrometers, PDA contains a high number of photodiodes on a silicon chip that allows a PDA to scan multiple wavelengths at once. The ability to quickly analyse the whole spectrum is essential for successful online coupling with common separation techniques. Due to its simplicity and versatility, PDA is the most common detector used in combination with main separation techniques such as liquid chromatography and capillary electrophoresis. Because all flavonoids absorb UV radiation (kaempferol A_{max} at 265 and 365 nm), they can be detected by PDA. However, it is not possible to effectively distinguish between different flavonoids and their derivatives by only considering their UV spectra. On the whole, the main disadvantage of PDA is not only low sensitivity but especially its low specificity when compared to either mass spectrometers or electrochemical detectors.

Flavonoids were analysed many times by HPLC–PDA which, for example, was successfully used to analyse kaempferol and quercetin in *Cissus quadrangularis* Linn, an ancient medicinal plant used as an analgesic and general tonic as well as for the treatment of bloody diarrhoea and skin disorders (Thakur et al., 2009). The method was validated and all of its parameters including LOD and linear range can be found in Table 3.

As an example of less common approaches, PDA was also used to perform derivative spectroscopy (DS) to analyse kaempferol, quercetin, myricetin and luteolin in two-compound mixtures (Baranowska and Rarog, 2001). With the use of zero-crossing DS they were able to determine the amounts of the above-described compounds from their two-compound mixtures. DS can be used as a supplementary method for HPLC as it does not require the separation of examined compounds but only a transfer of spectra recorded by PDA into spectrophotometric software.

It was suggested that this DS method may be used for direct analysis of samples, which contain only two dominating flavonols, for example after SPE elimination of interfering matrix.

Chemiluminiscence

These detectors measure light emitted from excited products of chemical reactions. With proper chemical reagents, this method is very sensitive and has almost no background noise compared to fluorimetric or photodiode detectors. Chemiluminiscence (CL) can be coupled with separation techniques such as HPLC or electrophoresis.

HPLC coupled with CL detector was used to analyse pharmaceuticals containing *Hippophae rhamnoides* L. extracts (Zhang and Cui, 2005). Three main flavonoids, including kaempferol, quercetin and isorhamnetin, were succesfully determined by this method. In sulfuric acid medium, the oxidation reaction between cerium(IV) and rhodamine 6G emitted weak CL. Flavonols including quercetin, kaempferol, and isorhamnetin were observed to enhance the CL intensity of the cerium(IV)-rhodamine 6G system. Considering not only good separation but also high intensity of CL, methanol-acetic acid-water (37:1:62 v/v/v) was chosen as an optimal mobile phase. The column effluent from PDA was mixed on-line with 0.1 M sulfuric acid, 5.06 mM cerium(IV), 5.06 μM rhodamine 6G using peristaltic pump at a flow rate of 1.5 mL min^{-1}. To summarize, HPLC-CL is a relatively selective method which reaches almost 30 times lower LOD of kaempferol than conventional HPLC-PDA but requires more complicated instrumentation and time-consuming optimisation of reagent concentrations.

Mass Spectrometry

Mass Spectrometry (MS) is a detection method widely used in combination with HPLC separation. It is based on the determination of mass-to-charge ratio of ionized molecules or their fragments. MS is a selective method which can distinguish ions based on their molecular weight and in the case of tandem mass spectrometry (MS/MS), also on molecular weight of their fragments. The detection of fragments (product ions) generated from a specific precursor ion by collision induced dissociation (CID) is a valuable technique to analyse complex matrices named Multiple-Reaction Monitoring (MRM). The critical part of MS is the ionization of analytes prior to their detection because it directly affects sensitivity. Ionization is affected by many factors, including compound structure and mobile phase composition, and therefore different ionization modules, polarities and applied voltages are used to optimize ionization effectivity. When liquid chromatography is coupled with MS, soft ionization techniques, such as Electrospray Ionization (ESI) and Atmospheric Pressure Chemical Ionization (APCI), are mostly used for ionization of polar and non-polar compounds, respectively. To determine kaempferol concentrations, ESI-MS can also be used in combination with gas chromatography, although such application is less common. Taken together, MS is detector of choice for analysis of kaempferol derivatives in plant samples due to its high sensitivity and selectivity.

Various MS instruments were used to analyse kaempferol. UPLC-MS/MS with ESI operating in positive ion mode was used for quantification of 26 flavonoids and other phenylpropanoids including kaempferol (Prokudina et al., 2012). All compounds were detected in selective MRM mode and their quantification was based on isotopically labeled internal standards to compensate for signal instability.

In general, ESI-MS signals often exhibits some drift and is substantially affected by the sample matrix which is the main reason for using internal standards (Chernetsova et al., 2010). If only kaempferol and its derivatives are analysed, negative ion mode is more frequently used due to its better performance in real samples. Negative ion mode commonly results in lower background noise, but it is not suitable for ionization of some compounds, such as methoxylated flavonoids.

Thus the selection of polarity mode also depends on ionization effectivity of other compounds quantified in the same chromatographic run. The dilemma of polarity selection may be solved by modern instruments that can quickly switch polarity even within short retention windows, thus giving the opportunity to optimise parameters for each analyte independently.

The main advantage of tandem and ion trap mass spectrometers is their ability to study specific fragmentation patterns. A common fragmentation of kaempferol glycosides is the loss of a glycan residue which occurs not only in the collision cell but also in the ionization source and results in generation of the aglycone fragment ion.

The aglycone ion $[M-H]^-$ (*m/z* 285) can be further fragmented by CID in tandem or ion trap MS to produce typical ions of *m/z* 151 and 153 (Figure 2). Described fragmentation patterns can be utilized in the development of MRM mode for quantification.

Besides routine quantification, MS could also be used for the identification of kaempferol derivatives. An interesting ion trap MS method based on the tendency of sodium ions to associate with the glycosyl group was developed to determine the distribution of sugars on the aglycone (Kite and Veitch, 2011).

Such approach is extremely useful, as glycosylated standards are not commercially available and isolation followed by NMR might be too complicated in many cases.

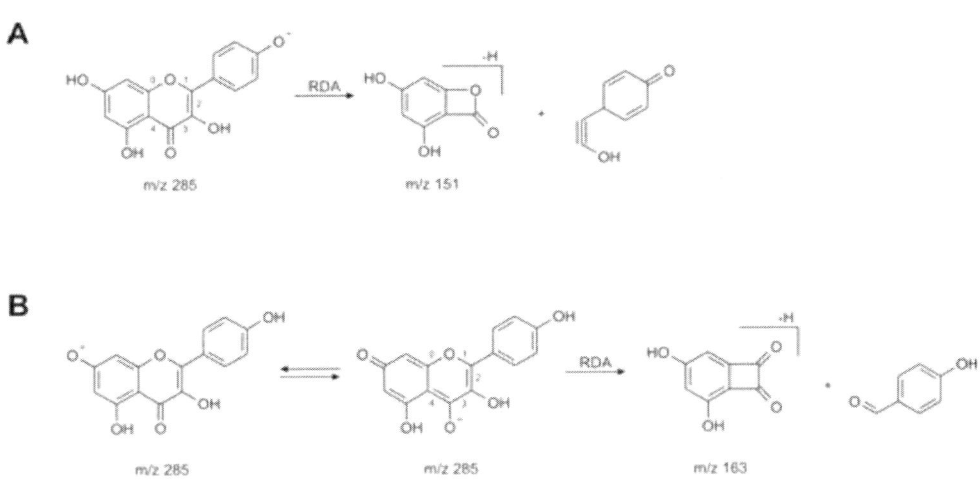

Figure 2. Proposed mechanisms (A, B) of kaempferol fragmentation after CID (Lu et al., 2010).

Nuclear Magnetic Resonance Spectroscopy

NMR is a phenomenon based on magnetically active nuclei (those that have non-zero spin, e.g. ^{1}H, ^{13}C) placed in strong magnetic field with electromagnetic radiation of radio waves.

Mutual interactions of observed nuclei and their chemical environments lead to characteristic shifts and splitting of signals in the spectrum. The resulting spectrum can then be interpreted as the structure of the observed molecule.

NMR is an indispensable technique for any structural identification of newly isolated or synthetized compounds.

There are various NMR experiments and each of them provides different information about structure. Combined information from those can provide complete information about the structure of an analyte. As a detection method, NMR can stand offline or can be online connected to a separation technique like LC.

As an illustration, two kaempferol glycosides were succesfully identified by NMR after their isolation from violet and purple flowers of *Aubrieta* × *cultorum*. The structural elucidation was allowed by various experiments, including 2D COSY, NOESY, ^{1}H-^{13}C HMQC, and ^{1}H-^{13}C HMBC (Tatsuzawa et al., 2012).

Voltammetry

Voltammetry is an electroanalytical method which measures current as a function of applied potential. There are some limitations of voltammetry. Analyzed substances must be oxidizable or reducible in the range where the solvent and electrode are electrochemically inert.

Furthermore, voltammetry provides relatively low information on analyte identity, so it is advantageous to use another complementary detection method for compound identification.

1. Cyclic Voltammetry

Cyclic Voltammetry (CV) is a potentiodynamic electrochemical method. Working electrode potential is ramped linearly versus time and when it reaches set potential, potential ramp is inverted.

CV was reported as a method for detection of kaempferol from a standard solution using saturated calomel electrode as the reference electrode, platinum electrode as a counter electrode and pyrolytic-graphite (PG) or hemoglobin/polysorbate 20 (HP) as a working electrode (Zheng et al., 2008). Starting with pyrolytic-graphite electrode, the authors studied voltammetric behaviour of kaempferol with different buffers at different pH values. The highest value of the peak current was obtained at pH 5.

The authors also find that the HP electrode, prepared by coating PG electrode with hemoglobin and polysorbate 20, showed significantly better results than the bare PG electrode.

2. Differential Pulse Voltammetry

Differential Pulse Voltammetry (DPV) uses a series of potential pulses of constant amplitude (10 to 100 mV) which are superimposed on a slowly changing base potential. Current is measured two times per pulse, at the begining and at the end of the pulse.

A validated method using DPV was developed to determine quercetin and kaempferol in phytopharmaceuticals made from *Ginkgo biloba* (Aguilar-Sanchez et al., 2005). By considering the validation parameters of reversed/normal phase HPLC coupled to PDA/DPV detectors, the authors decided to choose HPLC-DPV instrumentation as the optimal one for kaempferol and quercetin determination. It was also found that C-PVC used as measuring electrode provided two orders higher sensitivity than a carbon fiber electrode. Further method details are summarized in Table 3.

Coulometric Array

Coulometric Array Detection (CAD) measures electric charge needed for oxidation/reduction of analyte during its flow through a measuring cell. CAD is a relatively sensitive method but needs conductive mobile phase. This detection is almost exclusively combined with HPLC as a separation method. HPLC coupled with CAD was used to analyse 17 flavonoids including kaempferol in standard solutions. The method was suggested as an alternative for online screening of natural antioxidants. CAD consisted of 8 porous working graphite electrodes associated with palladium reference electrodes (Aaby et al., 2004).

Conclusion

Kaempferol is a relatively simple compound which can be succesfully analyzed by a number of methods. However, in real samples, various kaempferol derivatives with complicated chemical structures are present. While free kaempferol can be easily analyzed by most of the described analytical approaches, there are only a few methods that are suitable for the analysis of kaempferol derivatives. The main problem with kaempferol derivatives is the absence of commercial standards which greatly complicates method development. Thus only selective detectors (MS, NMR) with the ability to elucidate chemical structure can be recommended for determination of kaempferol derivatives. From this point of view, applied separation technique is at least partially predetermined by its compatibility with the respective detection system.

References

Aaby, K., Hvattum, E., Skrede, G. (2004) Analysis of flavonoids and other phenolic compounds using high-performance liquid chromatography with coulometric array detection: Relationship to antioxidant activity. *Journal of Agricultural and Food Chemistry* 52: 4595-4603

Aguilar-Sanchez, R., Ahuatl-Garcia, F., Davila-Jimenez, M. M., Elizalde-Gonzalez, M. P., Guevara-Villa, M. R. G. (2005) Chromatographic and electrochemical determination of quercetin and kaempferol in phytopharmaceuticals. *Journal of Pharmaceutical and Biomedical Analysis* 38: 239-249

Aguirre-Hernandez, E., Gonzalez-Trujano, M. E., Martinez, A. L., Moreno, J., Kite, G., Terrazas, T., Soto-Hernandez, M. (2010) HPLC/MS analysis and anxiolytic-like effect of quercetin and kaempferol flavonoids from Tilia americana var. mexicana. *Journal of Ethnopharmacology* 127: 91-97

Bajer, T., Adam, M., Galla, L., Ventura, K. (2007) Comparison of various extraction techniques for isolation and determination of isoflavonoids in plants. *Journal of Separation Science* 30: 122-127

Baranowska, I., Rarog, D. (2001) Application of derivative spectrophotometry to determination of flavonoid mixtures. *Talanta* 55: 209-212

Benyahia, S., Benayache, S., Benayache, F., Quintana, J., Lopez, M., Leon, F., Hernandez, J. C., Estevez, F., Bermejo, J. (2004) Isolation from Eucalyptus occidentalis and identification of a new kaempferol derivative that induces apoptosis in human myeloid leukemia cells. *Journal of Natural Products* 67: 527-531

Bhouri W, Ben Sghaier M, Kilani S, Bouhlel I, Dijoux-Franca MG, Ghedira K, Ghedira LC (2011) Evaluation of antioxidant and antigenotoxic activity of two flavonoids from Rhamnus alaternus L. (Rhamnaceae): Kaempferol 3-O-beta-isorhamninoside and rhamnocitrin 3-O-beta-isorhamninoside. *Food and Chemical Toxicology* 49: 1167-1173

Bloor, S. J (1995) An antimicrobial kaempferol-diacyl-rhamnoside from pentachondra-pumila. *Phytochemistry* 38: 1033-1035

Catalano, S., Bilia, A. R., Martinozzi, M., Morelli, I. (1994) kaempferol 3-o-beta-d-(6"-feruloylglucoside) from polylepis-incana. *Phytochemistry* 37: 1777-1778

Chernetsova, E. S., Kovaleva, A. S., Koryakova, A. G. (2010) Stability of operation of triple quadrupole mass spectrometers with electrospray ionization in the studies of pharmacokinetics and metabolism of new pharmaceutical entities. *Journal of Analytical Chemistry* 65: 1420-1422

Costa, S. S., Jossang, A., Bodo, B. (1996) 4""-acetylsagittatin A, a kaempferol triglycoside from Kalanchoe streptantha. *Journal of Natural Products* 59: 327-329

De Melo, G. O., Malvar, D. D., Vanderlinde, F. A., Rocha, F. F., Pires, P. A., Costa, E. A., de Matos, L. G., Kaiser, C. R., Costa, S. S. (2009) Antinociceptive and anti-inflammatory kaempferol glycosides from Sedum dendroideum. *Journal of Ethnopharmacology* 124: 228-232

Diaz, J. G., Carmona, A. J., Torres, F., Quintana, J., Estevez, F., Herz, W. (2008) Cytotoxic activities of flavonoid glycoside acetates from Consolida oliveriana. *Planta Medica* 74: 171-174

Dong, J. Z., Gao, W. S., Lu, D. Y., Wang, Y. (2011) simultaneous extraction and analysis of four polyphenols from leaves of lycium barbarum l. *Journal of Food Biochemistry* 35: 914-931

El-Sayed, N. H., Omara, N. M., Yousef, A.K., Farag, A. M., Mabry, T. J. (2001) Kaempferol triosides from Reseda muricata. *Phytochemistry* 57: 575-578

El-Sayed, N. H., Wojcinska, M., Drost-Karbowska, K., Matlawska, I., Williams, J., Mabry, T. J. (2002) Kaempferol triosides from Silphium perfoliatum. *Phytochemistry* 60: 835-838

Emiliani, G., Fondi, M., Fani, R., Gribaldo, S. (2009) A horizontal gene transfer at the origin of phenylpropanoid metabolism: a key adaptation of plants to land. *Biology Direct* 4: 12

Fiorentino, A., Ricci, A., D'Abrosca, B., Golino, A., Izzo, A., Pascarella, M. T., Piccolella, S., Esposito, A. (2009) Kaempferol Glycosides from Lobularia maritima and Their Potential Role in Plant Interactions. *Chemistry and Biodiversity* 6: 204-217

Fiorini, C., David, B., Fouraste, I., Vercauteren, J. (1998) Acylated kaempferol glycosides from Laurus nobilis leaves. *Phytochemistry* 47: 821-824

Fukumoto, H., Ishiguro, K., Murashima, T., Yamaki, M., Isoi, K. (1994) structure determination of a kaempferol 3-rhamnosyldiglucoside from impatiens-balsamina. *Phytochemistry* 37: 1486-1488

Gao, X., Williams, S. J., Woodman, O. L., Marriott, P. J. (2010) Comprehensive two-dimensional gas chromatography, retention indices and time-of-flight mass spectra of flavonoids and chalcones. *Journal of Chromatography A* 1217: 8317-8326

Gikas, E., Bazoti, F. N., Papadopoulos, N., Alesta, A., Economou, G., Tsarbopoulos, A. (2011) quantitation of the flavonols quercetin and kaempferol in the leaves of trigonella foenum-graecum by high-performance liquid chromatography - diode array detection. *Analytical Letters* 44: 1463-1472

Grayer, R. J., Kite, G. C., Abou-Zaid, M., Archer, L. J. (2000) The application of atmospheric pressure chemical ionisation liquid chromatography-mass spectrometry in the chemotaxonomic study of flavonoids: Characterisation of flavonoids from Ocimum gratissimum var. gratissimum. *Phytochemical Analysis* 11: 257-267

Halabalaki, M., Urbain, A., Paschali, A., Mitakou, S., Tillequin, F., Skaltsounis, A. L. (2011) Quercetin and Kaempferol 3-O- alpha-L-Rhamnopyranosyl-(1 -> 2)-alpha-L-arabinopyranoside -7-O-alpha-L-rhamnopyranosides from Anthyllis hermanniae: Structure Determination and Conformational Studies. *Journal of Natural Products* 74: 1939-1945

Ho, H. M., Chen, R. Y., Huang, Y., Chen, Z. Y. (2002) Vascular effects of a soy leaves (Glycine max) extract and kaempferol glycosides in isolated rat carotid arteries. *Planta Medica* 68: 487-491

Huang, J. L., Gu, M., Lai, Z. B., Fan, B. F., Shi, K., Zhou, Y. H., Yu, J. Q., Chen, Z. X. (2010) Functional Analysis of the Arabidopsis PAL Gene Family in Plant Growth, Development, and Response to Environmental Stress. *Plant Physiology* 153: 1526-1538

Huang, Y. L., Chen, C. C., Hsu, F. L., Chen, C. F. (1998) Tannins, flavonol sulfonates, and a norlignan from Phyllanthus virgatus. *Journal of Natural Products* 61: 1194-1197

Ibrahim, L. F., Kawashty, S. A., El-Hagrassy, A. M., Nassar, M. I., Mabry, T. J. (2008) A new kaempferol triglycoside from Fagonia taeckholmiana: cytotoxic activity of its extracts. *Carbohydrate Research* 343: 155-158

Imperato, F. (1995) flavonol glycosides from pteridium-aquilinum. *Phytochemistry* 40: 1801-1802

Imperato, F. (1996) Kaempferol 3-O-(5''-feruloylapioside) from Pteridium aquilinum. *Phytochemistry* 43: 1421-1423

Imperato, F. (1998) Kaempferol 7-O-rhamnoside-4 '-O-glucoside from Pteridium aquilinum. *Phytochemistry* 47: 911-913

Imperato, F. (2003) Kaempferol and quercetin 3-O-(2 '',3 ''-di-O-p-coumaroyl)-glucosides from Pteris vittata. *American Fern Journal* 93: 157-160

Imperato, F. (2006) Kaempferol 3-O-(acetylrutinoside), a new flavonoid and two new fern constituents, quercetin 3-O-(acetylglucoside) and 3-O-(acetylrutinoside) from Dryopteris villarii. *American Fern Journal* 96: 93-96

Imperato, F. (2007) Three new flavonoid glycosides, kaempferol 3-o-(caffeoylrhamnoside), apigenin 4 '-O-(caffeoylglucoside) and 4 '-O-feruloylglucoside) from Dryopteris villarii. *American Fern Journal* 97: 233-236

Imperato, F., Minutiello, P. (1997) Kaempferol 3-O-(6"-caffeoylglucoside) from Pteridium aquilinum. *Phytochemistry* 45: 199-200

Jin, H., Tanaka, T., Kouno, I., Ishimaru, K. (2007) A new kaempferol trioside from Solidago altissima L. *Journal of Natural Medicines* 61: 351-354

Johnson, E. L., Schmidt, W. F., Emche, S. D., Mossoba, M. M., Musser, S. M. (2003) Kaempferol (rhamnosyl) glucoside, a new flavonol from Erythroxylum coca. var.ipadu. *Biochemical Systematics and Ecology* 31: 59-67

Jung, H. A., Woo, J. J., Jung, M. J., Hwang, G. S., Choi, J. S. (2009) Kaempferol Glycosides with Antioxidant Activity from Brassica juncea. *Archives of Pharmacal Research* 32: 1379-1384

Jung, K. Y., Do, J. C., Son, K. H. (1993) kaempferol 3-o- 6"-o-(3-hydroxy-3-methylglutaroyl) glucoside from leaves of polygala-japonica. *Phytochemistry* 34: 1196-1197

Kaouadji, M., Morand, J. M., Garcia, J. (1993) polyphenols from platanus-acerifolia buds .8. Further acylated kaempferol rhamnosides from platanus-acerifolia buds. *Journal of Natural Products* 56: 1618-1621

Kim, J. E., Jung, M. J., Jung, H. A., Woo, J. J., Cheigh, H. S., Chung, H. Y., Choi, J. S. (2002) A new kaempferol 7-O-triglucoside from the leaves of Brassica juncea L. *Archives of Pharmacal Research* 25: 621-624

Kite, G. C., Veitch, N. C. (2011) Identification of common glycosyl groups of flavonoid O-glycosides by serial mass spectrometry of sodiated species. *Rapid Communications in Mass Spectrometry* 25: 2579-2590

Kovganko, N. V., Kashkan, Z. N., Krivenok, S. N. (2004) Bioactive compounds of the flora of Belarus. 3. Gymnocarpium dryopteris, a kaempferol source. *Chemistry of Natural Compounds* 40: 16-18

Kuo, Y. C., Lu, C. K., Huang, L. W., Kuo, Y. H., Chang, C., Hsu, F. L., Lee, T. H. (2005) Inhibitory effects of acylated kaempferol glycosides from the leaves of Cinnamomum kotoense on the proliferation of human peripheral blood mononuclear cells. *Planta Medica* 71: 412-415

Leal, P. F., Maia, N. B., Carmello, Q. A. C., Catharino, R. R., Eberlin, M. N., Meireles, M. A. A. (2008) Sweet Basil (Ocimum basilicum) Extracts Obtained by Supercritical Fluid Extraction (SFE): Global Yields, Chemical Composition, Antioxidant Activity, and Estimation of the Cost of Manufacturing. *Food and Bioprocess Technology* 1: 326-338

Lei, J., Xiao, Y. C., Huang, J., Tang, M., Deng, P. C. (2009) Two New Kaempferol Glycosides from Androsace umbellata. *Helvetica Chimica Acta* 92: 1439-1444

Leitao, G. G., Soares, S. S. V., Brito, T. D. M., Delle Monache, F. (2000) Kaempferol glycosides from Siparuna apiosyce. *Phytochemistry* 55: 679-682

Li, G. W., Zeng, X. L., Xie, Y., Cai, Z. Z., Moore, J. C., Yuan, X. R., Cheng, Z. H., Ji, G. (2012) Pharmacokinetic properties of isorhamnetin, kaempferol and quercetin after oral gavage of total flavones of Hippophae rhamnoides L. in rats using a UPLC-MS method. *Fitoterapia* 83: 182-191

Li, Y., Yang, Y., Wang, S. X., Wei, Y. M., Zheng, X. H. (2009) Simultaneous LC Determination of Quercetin, Kaempferol and Isorhamnetin in Rabbits after Intragastric Administration of an Ethanol Extract from Pollen Typhae. *Chromatographia* 69: 117-121

Liu, C. D., Chen, J., Wang, J. H. (2009) a novel kaempferol triglycoside from flower buds of Panax quinquefolium. *Chemistry of Natural Compounds* 45: 808-810

Liu, L., Liu, R. L., Zhang, J., Zhang, Z. Q. (2012) Study on the PEG-based microwave-assisted extraction of flavonoid compounds from persimmon leaves. *Journal of Separation Science* 35: 3412-3420

Lu, L., Song, F. R., Tsao, R., Jin, Y. R., Liu, Z. Q., Liu, S. Y. (2010) Studies on the homolytic and heterolytic cleavage of kaempferol and kaempferide glycosides using electrospray ionization tandem mass spectrometry. *Rapid Communications in Mass Spectrometry* 24: 169-172

Males, Z., Medic-Saric, M. (2001) Optimization of TLC analysis of flavonoids and phenolic acids of Helleborus atrorubens Waldst. et Kit. *Journal of Pharmaceutical and Biomedical Analysis* 24: 353-359

Marzouk, M. M., Kawashty, S. A., Saleh, N. A. M., Al-Nowaihi, A. S. M. (2009) A new kaempferol trioside from Farsetia aegyptia. *Chemistry of Natural Compounds* 45: 483-486

Min, B. S., Tomiyama, M., Ma, C. M., Nakamura, N., Hattori, M. (2001) Kaempferol acetylrhamnosides from the rhizome of Dryopteris crassirhizoma and their inhibitory effects on three different activities of human immunodeficiency virus-1 reverse transcriptase. *Chemical and Pharmaceutical Bulletin* 49: 546-550

Mizuno, M., Iinuma, M., Tanaka, T., Yamamoto, H., Tu, Z. B. (1991) sutchuenoside-a - a new kaempferol glycoside from the aerial parts of epimedium-sutchuenense. *Journal of Natural Products* 54: 1427-1429

Moon, S. S., Rahman, M. A. A., Manir, M. M., Ahamed, V. S. J. (2010) Kaempferol Glycosides and Cardenolide Glycosides, Cytotoxic Constituents from the Seeds of Draba nemorosa (Brassicaceae). *Archives of Pharmacal Research* 33: 1169-1173

Moriyama, H., Iizuka, T., Nagai, M., Murata, Y. (2003) HPLC quantification of kaempferol-3-O-gentiobioside in Cassia alata. *Fitoterapia* 74: 425-430

Muminova, B. A., Batirov, E. K., Yuldashev, M. P., Inamova, Z. G. (2006) Kaempferol glycosedes from Allium cepa and Raphanus sativus. *Chemistry of Natural Compounds* 42: 110-111

Nielsen, J. K., Norbaek, R., Olsen, C. E. (1998) Kaempferol tetraglucosides from cabbage leaves. *Phytochemistry* 49: 2171-2176

Olszewska, M. (2008) Separation of quercetin, sexangularetin, kaempferol and isorhamnetin for simultaneous HPLC determination of flavonoid aglycones in inflorescences, leaves and fruits of three Sorbus species. *Journal of Pharmaceutical and Biomedical Analysis* 48: 629-635

Ozden, S., Durust, N., Toki, K., Saito, N., Honda, T. (1998) Acylated kaempferol glycosides from the flowers of Delphinium formosum. *Phytochemistry* 49: 241-245

Porgali, E., Buyuktuncel, E. (2012) Determination of phenolic composition and antioxidant capacity of native red wines by high performance liquid chromatography and spectrophotometric methods. *Food Research International* 45: 145-154

Prokudina, E. A., Havlicek, L., Al-Maharik, N., Lapcik, O., Strnad, M., Gruz, J. (2012) Rapid UPLC-ESI-MS/MS method for the analysis of isoflavonoids and other phenylpropanoids. *Journal of Food Composition and Analysis* 26: 36-42

Rao, L. J. M., Yada, H., Ono, H., Yoshida, M. (2002) Acylated and non-acylated flavonol monoglycosides from the Indian minor spice nagkesar (Mammea longifolia). *Journal of Agricultural and Food Chemistry* 50: 3143-3146

Rodriguez, P., Gonzalez-Mujica, F., Bermudez, J., Hasegawa, M. (2010) Inhibition of glucose intestinal absorption by kaempferol 3-O-alpha-rhamnoside purified from Bauhinia megalandra leaves. *Fitoterapia* 81: 1220-1223

Selyutina, I. Y., Tankhaeva, L. M., Olennikov, D. N (2008) Acylated kaempferol diglycoside from Allium senescens. *Chemistry of Natural Compounds* 44: 769-770

Shen, C. C., Lin, T. W., Huang, Y. L., Wan, S. T., Shien, B. J., Chen, C. C. (2006) Phenolic constituents of the roots of Sophora flavescens. *Journal of Natural Products* 69: 1237-1240

Singh, B., Mungara, P., Nivsarkar, M., Anandjiwala, S. (2009) HPTLC Densitometric Quantification of Glycyrrhizin, Glycyrrhetinic Acid, Apigenin, Kaempferol and Quercetin from Glycyrrhiza glabra. *Chromatographia* 70: 1665-1672

Singh, R., Singh, B., Singh, S., Kumar, N., Kumar, S., Arora, S. (2008) Anti-free radical activities of kaempferol isolated from Acacia nilotica (L.) Willd. Ex. Del. *Toxicology in Vitro* 22: 1965-1970

Soliman, F. M., Shehata, A. H., Khaleel, A. E., Ezzat, S. M. (2002) An acylated kaempferol glycoside from flowers of Foeniculum vulgare and F-Dulce. *Molecules* 7: 245-251

Sun, P. Y., Chen, Y. J., Shimizu, N., Takeda, T. (1998) Studies on the constituents of Epimedium koreanum. III. *Chemical and Pharmaceutical Bulletin* 46: 355-358

Tang, L., Wu, X. J., Feng, B. M., Shi, L. Y., Fu, X. Y., Wang, Y. Q., Liu, M. F. (2011) Two new kaempferol glycosides from the seeds of Camellia semiserrata Chi. *Chinese Chemical Letters* 22: 835-838

Tanikawa, N., Yoshida, K., Kondo, T., Mizuno, T., Iwashina, T., Nakayama, M. (2011) Urakunoside, a New Tetraglycosyl Kaempferol from Petals of the Wabisuke Camellia cv. Tarokaja. *Bioscience Biotechnology and Biochemistry* 75: 2046-2048

Tanjung, M., Hakim, E. H., Mujahidin, D., Hanafi, M., Syah, Y. M. (2009) Macagigantin, a farnesylated flavonol from Macaranga gigantea. *Journal of Asian Natural Products Research* 11: 929-932

Tatsuzawa, F., Aiba, Y., Morino, T., Saito, N., Shinoda, K., Kato, K., Toki, K., Honda, T. (2012) Copigmentation with Acylated Anthocyanin and Kaempferol Glycosides in Violet and Purple Flower Cultivars of Aubrieta x cultorum (Brassicaceae). *Journal of the Japanese Society for Horticultural Science* 81: 275-284

Thakur, A. K., Jain, V., Hingorani, L., Laddha, K. S. (2009) Improved High-Performance Liquid Chromatography-DAD Method for the Simultaneous Analysis of Quercetin and Kaempferol in the Stems of Cissus quadrangularis Linn. *Acta Chromatographica* 21: 95-103

Tsiklauri, L. K., An, G., Alania, M. D., Kemertelidze, E. P., Morris, M. E. (2012) Optimum HPLC parameters for simultaneous determination of Robinin and Kaempferol. *Pharmaceutical Chemistry Journal* 46: 64-67

Van Beek, T. A. (2002) Chemical analysis of Ginkgo biloba leaves and extracts. *Journal of Chromatography* A 967: 21-55

Veitch, N. C., Bristow, J. M., Kite, G. C., Lewis, G. P. (2005) Mildbraedin, a novel kaempferol tetraglycoside from the tropical forest legume Mildbraediodendron excelsum. *Tetrahedron Letters* 46: 8595-8598

Vitalini, S., Flamini, G., Valaguzza, A., Rodondi, G., Iriti, M., Fico, G. (2011) Primula spectabilis Tratt. aerial parts: Morphology, volatile compounds and flavonoids. *Phytochemistry* 72: 1371-1378

Wang, S. P., Huang, K. J. (2004) Determination of flavonoids by high-performance liquid chromatography and capillary electrophoresis. *Journal of Chromatography A* 1032: 273-279

Wei, Y., Xie, Q. Q., Fisher, D., Sutherland, I. A. (2011) Separation of patuletin-3-O-glucoside, astragalin, quercetin, kaempferol and isorhamnetin from Flaveria bidentis (L.) Kuntze by elution-pump-out high-performance counter-current chromatography. *Journal of Chromatography A* 1218: 6206-6211

Wei, Y., Xie, Q. Q., Ito, Y. C. (2008) Preparative isolation of kaempferol pyranosides from a traditional Chinese herb using high speed countercurrent chromatography. *Journal of Liquid Chromatography and Related Technologies* 31: 443-451

Xiao, Z. P., Wu, H. K., Wu, T., Shi, H., Hang, B., Aisa, H. A. (2006) Kaempferol and quercetin flavonoids from Rosa rugosa. *Chemistry of Natural Compounds* 42: 736-737

Yang, J. H., Kondratyuk, T. P., Marler, L. E., Qiu, X., Choi, Y., Cao, H. M., Yu, R., Sturdy, M., Pegan, S., Liu, Y., Wang, L. Q., Mesecar, A. D., Van Breemen, R. B., Pezzuto, J. M., Fong, H. H. S., Chen, Y. G., Zhang, H. J. (2010) Isolation and evaluation of kaempferol glycosides from the fern Neocheiropteris palmatopedata. *Phytochemistry* 71: 641-647

Yi, Y., Zhang, Q. W., Li, S. L., Wang, Y., Ye, W. C., Zhao, J., Wang, Y. T (2012) Simultaneous quantification of major flavonoids in "Bawanghua", the edible flower of Hylocereus undatus using pressurised liquid extraction and high performance liquid chromatography. *Food Chemistry* 135: 528-533

Yu, S. F., Shun, C. T., Chen, T. M., Chen, Y. H. (2006) 3-O-beta-D-glucosyl-(1 -> 6)-beta-D-glucosyl-kaempferol isolated from Sauropus androgenus reduces body weight gain in Wistar rats. *Biological and Pharmaceutical Bulletin* 29: 2510-2513

Zang, Y. Q., Sato, H., Igarashi, K. (2011) Anti-Diabetic Effects of a Kaempferol Glycoside-Rich Fraction from Unripe Soybean (Edamame, Glycine max L. Merrill. 'Jindai') Leaves on KK-A(y) Mice. *Bioscience Biotechnology and Biochemistry* 75: 1677-1684

Zhang, Q. L., Cui, H. (2005) Simultaneous determination of quercetin, kaempferol, and isorhamnetin in phytopharmaceuticals of Hippophae rhamnoides L. by high-performance liquid chromatography with chemiluminescence detection. *Journal of Separation Science* 28: 1171-1178

Zheng, X. F., Zao, J., Chen, G. F., Liu, S. L., Li, G. X. (2008) Electroanalysis of kaempferol using pyrolytic graphite and a hemoglobin/polysorbate-20 modified electrodes. *Journal of Analytical Chemistry* 63: 381-385

Zhu, X. Y., Lin, H. M., Chen, X., Xie, J., Wang, P. (2011) Mechanochemical-Assisted Extraction and Antioxidant Activities of Kaempferol Glycosides from Camellia oleifera Abel. Meal. *Journal of Agricultural and Food Chemistry* 59: 3986-3993

In: Kaempferol
Editors: Garion Villers and Yves Fougere
ISBN: 978-1-62618-515-9

Chapter 5

Chemoprotective Effects of Kaempferol in Colorectal Tumorigenesis

Gerd Bobe*[1,2], Matthew Young[3], Elaine Lanza[3], Amanda J. Cross[4] and Nancy H. Colburn[3]
[1]Linus Pauling Institute, Oregon State University, Corvallis, OR, US
[2]Department of Animal and Rangeland Sciences, Oregon State University, Corvallis, OR, US
[3]Laboratory of Cancer Prevention, Center for Cancer Research, National Cancer Institute, Frederick, MD, US
[4]Nutritional Epidemiology Branch, Division of Cancer Epidemiology and Genetics, National Cancer Institute, National Institutes of Health, Rockville, MD, US

Abstract

Kaempferol is a bioactive polyphenol that has been shown to have chemoprotective properties in cell culture. The objective of this study was to evaluate the potential chemoprotective effects of kaempferol against colorectal tumorigenesis in humans and animal models. The association between kaempferol intake and colorectal adenoma recurrence was examined in the Polyp Prevention Trial (PPT), a 4-year randomized, clinical intervention trial which examined the effect of a high-fruit and vegetable, high-fiber, and low-fat diet on colorectal adenoma recurrence. Kaempferol consumption was estimated annually by self-administered food frequency questionnaires that were connected to the U.S. Department of Agriculture flavonoid database for foods.

Kaempferol intake was inversely associated with high-risk adenoma [odds ratio (OR) for individuals in the highest compared to those in the lowest quartile: 0.62; 95% confidence interval (CI): 0.40, 0.97; $P_{\text{Linear}} = 0.04$) and advanced adenoma recurrence (OR: 0.49; 95% CI: 0.26, 0.91; $P_{\text{Linear}} = 0.03$).

* Correspondence and current affiliation: Gerd Bobe, 307 Linus Pauling Science Center, Oregon State University, Corvallis, OR 97331; e-mail: gerd.bobe@oregonstate.edu; phone: 541-737-1898; fax: 541-737-4174.

To examine whether the chemoprotective effect of kaempferol could be in part explained by its anti-inflammatory properties, we measured serum concentrations of cytokines at baseline, year 1, and year 3 in the intervention arm of the PPT. High intake of kaempferol was associated with lower serum interleukin 6 (IL-6) concentrations in year 1 and 3 (highest versus lowest kaempferol intake quartile: 1.86 versus 2.09 pg/mL; $P_{Kaempferol} = 0.05$). The greatest decrease in serum IL-6 concentrations was observed in participants with elevated serum IL-6 concentrations (> 2.37 pg/mL) at baseline (highest versus lowest intake quartile: 2.63 versus 3.23 pg/mL; $P_{Kaempferol} = 0.05$). Individuals with above median kaempferol consumption and equal or below median change in IL-6 concentrations had the lowest risk of advanced adenoma recurrence.

To examine whether the beneficial effects of kaempferol on colorectal tumorigenesis were associated with other compounds enriched in kaempferol-containing foods, we conducted a study in male *FVB/NCr* mice fed a diet with or without 500 ppm purified kaempferol for 85 days after colorectal tumor induction and promotion by the chemical carcinogen azoxymethane and the colonic irritant dextran sodium sulfate. In comparison to mice on the control diet, mice consuming kaempferol had fewer tumors (16 vs. 21 tumors; $P = 0.04$) and smaller tumors (11.6 vs. 7.4 mm^3/tumor; $P = 0.09$).

Our results suggest kaempferol is a potential chemoprotective compound against colorectal tumorigenesis.

Keywords: Cancer prevention, colorectal carcinogenesis, inflammation, kaempferol

Abbreviations

AOM	azoxymethane
AP-1	activating protein-1
BMI	body mass index
CI	confidence interval
CRC	colorectal cancer
CV	coefficient of variation
CXCL	chemokine ligand
DSS	dextran sodium sulfate
FFQ	food frequency questionnaire
GMCSF	granulocyte macrophage colony stimulating factor
IL	interleukin
IFN	interferon
IQR	interquartile range
MAPK	mitogen activated protein kinase
NFκB	nuclear factor κB
NSAID	nonsteroidal anti-inflammatory drug
OR	odds ratio
PPT	Polyp Prevention Trial
TNF	tumor necrosis factor
USDA	U.S. Department of Agriculture

Introduction

Colorectal cancer (CRC) is an important public health problem as it is the fourth most common cancer (143,460 new cases/year) and the second leading cause of cancer mortality (51,690 people/year) in the U.S. (American Cancer Society, 2012). These numbers do not include the even greater number of people that are diagnosed and treated each year for colorectal adenoma, a pre-carcinogenic stage of CRC. Individuals with advanced adenomas, defined as those that are large ($\geq$ 1 cm) or have atypical histological features (high-grade dysplasia or villous characteristics), or individuals with 3 or more adenomas are at increased risk for developing CRC (Laiyemo *et al.*, 2009; Leung *et al.*, 2010). Treatment costs for CRC are high with approximately $6.5 billion per year in the U.S. (Gill and Sinicrope, 2005). Growing evidence suggest that the quantity and type of food we eat plays an important role in the induction and progression of colorectal neoplasms. Dietary change or supplementation with components naturally found in our diet represents a safe and cost-efficient strategy for CRC prevention.

Inflammatory responses are crucial to tissue repair and host defense; however, they also play a central role in cancer initiation and progression (Coussens and Werb, 2002). The association between inflammation and CRC is exemplified by the fact that patients with inflammatory bowel diseases have an increased CRC risk (Eaden *et al.*, 2001; Canavan *et al.*, 2007) and that regular use of anti-inflammatory drugs is known to decrease CRC risk (Asano and McLeod, 2007). The inflammatory process is initiated by the synthesis and secretion of pro-inflammatory cytokines such as interleukin-1β (IL-1 β), IL-6, IL-12, chemokine ligand 1 (CXCL-1 or RANTES), interferon-γ (IFN- γ), and tumor necrosis factor-α (TNF-α) in macrophages in response to an inflammatory insult (Lin and Karin, 2007). These pro-inflammatory cytokines bind to their receptors on the cell surface of other cells and induce the mitogen activated protein kinase (MAPK) and IκB kinase signaling pathways, which amplify the signal and activate two redox-sensitive transcription factors, nuclear factor κB (NFκB) and the c-Jun part of activating protein-1 (AP-1). Nuclear factor κB and AP-1 induce gene expression of a wide variety of genes, including genes that facilitate cancer cell growth, survival, and progression (Bode and Dong, 2006; Nam *et al.*, 2006). This pathway can be attenuated by the secretion of anti-inflammatory cytokines, including IL-4 and IL-10, and by bioactive food components such as kaempferol (Aggarwal and Shishodia, 2006; Hämäläinen *et al.*, 2007; Lee *et al.*, 2010; Chen *et al.*, 2012; Gong *et al.*, 2012).

Kaempferol is a bioactive polyphenol out of the flavonol subgroup that is present in many plant-based foods and beverages, in particular tea, dry beans, and, in lower concentrations, in other vegetables (U.S. Department of Agriculture, 2007). Kaempferol may inhibit colorectal inflammation and tumorigenesis, as kaempferol induces death of colon cancer cells (Li *et al.*, 2009; Niestroy *et al.*, 2011; Wang and Zhang, 2012) and decrease dextran sodium sulfate (DSS)-induced colorectal inflammation and cytokine expression in mice (Park *et al.*, 2012).

As a first step for establishing kaempferol as a potential dietary CRC preventive compound, we examined the association between dietary kaempferol and colorectal adenoma recurrence in humans in the Polyp Prevention Trial (PPT). We previously estimated that U.S. adults consume on average approximately 4 to 5 mg/d of kaempferol from food and beverages (Bobe *et al.*, 2010a) and that, at 3 to 4-fold greater concentrations, a kaempferol-enriched diet may decrease the risk of advanced adenoma recurrence (Bobe et al., 2008).

Other human cohort studies, however, did not find an association between kaempferol intake and CRC (Knekt *et al.*, 2002; Lin *et al.*, 2006; Simons *et al.*, 2009; Kyle *et al.*, 2010).

As a second step, we examined the association between dietary kaempferol, colorectal adenoma recurrence, and serum cytokine concentrations as potential indicators of a beneficial kaempferol response. Previously, we reported that a decrease in serum cytokine concentrations during the PPT was associated with a decreased risk of advanced adenoma recurrence and that increased flavonol consumption and decreased serum cytokine concentrations during the trial act synergistically to inhibit advanced adenoma recurrence (Bobe *et al.*, 2010a; 2010b; 2011). We hypothesized that the kaempferol-induced decrease in advanced colorectal adenoma recurrence might be mediated in part by attenuated serum cytokine concentrations during the PPT.

As a third step, we examined whether purified kaempferol at concentrations that could be used as a dietary supplement may inhibit inflammation-associated colorectal tumorigenesis. Because human clinical trials at this phase of the investigation are cost-prohibitive, we used a well-established two-stage inflammation-associated CRC mouse model; tumors are initiated by the chemical carcinogen azoxymethane (AOM), and tumor growth is promoted by the colonic irritant DSS (Tanaka *et al.*, 2003; Kohno *et al.*, 2006, 2007). We hypothesized that purified kaempferol inhibits inflammation-associated colorectal tumorigenesis. To our knowledge, the effects of dietary kaempferol on colorectal tumorigenesis have not been studied.

Materials and Methods

Human Study Design and Outcome

The PPT was a 4-year multi-center, randomized, nutritional intervention study with the primary objective to evaluate whether increasing consumption of fiber, fruit and vegetables, and decreasing the amount of fat in the diet could reduce the risk of adenoma recurrence. The study has previously been described in detail (Schatzkin et al., 2000; Lanza et al., 2001). To be eligible for this study, participants had to have at least 1 histologically confirmed colorectal adenoma identified by complete colonoscopy in the 6 months before study entry. Our study included individuals from both the control (n = 930) and the intervention arm (n = 929) that had a colonoscopy at baseline (T0) and at the end of the four year trial (T4), as well as dietary data for any of the first 3 years of the study (T1, T2, or T3). For analysis of serum cytokines, we included 872 participants from the intervention arm with available dietary data for any of the first 3 years of the trial and serum from baseline (T0) and either from year 1 (T1) or year 3 (T3). Two pathologists independently examined all lesions for histological features and degree of atypia. Adenoma recurrence was defined as: any ($\geq$ 1 adenoma, n = 736 for complete study, n = 348 for cytokine study), high-risk ($\geq$ 3 adenomas or $\geq$ 1 advanced adenoma, n = 226 for complete study, n = 100 for cytokine study) or advanced ($\geq$ 1 adenoma of $\geq$ 1 cm in size, having $\geq$ 25% villous component, or exhibiting high-grade dysplasia, n = 123 for complete study, n = 49 for cytokine study). The institutional review boards of the National Cancer Institute and each participating center approved the study, and all participants provided written, informed consent.

Lifestyle and Kaempferol Data

At T0 and at each of the annual follow-up visits (T1, T2, T3, T4), participants were asked to complete an interviewer-administered questionnaire about demographics, family history, and use of medication or supplements (including name and dosage), as well as a self-administered food frequency questionnaire (FFQ) that was reviewed with a certified nutritionist. The FFQ was specifically designed and validated to accurately measure fat, fiber, fruit, and vegetable consumption (Block *et al.*, 1990). Relative to 24-hour dietary recall and four-day food record data, the FFQ slightly overestimated fat and underestimated fiber, fruit and vegetable intake and had acceptable correlations of macro- and micronutrients (Caan *et al.*, 1999; Lanza *et al.*, 2001). The average kaempferol intake was estimated using 30 of the 119 questions on the FFQ using the 2007 U.S. Department of Agriculture (USDA) flavonoid database (USDA, 2007). The dietary sources of kaempferol that accounted each for over 5% of the total kaempferol consumption during the trial were dry beans (69%), chili (7%), and tea (6%).

Human Serum Data

At each annual visit, participants provided an overnight fasting blood sample, the serum from which was stored at -70°C until analysis. Among the 872 participants, 23 and 69 had no available samples at T1 and T3, respectively. Serum concentrations of IL-1β, IL-2, IL-6, IL-8, IL-10, IL-12p70, GMCSF, IFNγ, and TNFα were measured at T0, T1, and T3 by the Clinical Support Laboratory of SAIC Frederick, Inc. (Frederick, MD) using a commercially available multiplex 96-well enzyme-linked immunoabsorbent assay kit (MS6000 Human Pro-Inflammatory 9-Plex Ultra-Sensitive Kit K11007; Meso Scale Diagnostics, Gaithersburg, MD) on a Sector^{TM} Imager 6000 according to the manufacturer's recommendation (Meso Scale Diagnostics, Gaithersburg, MD). Study samples were run with 2 pooled serum samples and 3 assay specific standards in duplicate and the average of the duplicate was used. Fewer than 1% of the samples were below the detection limit and the interassay coefficient of variation (CV) was below 15%.

Animal Study Design and Outcome

Mice were housed in individual cages in animal research facilities overseen by the In Vivo Carcinogenesis Program of SAIC Frederick, Inc, National Cancer Institute – Frederick. All mice were maintained in temperature (20-22°C) and humidity (50%) controlled rooms with a twelve-hour light:dark cycle. Animal research approval had been obtained for feeding studies in the AOM/DSS mouse colon cancer model from the National Cancer Institute - Frederick Animal Use and Care Committee.

Experimental diets were AIN-93G rodent diets (Reeves *et al.*, 1993) that contained either 0 (control) or 500 mg/kg kaempferol. To keep diets isocaloric and isonitrogenous, kaempferol replaced part of the corn starch in the diet. Kaempferol (98% purity) was purchased from Jinan Haohua Industry Co., Ltd (Jinan Shandong, China) and was verified for purity using HPLC/MS and NMR by the Natural Products Support Group of Applied Developmental Research Program of SAIC Frederick, Inc. (Frederick, MD). Purified diets were prepared by Harlan Laboratories (Madison, WI). The kaempferol dosage was based on effective dietary dosages of flavonols for attenuating inflammatory responses (Kwon *et al.*, 2005). Assuming a daily food intake of 3.5 gram per 25-gram weighing mouse, 500 ppm kaempferol would be the human equivalent of 8.4 mg kaempferol/kg of body weight per day or 504 mg of kaempferol per day for a 60 kg human, a dose that could be consumed by a supplement (Reagan-Shaw *et al.*, 2007).

Pathogen-free male *FVB/NCr* mice were purchased at 5 weeks of age from the NCI-Frederick Animal Production Area (Frederick, MD) and were fed an AIN-93G purified diet from Harlan Teklad (Madison, WI) and drinking water *ad libitum*. After 1 week of acclimatization, mice were injected intraperitoneally with the colon carcinogen AOM (Sigma St. Louis, MO) dissolved in 0.1 mL of saline at 10 mg AOM per kg of body weight to induce colon tumors. One week after AOM injection, mice received 2% of the colon irritant DSS (weight/vol; molecular weight: 36,000-250,000 kD; MP Biomedicals LL, Solon, OH) dissolved in normal drinking water (reverse osmosis-purified water) for 1 week to promote colon tumorigenesis. Response to DSS was monitored by weighing mice at days 3, 5, 7, and 10 after the start of DSS exposure. Mice that gained weight throughout the weighing period, indicating no DSS response, received another week of 2% DSS in their drinking water 14 days after the end of the first DSS dosage. Mice that did not gain weight during and shortly after DSS exposure, indicating a DSS response, were sorted 3 days after the end of their last DSS dosage by body weight loss and were alternately assigned to remain on the AIN-93G diet or immediately switched to a AIN-93G diet containing 500 ppm kaempferol. Mice were assigned to diets according to body weight loss to ensure that both groups were similar in initial tumor initiation and promotion response because greater weight loss during and shortly after DSS exposure is associated with greater inflammation and tumor number. A total of 32 mice per treatment group started the dietary treatments and remained on the diets for 85 days, until the mice lost more than 10% of their maximum body weight, or had an intestinal prolapse, whichever came first. Mice had *ad libitum* access to feed and water during the treatment period. The experimental protocol is based on previous research of our laboratory group and published data (Tanaka *et al.*, 2003; Kohno *et al.*, 2006, 2007).

After sacrifice by CO_2 asphyxiation, the colorectum was removed, washed, weighed, opened longitudinally, and its length measured. The colorectum was briefly held in phosphate-buffered saline solution (pH 7.4) until pinned flat on cardboard, and then was fixed for one day in 10% neutral-buffered formalin solution (vol/vol; pH 7.4). Until further analysis, the colorectum was stored at room temperature in 1% neutral buffered formalin solution (vol/vol). To facilitate tumor identification and quantification, tissues were stained in phosphate-buffered saline solution (pH 7.4) containing 0.02% (weight/vol) methylene blue. Using a stereomicroscope and a measuring grid, tumor number and dimensions of individual tumors were determined for each colorectum. The areas of flat adenomas were calculated using the formula: adenoma size = $(\pi * d_1 * d_2) / 4$. The areas of papillary tumors were calculated by using the formula: adenoma size = $(\pi * d_1 * d_2 * d_3) / 6$.

Animal Serum Data

After 85 day on the treatment diets, an overnight fasting blood sample was taken. Serum was stored at -70°C until analysis. Of the 32 mice per group, blood was only taken off mice that completed the 85 day diet treatment, which was 17 mice in the control group and 18 mice in the kaempferol group. Serum concentrations of IL-1β, IL-6, IL-10, IL-12p70, CXCL1, IFNγ, and TNFα were measured at the same location as the human serum samples under the same conditions described above using a commercially available multiplex 96-well enzyme-linked immunoabsorbent assay kit (MS6000 Mouse Pro-Inflammatory 7-Plex Ultra-Sensitive Kit K15012C; Meso Scale Diagnostics).

Statistical Analyses

Statistical analyses were performed using SAS, version 9.2 (SAS, 2009; SAS, Inc., Cary, NC) software. For the human study, baseline characteristics, average dietary intake for the first 3 years of the trial, and serum cytokine concentrations were evaluated by adenoma recurrence at T4 (no versus any, high-risk, or advanced adenoma recurrence) using Wilcoxon rank-sum test for continuous variables and Fisher's exact test for categorical variables and are shown as medians and interquartile ranges (IQRs). The association between serum cytokine concentrations and kaempferol consumption during the first 3 years of the trial was evaluated with the Kruskal Wallis test and multiple linear regression models.

We defined trial cytokine concentrations as the geometric mean of T1 and T3. Cytokine concentration changes during the trial were defined as the geometric mean of T1 and T3 minus the baseline values. The association between cytokine change and colorectal adenoma recurrence was estimated by odds ratios (ORs) and 95% confidence intervals (CIs) using logistic regression. A trend test was performed using the natural log-transformed median values of each quartile as a continuous variable in a logistic regression model. The median values of both kaempferol intake and cytokine change were used as cut-offs (≤ median, > median) to examine the combined effect of kaempferol intake and cytokine change on colorectal adenoma recurrence. Potential confounders were added to the models in a stepwise fashion and remained in the model if they changed the association by >10%, were associated with both study variables, and had a χ^2 *p*-value ≤ 0.20.

In mice, the effect of kaempferol treatment on serum cytokine concentrations and tumor number, burden, and size was evaluated with the non-parametric Kruskal Wallis test (PROC NONPAR1WAY) and the Wilcoxon rank sum test (PROC LIFETEST), and analysis of variance (PROC GLM) models. Beside diet, number of DSS rounds (1 or 2) was included in the analysis of variance model. To achieve normality for the analysis of variance procedure, all dependent variables were natural log-transformed or for tumor burden and size twice natural log-transformed, after adding a 1 to prevent values below zero. All statistical tests were two-sided. The *P* values were not adjusted for multiple comparisons. Significance was declared at $P \leq 0.05$ and a tendency at 0.05 to 0.10.

Results

At the end of the 4-year trial, 40% of participants had at least 1 adenoma, 11% (intervention arm) or 12% (control and intervention arm) had high-risk adenoma, and 6% (intervention arm) or 7% (control and intervention arm) had an advanced adenoma recurrence. The association between participant demographics, food intake, and adenoma recurrence has previously been described in detail (Bobe *et al.*, 2008; 2010a; 2010b). Individuals who had recurrence of a high-risk or advanced adenoma consumed less fiber (limited to individuals with a high-risk adenoma), fruits and vegetables, flavonols, dry beans, and other vegetables (Bobe *et al.*, 2008; 2010a; 2010b).

Table 1. Association between kaempferol consumption and colorectal adenoma recurrence overall (n = 1858) and in the intervention arm of the Polyp Prevention Trial (n = 872)

Kaempferol[1] Change Quartiles[1]	Adenoma Recurrence (T4)						
	None	Any		High-Risk		Advanced	
(in mg/d)	*n* (%)	*n* (%)	OR (95% CI)[2]	*n* (%)	OR (95% CI)[2]	*n* (%)	OR (95% CI)[2]
Overall (T1,2,3)							
Q1: < 4.61	268 (58.8)	188 (41.2)	1.00	67 (14.7)	1.00	34 (7.5)	1.00
Q2: 4.61-8.23	280 (61.0)	179 (39.0)	0.88 (0.67-1.15)	59 (12.9)	0.79 (0.53-1.18)	38 (8.3)	0.99 (0.60-1.65)
Q3: 8.24-15.8	278 (59.4)	190 (40.6)	0.93 (0.71-1.22)	57 (12.2)	0.79 (0.52-1.18)	33 (7.1)	0.88 (0.52-1.48)
Q4: > 15.8	297 (62.4)	179 (37.6)	0.88 (0.66-1.15)	43 (9.0)	0.62 (0.40-0.97)	18 (3.8)	0.49 (0.26-0.91)
P for trend[3]			0.44		0.04		0.03
Intervention (T1,2,3)							
Q1: < 7.84	132 (60.6)	86 (39.4)	1.00	38 (17.4)	1.00	24 (11.0)	1.00
Q2: 7.84-13.8	128 (58.7)	90 (41.3)	1.08 (0.73-1.60)	20 (9.2)	0.50 (0.27-0.94)	10 (4.6)	0.35 (0.15-0.80)
Q3: 13.9-21.4	130 (59.6)	88 (40.4)	1.14 (0.77-1.70)	20 (9.2)	0.60 (0.32-1.11)	9 (4.1)	0.40 (0.17-0.91)
Q4: > 21.4	134 (61.5)	84 (38.5)	1.11 (0.73-1.67)	22 (10.1)	0.67 (0.36-1.26)	6 (2.8)	0.25 (0.09-0.67)
P for trend[3]			0.56		0.16		0.002

[1] Kaempferol intake in quartiles (Q1-Q4) by the mean consumption during the first 3 trial years.

[2] Multivariate odds ratio (OR) and 95% confidence interval (CI) models were adjusted for age tertiles (<58, 58-66, >66 yrs), sex, average BMI (<25, 25.0-29.9, ≥30 kg/m^2), and current smoking status during the first 3 trial years.

[3] The natural log-transformed median intakes in mg/d of each quartile were used as a continuous score variable to determine the *p* value for trend.

Table 2. Medians (interquartile ranges) of serum cytokine concentrations (overall and stratified by baseline concentrations) and kaempferol consumption during the Polyp Prevention Trial (n = 872)

Cytokine[1] (in pg/mL)	Kaempferol Intake Quartiles (in mg/d)[2]					
	Q1: < 7.84	Q2: 7.84-13.8	Q3: 13.9-21.4	Q4: > 21.4	*P* Non-	*P* for
	Median (IQR)	Median (IQR)	Median (IQR)	Median (IQR)	Param.[3]	Trend[4]
Sample size	218	218	218	218		
Interleukin 1β						
Overall	0.38 (0.25-0.64)	0.38 (0.23-0.61)	0.35 (0.22-0.58)	0.34 (0.22-0.55)	0.36	0.15
Low (<0.26)	0.28 (0.16-0.36)	0.23 (0.15-0.40)	0.22 (0.16-0.34)	0.24 (0.12-0.33)	0.66	0.46
Medium (0.26-0.49)	0.32 (0.25-0.46)	0.34 (0.24-0.47)	0.35 (0.26-0.51)	0.36 (0.25-0.49)	0.82	0.79
High (>0.49)	0.67 (0.44-1.06)	0.68 (0.44-1.05)	0.58 (0.37-0.97)	0.60 (0.35-1.16)	0.51	0.44

Cytokine[1] (in pg/mL)	Kaempferol Intake Quartiles (in mg/d)[2]					
	Q1: < 7.84	Q2: 7.84-13.8	Q3: 13.9-21.4	Q4: > 21.4	*P* Non-	*P* for
	Median (IQR)	Median (IQR)	Median (IQR)	Median (IQR)	Param.[3]	Trend[4]
Interleukin 2						
Overall	0.76 (0.41-1.64)	0.80 (0.33-1.43)	0.80 (0.46-1.58)	0.79 (0.39-1.60)	0.67	0.16
Low (<0.46)	0.41 (0.28-0.67)	0.32 (0.17-0.67)	0.44 (0.23-0.67)	0.42 (0.21-0.80)	0.73	0.65
Medium (0.46-1.10)	0.68 (0.41-1.06)	0.75 (0.44-1.00)	0.76 (0.54-1.19)	0.67 (0.45-1.03)	0.44	0.35
High (>1.10)	1.84 (1.00-3.39)	2.00 (1.15-3.38)	2.09 (1.15-3.84)	1.83 (1.08-3.36)	0.74	0.86
Interleukin 8						
Overall	10.9 (8.52-15.3)	10.8 (8.48-15.2)	10.5 (7.79-15.1)	10.8 (8.18-15.1)	0.74	0.84
Low (<8.66)	8.42 (6.59-11.2)	8.45 (6.40-10.1)	7.68 (6.48-9.19)	7.93 (6.36-10.2)	0.38	0.16
Medium (8.66-12.6)	10.6 (9.10-13.8)	10.9 (8.72-13.6)	10.6 (9.02-13.2)	10.6 (8.76-12.8)	0.85	0.16
High (>12.6)	14.6 (11.1-19.4)	15.2 (11.7-22.4)	16.4 (12.9-21.5)	15.1 (11.5-20.2)	0.46	0.30
Interleukin 10						
Overall	3.30 (2.14-6.09)	3.22 (2.22-5.72)	3.01 (2.10-5.19)	3.21 (2.22-5.65)	0.78	0.87
Low (<2.46)	1.89 (1.52-2.42)	1.98 (1.58-2.39)	1.98 (1.57-2.29)	1.91 (1.60-2.37)	0.97	0.87
Medium (2.46-4.56)	3.23 (2.67-4.29)	3.40 (2.60-4.08)	3.20 (2.54-4.15)	3.01 (2.44-3.71)	0.34	0.34
High (>4.56)	7.34 (4.68-20.4)	9.29 (5.31-42.8)	7.42 (4.44-16.6)	12.1 (5.46-46.2)	0.14	0.12
Interleukin 12p70						
Overall	3.13 (1.69-7.22)	3.02 (1.52-7.67)	2.80 (1.56-6.07)	3.09 (1.61-7.81)	0.92	0.76
Low (<1.98)	1.31 (0.97-1.80)	1.33 (1.03-1.77)	1.33 (0.90-1.76)	1.21 (0.89-1.68)	0.76	0.78
Medium (1.98-5.20)	2.93 (2.20-4.35)	2.79 (2.17-3.81)	2.84 (2.19-4.18)	2.91 (2.28-3.88)	0.84	0.53
High (>5.20)	9.26 (5.41-30.1)	13.3 (6.69-45.0)	11.5 (5.60-28.9)	17.7 (6.43-65.1)	0.34	0.27
GMCSF[5]						
Overall	0.77 (0.42-1.77)	0.80 (0.39-1.58)	0.68 (0.37-1.88)	0.84 (0.41-2.03)	0.35	0.86
Low (<0.48)	0.35 (0.21-0.54)	0.35 (0.26-0.44)	0.30 (0.21-0.45)	0.29 (0.20-0.45)	0.53	0.82
Medium (0.48-1.29)	0.77 (0.51-1.19)	0.97 (0.61-1.22)	0.66 (0.47-1.01)	0.84 (0.54-1.08)	0.09	0.65
High (>1.29)	2.27 (1.63-5.04)	2.55 (1.29-6.56)	3.02 (1.80-6.36)	3.96 (1.67-14.4)	0.17	0.07
Tumor necrosis factor α						
Overall	8.20 (6.76-10.2)	8.25 (6.97-10.1)	8.21 (6.92-9.82)	8.14 (6.95-9.68)	0.87	0.70
Low (<7.19)	6.59 (3.12-7.34)	6.56 (5.64-7.51)	6.40 (5.63-7.16)	6.47 (6.07-7.34)	0.60	0.89
Medium (7.19-9.37)	8.15 (7.35-9.24)	8.19 (7.30-9.02)	8.38 (7.58-9.25)	8.00 (7.35-8.87)	0.64	0.77
High (>9.37)	11.1 (9.46-13.0)	10.9 (9.24-12.4)	10.2 (9.32-12.1)	10.2 (9.27-11.8)	0.24	0.31

[1] Geometric mean of year 1 and 3 cytokine values [Trial (T1,3)].

[2] Participants were grouped in quartiles (Q1-Q4) by mean kaempferol intake during the first 3 trial years.

[3] *P* values for differences in medians among the kaempferol intake quartiles were calculated based on the Kruskal-Wallis test.

[4] The natural log-transformed median intakes in mg/d of each kaempferol quartile were used to determine *P* for trend of the cytokine concentrations using a multiple regression model adjusting for age tertiles (<58, 58-66, >66 yrs), sex, average BMI (<25, 25.0-29.9, ≥30 kg/m^2), smoking status, and average energy intake (continuous) during the first 3 trial years.

[5] GMCSF: Granulocyte macrophage colony stimulating factor.

Table 3. Association between kaempferol intake by change in serum cytokine concentrations and colorectal adenoma recurrence among intervention group participants of the Polyp Prevention Trial (n=872)

Kaempferol[1] (in mg/d)	Cytokine[2] (in pg/mL)	Adenoma Recurrence (T4) No	Any		High-Risk		Advanced	
		n (%)	*n* (%)	OR (95% CI)[3]	*n* (%)	OR (95% CI)[3]	*n* (%)	OR (95% CI)[3]
Mean (T1,2,3)	*Interleukin (IL)1β*							
Low ≤ 13.9	High > 0.01	124 (57.4)	92 (42.6)	1.00	36 (16.7)	1.00	19 (8.8)	1.00
	Low ≤ 0.01	135 (61.9)	83 (38.1)	0.83 (0.56-1.23)	22 (10.1)	0.51 (0.28-0.95)	15 (6.9)	0.66 (0.31-1.40)
High > 13.9	High > 0.01	137 (66.2)	70 (33.8)	0.79 (0.53-1.19)	14 (6.8)	0.42 (0.21-0.83)	8 (3.9)	0.44 (0.18-1.06)
	Low ≤ 0.01	125 (55.3)	101 (447)	1.20 (0.81-1.77)	28 (12.4)	0.88 (0.49-1.58)	4 (3.1)	0.40 (0.16-1.02)
Mean (T1,2,3)	*Interleukin 2*							
Low ≤ 13.9	High > 0.11	121 (61.7)	75 (38.3)	1.00	24 (12.2)	1.00	13 (6.6)	1.00
	Low ≤ 0.11	136 (58.1)	98 (41.9)	1.14 (0.77-1.70)	33 (14.1)	1.21 (0.66-2.21)	20 (8.5)	1.38 (0.64-2.98)
High > 13.9	High > 0.11	135 (63.4)	78 (36.6)	1.03 (0.68-1.56)	19 (8.9)	0.84 (0.43-1.65)	9 (4.2)	0.86 (0.29-1.83)
	Low ≤ 0.11	124 (57.7)	91 (42.3)	1.30 (0.86-1.95)	23 (10.7)	1.06 (0.55-2.04)	6 (2.8)	0.49 (0.17-1.36)
Mean (T1,2,3)	*Interleukin 6*							
Low ≤ 13.9	High > 0.08	127 (56.2)	99 (43.8)	1.00	35 (15.5)	1.00	21 (9.3)	1.00
	Low ≤ 0.08	133 (63.3)	77 (36.7)	0.77 (0.52-1.14)	23 (11.0)	0.63 (0.34-1.15)	13 (6.2)	0.61 (0.28-1.29)
High > 13.9	High > 0.08	121 (59.9)	81 (40.1)	0.98 (0.65-1.46)	20 (9.9)	0.75 (0.39-1.41)	9 (4.5)	0.55 (0.23-1.30)
	Low ≤ 0.08	143 (61.1)	91 (38.9)	0.93 (0.63-1.37)	22 (9.4)	0.65 (0.35-1.20)	6 (2.6)	0.29 (0.11-0.76)
Mean (T1,2,3)	*Interleukin 8*							
Low ≤ 13.9	High > 0.27	137 (62.8)	81 (37.2)	1.00	24 (11.0)	1.00	14 (6.4)	1.00
	Low ≤ 0.27	123 (56.4)	95 (43.6)	1.23 (0.83-1.81)	34 (15.6)	1.45 (0.79-2.63)	20 (9.2)	1.42 (0.67-2.99)
High > 13.9	High > 0.27	127 (59.1)	88 (40.9)	1.26 (0.85-1.88)	17 (7.9)	0.86 (0.43-1.72)	6 (2.8)	0.48 (0.18-1.32)
	Low ≤ 0.27	137 (62.0)	84 (38.0)	1.14 (0.76-1.69)	25 (11.3)	1.21 (0.64-2.29)	9 (4.1)	0.73 (0.30-1.80)
Mean (T1,2,3)	*Interleukin 10*							
Low ≤ 13.9	High > -0.02	129 (57.6)	95 (42.4)	1.00	34 (15.2)	1.00	21 (9.4)	1.00
	Low ≤ -0.02	131 (51.8)	81 (38.2)	0.82 (0.55-1.22)	24 (11.3)	0.59 (0.32-1.07)	13 (6.1)	0.52 (0.24-1.12)
High > 13.9	High > -0.02	127 (62.9)	75 (37.1)	0.87 (0.58-1.31)	18 (8.9)	0.57 (0.30-1.10)	9 (4.5)	0.45 (0.19-1.07)
	Low ≤ -0.02	137 (58.5)	97 (41.5)	1.08 (0.74-1.59)	24 (10.3)	0.76 (0.41-1.40)	6 (2.6)	0.30 (0.12-0.79)
Mean (T1,2,3)	*Interleukin 12p70*							
Low ≤ 13.9	High > -0.09	125 (59.0)	87 (41.0)	1.00	32 (15.1)	1.00	21 (9.9)	1.00
	Low ≤ -0.09	135 (60.3)	89 (39.7)	0.92 (0.63-1.37)	26 (11.6)	0.68 (0.37-1.23)	13 (5.8)	0.55 (0.26-1.18)
High > 13.9	High > -0.09	131 (60.9)	84 (39.1)	1.01 (0.68-1.52)	23 (10.7)	0.76 (0.41-1.41)	10 (4.7)	0.50 (0.22-1.15)
	Low ≤ -0.09	133 (60.2)	88 (39.8)	1.06 (0.71-1.58)	19 (8.6)	0.66 (0.35-1.26)	5 (2.3)	0.27 (0.10-0.75)
Mean (T1,2,3)	*Granulocyte macrophage colony stimulating factor (GMCSF)*							
Low ≤ 13.9	High > 0.00	131 (57.0)	99 (43.0)	1.00	34 (14.8)	1.00	22 (9.6)	1.00
	Low ≤ 0.00	129 (62.6)	77 (37.4)	0.78 (0.52-1.15)	24 (11.7)	0.65 (0.35-1.19)	12 (5.8)	0.50 (0.23-1.09)
High > 13.9	High > 0.00	128 (62.1)	78 (37.9)	0.89 (0.60-1.33)	18 (8.7)	0.60 (0.32-1.16)	9 (4.4)	0.45 (0.20-1.05)

Kaempferol[1] (in mg/d)	Cytokine[2] (in pg/mL)	Adenoma Recurrence (T4)						
		No	Any		High Risk		Advanced	
		n (%)	*n* (%)	OR (95% CI)[3]	*n* (%)	OR (95% CI)[3]	*n* (%)	OR (95% CI)[3]
	Low ≤ 0.00	136 (59.4)	93 (40.6)	1.02 (0.69-1.50)	24 (10.5)	0.80 (0.44-1.47)	6 (2.6)	0.30 (0.12-0.80)
Mean (T1,2,3)	*Interferon (IFN)γ*							
Low ≤ 13.9	High > 0.27	132 (59.2)	91 (40.8)	1.00	36 (16.1)	1.00	20 (9.0)	1.00
	Low ≤ 0.27	128 (60.1)	85 (39.9)	1.03 (0.69-1.52)	22 (10.3)	0.67 (0.36-1.23)	14 (6.6)	0.77 (0.36-1.63)
High > 13.9	High > 0.27	123 (62.1)	75 (37.9)	1.00 (0.66-1.50)	21 (10.6)	0.74 (0.40-1.39)	10 (5.1)	0.62 (0.27-1.43)
	Low ≤ 0.27	141 (59.2)	97 (40.8)	1.18 (0.80-1.75)	21 (8.8)	0.70 (0.37-1.29)	5 (2.1)	0.29 (0.10-0.82)
Mean (T1,2,3)	*Tumor necrosis factor (TNF)α*							
Low ≤ 13.9	High > 0.04	128 (57.1)	96 (42.9)	1.00	35 (15.6)	1.00	21 (9.4)	1.00
	Low ≤ 0.04	132 (62.3)	80 (37.7)	0.81 (0.55-1.20)	23 (10.8)	0.61 (0.33-1.11)	13 (6.1)	0.58 (0.27-1.23)
High > 13.9	High > 0.04	129 (62.0)	79 (38.0)	0.91 (0.61-1.36)	15 (7.2)	0.48 (0.24-0.96)	6 (2.9)	0.32 (0.12-0.84)
	Low ≤ 0.04	135 (59.2)	93 (40.8)	1.03 (0.70-1.52)	27 (11.8)	0.88 (0.49-1.59)	9 (3.9)	0.47 (0.20-1.10)

[1] Kaempferol intake below or above the median of the mean intake during the first 3 trial years (T1,2,3).

[2] Median change in cytokine concentrations from baseline [Change (T1,3-T0)].

[3] Multivariate odds ratios (OR) and 95% confidence intervals (CI) models were adjusted for age tertiles (<58, 58-66, >66 yrs), sex, average BMI (<25, 25.0-29.9, ≥30 kg/m^2), current smoking status, and average energy intake (continuous) during the first 3 trial years. The interaction term between kaempferol intake and serum cytokine concentrations for advanced adenoma recurrence was significant for IL-8, -10, -12p70, and GMCSF. The combination of high kaempferol intake and low cytokine concentrations had a significantly lower risk estimate of advanced adenoma recurrence for IFNγ *vs.* each of the other 3 combinations, for IL-1β, -2, -8, and -10 *vs.* both lower flavonol intake combinations, and for all 8 cytokines *vs.* lower kaempferol intake and low trial cytokine concentrations.

Kaempferol intake was not associated with any adenoma recurrence but was inversely associated with high-risk and advanced adenoma recurrence (Table 1; Figure 1). Overall, kaempferol intake was inversely associated with advanced adenoma (4th versus 1st quartile, OR = 0.49, 95% CI: 0.26-0.91, P_{Linear} = 0.03) and high-risk adenoma recurrence (OR = 0.62, 95% CI: 0.40-0.97, P_{Linear} = 0.04; Table 1; Figure 1). Compared to baseline intakes, kaempferol consumption increased 2.9-fold from 4.8 to 13.9 mg/d during the first three years of the trial in the intervention group. In the intervention arm, participants with kaempferol intakes in the higher three versus the lowest quartile had a lower risk of advanced adenoma (OR for Q2 to Q4 versus Q1 = 0.33, 95% CI: 0.17-0.63; P = 0.0008) and high risk adenoma recurrence (OR = 0.59, 95% CI: 0.36-0.95, P = 0.03). Our results suggest that a kaempferol-enriched diet may decrease the risk of the more progressive forms of colorectal adenomas.

One of the potential pathways by which kaempferol may inhibit colorectal adenoma progression is by attenuating inflammation. Therefore, we examined whether kaempferol consumption was associated with serum cytokine concentrations. Kaemferol consumption was inversely associated with serum concentrations of IL-6 (highest versus lowest kaempferol intake quartile: 1.86 vs. 2.09 pg/mL; $P_{Kaempferol}$ = 0.05) and IFNγ (highest versus lowest kaempferol intake quartile: 1.74 versus 1.52 pg/mL; $P_{Kaempferol}$ = 0.008) during the trial; the differences in serum concentrations were largest in participants with the highest IL-6 (highest versus lowest intake quartile: 2.63 versus 3.23 pg/mL; $P_{Kaempferol}$ = 0.05) and IFNγ concentrations (highest versus lowest intake quartile: 2.71 versus 2.16 pg/mL; $P_{Kaempferol}$ = 0.12) at baseline (Figure 2). In contrast, serum concentrations of other cytokines were not associated with kaempferol intake (Table 2). Our results indicate that kaempferol consumption may be associated with a decrease in some but not all inflammatory markers.

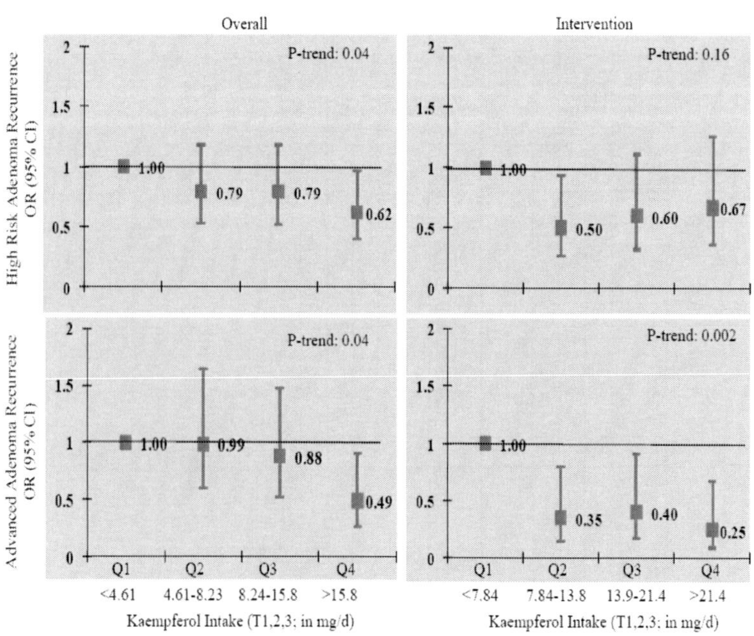

Figure 1. The association between kaempferol consumption and high-risk or advanced colorectal adenoma recurrence overall (n = 1858) and in the intervention arm of the Polyp Prevention Trial (n = 872).

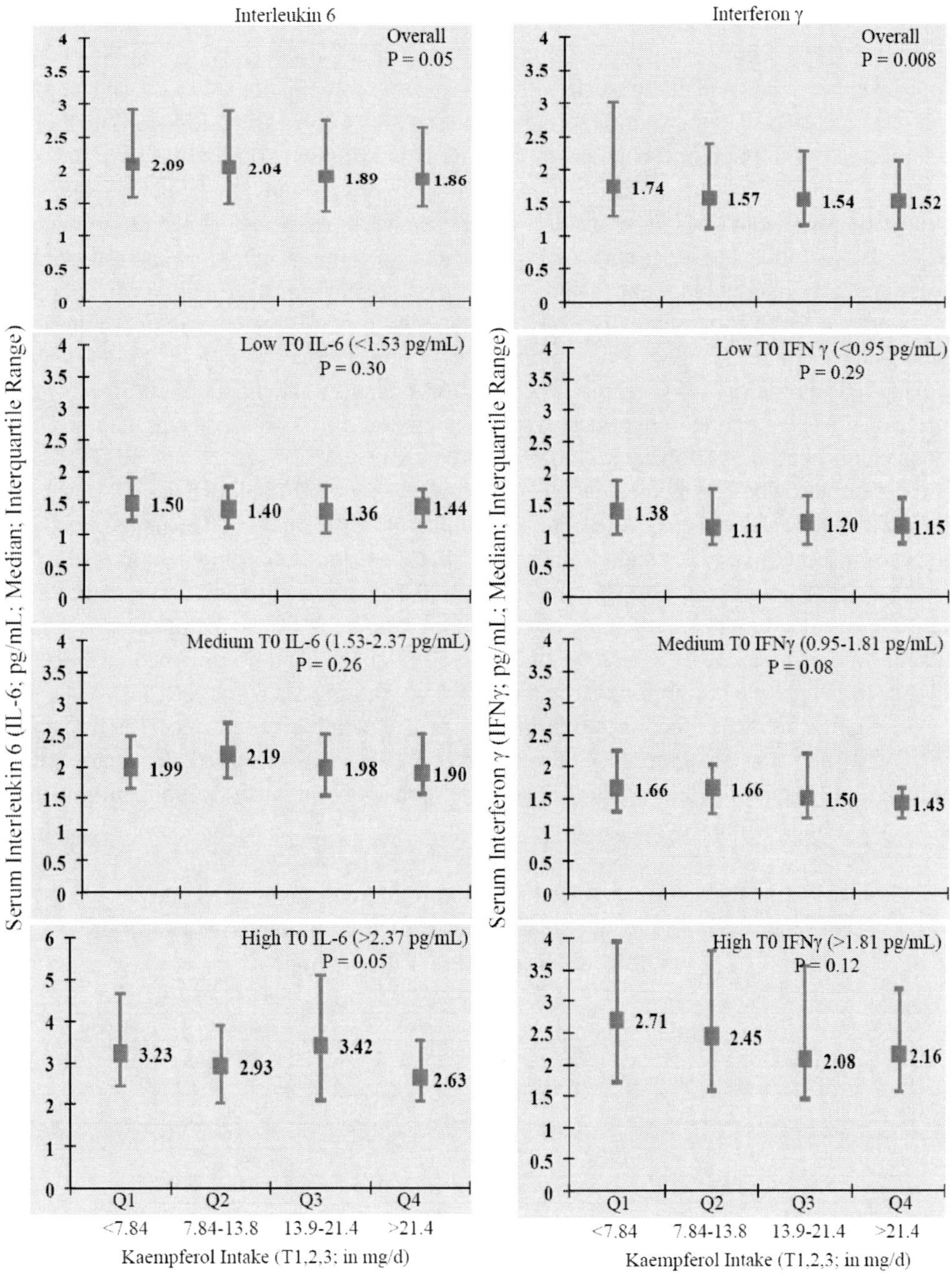

Figure 2. The association between kaempferol consumption and serum IL-6 and IFNγ concentrations (overall and stratified by baseline concentrations) during the Polyp Prevention Trial (n = 872). The *P* value indicates group differences using the Wilcoxon rank sum test.

Previously, we reported that a decrease in IL-6 concentrations during the trial was inversely associated with colorectal adenoma recurrence (any, high-risk, and advanced), as was for IL-1β or IL-10 and advanced adenoma recurrence (Bobe *et al.*, 2010a; 2010b). In contrast, a decrease in IL-2 concentrations was associated with increased risk of any adenoma recurrence (Bobe *et al.*, 2010a). To examine whether kaempferol consumption and a decrease in serum cytokine concentrations during the trial act synergistically to inhibit colorectal neoplasms, we examined the association between the combination of both and colorectal adenoma recurrence. No significant associations were observed for combinations of kaempferol consumption and changes in cytokine concentrations during the trial and any or high-risk adenoma recurrence (Table 3). However, individuals with above median kaempferol consumption and equal or below median change in serum cytokines concentrations had the lowest risk of advanced adenoma recurrence for 7 of the 9 cytokines investigated (the exceptions were interleukin 8 and TNFα) with the risk reduction being statistically significant for changes in concentrations of IL-6, IL-10, IL-12p70, and IFNγ (Figure 3, Table 3). Our results indicate that kaempferol consumption may act in concert with decreases in serum cytokine concentrations to decrease the risk of advanced adenoma recurrence. Similar changes have been reported previously for a combination of flavonol consumption and serum cytokine concentrations (Bobe *et al.*, 2010a; 2010b; 2011).

To examine whether feeding purified kaempferol induced similar beneficial effects on colorectal tumorigenesis, we used an inflammation-associated CRC mouse model. In comparison to mice on the standard diet, mice that were fed 500 ppm kaempferol had 24% fewer tumors (median 16 versus 21 tumors; $P = 0.04$ for parametric and non parametric test) and 36% smaller tumors (median 7 versus 12 mm^3/tumor; $P = 0.09$ for parametric and $P = 0.13$ for nonparametric test), which resulted in a 53% smaller tumor area (median 115 versus 247 mm^3/mouse; $P = 0.03$ for parametric and $P = 0.06$ for nonparametric test; Figure 4).

No significant differences in morbidity (14 of 32 kaempferol-fed mice versus 16 of 32 control-fed mice) or cytokine concentrations (Table 4) were observed between treatment groups. Our results indicate that kaempferol consumption may inhibit inflammation-associated colorectal tumorigenesis.

Table 4. The effect of feeding 500 ppm kaempferol for 85 days on serum cytokine concentrations of azoxymethane/dextran sodium sulfate-induced male *FVB/NCr* mice

Cytokine (pg/mL)	Control	Kaempferol	P-value	P-value
	N = 17	N = 18	Nonpar.[1]	Parametric[2]
	Median (Interquartile Range)			
Interleukin 1β	4.8 (4.1-5.6)	5.0 (4.0-7.0)	0.60	0.75
Interleukin 6	26.2 (20.7-31.5)	27.3 (13.2-32.2)	0.97	0.80
Interleukin 10	25.8 (16.4-36.9)	23.5 (12.2-41.9)	0.59	0.72
Interleukin 12p70	15.7 (0 -45.0)	16.9 (4.7-31.7)	0.92	0.65
Chemokine ligand 1	102 (79-130)	86 (67-105)	0.22	0.27
Interferon γ	8.1 (5.2-13.0)	7.5 (3.6-9.7)	0.18	0.25
Tumor necrosis factor α	1.7 (1.2-2.6)	1.9 (1.1-2.1)	0.95	0.76

[1] *P* value for difference in medians was calculated based on the Kruskal-Wallis test.

[2] *P* value for difference of natural log-transformed cytokine concentrations was calculated using generalized least squares means after adjusting for number of dextran sodium sulfate treatments (1, 2).

Discussion

The objective of this study was to evaluate the potential chemoprotective effects of kaempferol against colorectal tumorigenesis in humans and a mouse CRC model. In the human PPT, consumption of a diet rich in kaempferol-containing foods and beverages decreased the risk of high-risk and advanced adenoma recurrence. A combination of high kaempferol intake and decrease in serum cytokine concentrations during the trial resulted in the lowest risk of advanced adenoma recurrence. The role of kaempferol as a chemopreventive compound was supported by the lower colorectal tumor number and size in mice fed a diet containing 500 versus 0 ppm purified kaempferol. Our results suggest dietary kaempferol may be a potential chemoprotective compound for inhibiting colorectal tumorigenesis.

To our knowledge, the association between kaempferol intake and colorectal adenoma recurrence has not been reported for other human cohorts. In contrast, other human cohort studies examined the association kaempferol intake and CRC and did not find significant associations (Knekt *et al.*, 2002; Lin *et al.*, 2006; Simons *et al.*, 2009; Kyle *et al.*, 2010). Differences in outcomes could be the result of multiple factors, including dietary source of kaempferol, tumor endpoint, and intake level of kaempferol. Dry beans were the primary dietary source of kaempferol in this study, while the primary dietary source of kaempferol in the CRC studies was either not specified (Knekt *et al.*, 2002; Lin *et al.*, 2006; Simons *et al.*, 2009) or was tea (Kyle *et al.*, 2010). Colorectal adenoma recurrence was the endpoint in this study, while the more progressed form of colorectal adenocarcinoma was the endpoint in the other studies, suggesting that kaempferol intake may be more protective at earlier stages of CRC. We did not observe a chemoprotective effect of kaempferol at the kaempferol intake levels reported from the CRC studies (Knekt *et al.*, 2002; Lin *et al.*, 2006; Simons *et al.*, 2009; Kyle *et al.*, 2010), indicating that higher dietary kaempferol intake may be required to observe a chemoprotective effect of kaempferol.

In an effort to identify a non-invasive indicator of kaempferol response, we examined the association between dietary kaempferol, serum cytokine concentrations, and colorectal adenoma recurrence. Cell culture and animal model studies indicated that kaempferol may decrease lipopolysaccharide- or DSS-induced cytokine gene expression and cytokine concentrations in blood (Lee *et al.*, 2009; Chen *et al.*, 2012; Park *et al.*, 2012) and the release of cytokines from cancer cells (Sharma *et al.*, 2007). As anticipated from the cell culture and animal model data, dietary kaempferol intake was inversely associated with serum concentrations of IL-6 and IFNγ with greater cytokine decreases being observed at higher baseline cytokine concentrations. To our knowledge, the association between kaempferol intake and cytokine concentrations has not been reported for other human cohorts. Flavonol supplementation studies usually did not detect changes in cytokine concentrations unless individuals had elevated baseline values and received an inflammatory challenge (Nieman *et al.*, 2007; Boots *et al.*, 2008, 2009). Kaempferol-induced changes in inflammatory markers may be too small to be detected when baseline cytokine concentrations are close to the detection limit or in the absence of inflammation. Similar to our previous results for flavonol intake (Bobe *et al.*, 2010a; 2010b; 2011), the combination of high kaempferol consumption and decrease in serum cytokine concentrations resulted in the lowest risk for advanced adenoma recurrence for most serum cytokines. Our results suggest that the chemoprotective effect of kaempferol may in part be mediated by attenuating inflammation.

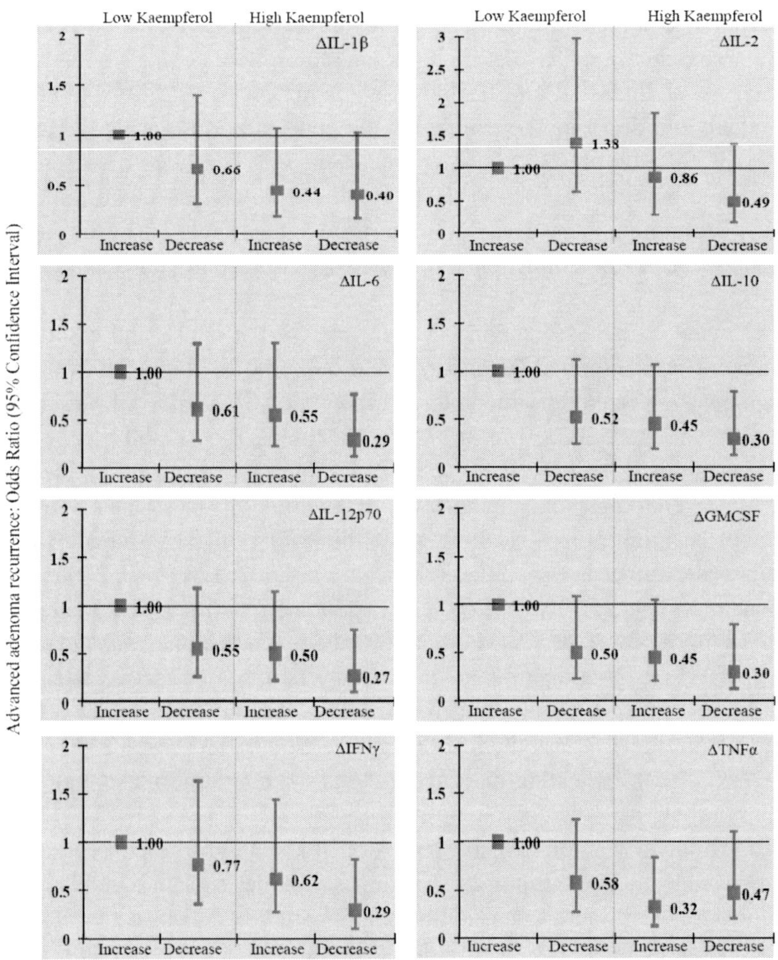

Figure 3. The association between the combination of high (>13.9 mg/d) or low (≤13.9 mg/d) kaempferol intake during the trial and change in serum concentration of cytokines (defined as the geometric mean of T1 and T3 minus baseline values) on advanced colorectal adenoma recurrence among participants in the intervention arm of the Polyp Prevention Trial. The cut-off values for an increase or decrease in serum cytokine concentrations are as follows (in pg/mL): ΔIL-1β: >0.01 (increase), ≤0.01 (decrease); ΔIL-2: >0.11 (increase), ≤0.11 (decrease); ΔIL-6: >0.08 (increase), ≤0.08 (decrease); ΔIL-10: >-0.02 (increase), ≤-0.02 (decrease); ΔIL-12p70: >-0.09 (increase), ≤-0.09 (decrease); ΔGMCSF: >0.00 (increase), ≤0.00 (decrease); ΔIFNγ: >0.27 (increase), ≤0.27 (decrease); TNFα: >0.04 (increase), ≤0.04 (decrease). The reference group is the combination of low kaempferol intake and increase in cytokine concentrations.

Other potential chemoprotective effects of kaempferol include its antioxidative properties (Choe *et al.*, 2012; Kim *et al.*, 2012), inhibition of β-catenin/Tcf signaling (Park and Choi, 2010) and COX-2 and Src kinase activity (Lee *et al.*, 2010), and induction of apoptosis or differentiation of cancer cells (Leung *et al.*, 2007; Sharma *et al.*, 2007; Kim *et al.*, 2008; Jeong *et al.*, 2009; Li *et al.*, 2009; Wang *et al.*, 2011).

To establish that in fact kaempferol and not another compound, enriched in kaempferol-containing foods, attenuates inflammation and recurrence of more progressed forms of adenomas, we tested purified kaempferol at concentrations that could be used as dietary supplement in a two-stage inflammation-associated CRC mouse model. Our results document that feeding 500 ppm of kaempferol, which would be equivalent to 504 mg of kaempferol per day for a 60 kg human (Reagan-Shaw *et al.*, 2007), decreases colorectal tumor number and size in AOM/DSS-induced mice (Figure 4).

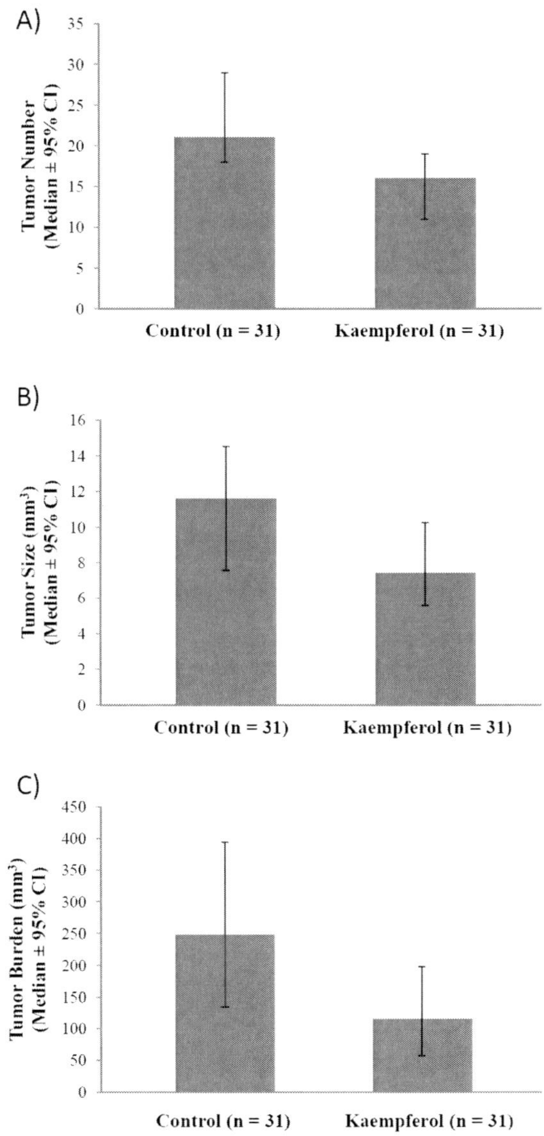

Figure 4. The effect of feeding 500 ppm kaempferol for 85 days on tumor number, size, and burden of azoxymethane/dextran sodium sulfate-induced male *FVB/NCr* mice.

To our knowledge, the effect of purified kaempferol on colorectal tumorigenesis has not been tested in animal models for CRC. The results from the mouse study support our results from the observational human study that dietary kaempferol may inhibit colorectal tumorigenesis.

A strength of this study was the combination of human observational and mouse model studies to document a potential chemoprotective effect of a dietary compound, which is novel in the literature. Additional strengths of the human study were the detailed adenoma endpoint information, the prospective and repeated collection of validated dietary exposure measures, and the repeated collection of serum, all of which decreased the risk of misclassification of exposure and endpoint measures. Additional strengths of the mouse study were the high purity of the dietary compound (98%) and the dietary dosage being similar to what humans could consume from dietary supplements.

Potential limitations include the fact that the human study was not designed to evaluate the effect of dietary kaempferol on adenoma recurrence. The PPT consisted of individuals with increased adenoma risk that mostly engaged in a relatively healthy lifestyle and those in the intervention arm consumed relatively high amounts of dry beans. Random as well as systematic measurement errors related to exposure measurements, including the FFQ, the kaempferol database, and the participants response, may be present and could bias the results. The possibility of a chance findings cannot be excluded because of the small number of advanced adenoma cases and the multiple testing.

Serum cytokine measurements have multiple limitations including low abundance, limited dynamic range, high CVs, daily fluctuations, short half-lives, and lack of specificity for location, strength and type of inflammation. Tumor number and size are relatively high and variable in the CRC mouse model compared to human adenoma numbers.

The histology and invasiveness of mouse tumors were not examined because of the large number of mice and tumors. Only male mice were used because the flavonol effects in the PPT were greater in males than females. To account for the variability in tumor number and size in the CRC mouse model, we used a relatively large number of mice per treatment, which also decreased the possibility of a chance finding. The number of mouse per treatment may be still insufficient to detect significant treatment differences in cytokine concentrations. Finally, the chemoprotective effect of kaempferol may only applicable to non-invasive forms of colorectal neoplasms.

Conclusion

In conclusion, our results indicate that consumption of a diet rich in kaempferol-containing foods and beverages may prevent recurrence of the more progressive forms of colorectal adenomas in humans, which is in part mediated by a decrease in serum cytokine concentrations. The role of kaempferol as a chemopreventive compound was supported by the lower tumor number and size of mice fed a diet containing purified kaempferol at concentrations similar to what can be consumed by dietary supplements. Our results suggest dietary kaempferol may be a potential chemoprotective compound for inhibiting colorectal tumorigenesis.

Acknowledgments

We would like to thank the Polyp Prevention Trial Study Group for their outstanding contribution to this project. In addition, the authors thank Helen Rager and Yanyu Wang from the Clinical Support Laboratory of SAIC Frederick Inc. (Frederick, MD) for cytokine analysis of the serum samples; Jen Wise and Craig Driver of the Laboratory Animal Sciences Program of SAIC Frederick Inc. for taking care of the mice; Darlene Green and Tammy Beachley of the Pathology/Histotechnology Laboratory of SIAC Frederick Inc. for tissue necropsy and preservation; and Thomas G. McCloud from the Natural Products Support Group of Applied Developmental Research Program of SAIC Frederick Inc. for verifying the purity of kaempferol.

Funding

This study was funded by the Office of Complementary and Alternative Medicine, Office of Dietary Supplements, and the Intramural Research Program, National Cancer Institute, NIH, DHHS, Bethesda, MD.

Conflict of Interest

None

References

Aggarwal BB and Shishodia S. Molecular targets of dietary agents for prevention and therapy of cancer. *Biochem Pharmacol.* 2006;71, 1397-1421.

American Cancer Society. *Cancer Facts and Figures 2012.* Atlanta, GA: American Cancer Society; 2012.

Asano TK and McLeod RS. Nonsteroidal anti-inflammatory drugs (NSAID) and aspirin for preventing colorectal adenomas and carcinomas (review). *Cochrane Database of Syst Rev.* 2004;1, 1-28.

Block G, Hartman AM, Naughton D. A reduced dietary questionnaire: development and validation. *Epidemiology.* 1990;1, 58-64.

Bobe G, Sansbury LB, Albert PS, Cross AJ, Kahle L, Ashby J, Slattery ML, Caan B, Paskett E, Iber F, Kikendall JW, Lance P, Daston C, Marshall JR, Schatzkin A, Lanza E. Dietary flavonoids and colorectal adenoma recurrence in the Polyp Prevention Trial. *Cancer Epidemiol Biomarkers Prev.* 2008;17, 1344-1353.

Bobe G, Albert PS, Sansbury LB, Lanza E, Schatzkin A, Colburn NH, Cross AJ. Interleukin-6 as a potential indicator for prevention of high-risk adenoma recurrence by dietary flavonols in the polyp prevention trial. *Cancer Prev Res (Phila).* 2010a;3, 764-775.

Bobe G, Murphy G, Albert PS, Sansbury LB, Lanza E, Schatzkin A, Colburn NH, Cross AJ. Serum cytokine concentrations, flavonol intake and colorectal adenoma recurrence in the Polyp Prevention Trial. *Br. J. Cancer.* 2010b;103, 1453-1461.

Bobe G, Murphy G, Albert PS, Sansbury LB, Young, MR, Lanza E, Schatzkin A, Colburn NH, Cross AJ. Do interleukin polymorphisms play a role in the prevention of adenoma recurrence by dietary flavonols? *Eur J Cancer Prev.* 2011;20, 86-95.

Bode AM and Dong Z. Molecular and cellular targets. *Mol Carcinog.* 2006;46, 422-30.

Boots AW, Wilms LC, Swennen EL, Kleinjans JC, Bast A, Haenen GR. In vitro and ex vivo anti-inflammatory activity of quercetin in healthy volunteers. *Nutrition.* 2008;24, 703-710.

Boots AW, Drent M, Swennen EL, Moonen HJ, Bast A, Haenen GR. Antioxidant status associated with inflammation in sarcoidosis: a potential role for antioxidants. *Respir Med.* 2009;103, 364-372.

Caan BJ, Lanza E, Schatzkin A, Coates AO, Brewer BK, Slattery ML, Marshall JR, Bloch A. Does nutritionist review of a self-administered food frequency questionnaire improve data quality? *Public Health Nutr.* 1999;2, 565-569.

Canavan C, Abrams KR, Mayberry J. Meta-analysis: colorectal and small bowel cancer isk in patients with Crohn's disease. *Aliment Pharmacol Ther.* 2006;23, 1097-1104.

Chen X, Yang X, Liu T, Guan M, Feng X, Dong W, Chu X, Liu J, Tian X, Ci X, Li H, Wei J, Deng Y, Deng X, Chi G, Sun Z. Kaempferol regulates MAPKs and NF-κB signaling pathways to attenuate LPS-induced acute lung injury in mice. *Int Immunopharmacol.* 2012;14, 209-216.

Choe KI, Kwon JH, Park KH, Oh MH, Kim MH, Kim HH, Cho SH, Chung EK, Ha SY, Lee MW. The antioxidant and anti-inflammatory effects of phenolic compounds isolated from the root of *Rhodiola sachalinensis* A. BOR. *Molecules* 2012;17, 11484-11494.

Coussens LM and Werb Z. Inflammation and cancer. *Nature.* 2002;420, 860-867.

Eaden JA, Abrams KR, Mayberry JF. The risk of colorectal cancer in ulcerative colitis: a meta analysis. *Gut.* 2001;48, 526-35.

Gill S and Sinicrope FA. Colorectal cancer prevention: is an ounce of prevention worth a pound of cure? *Semin Oncol.* 2005;32, 24-34.

Gong JH, Shin D, Han SY, Kim JL, Kang YH. Kaempferol suppresses eosionphil infiltration and airway inflammation in airway epithelial cells and in mice with allergic asthma. *J Nutr.* 2012;142,47-56.

Hämäläinen M, Nieminen R, Vuorela P, Heinonen M, Moilanen E. Anti-ninflammatory effects of flavonoids: genistein, kaempferol, quercetin, and daidzein inhibit STAT-1 and NF-κB actvations, whereas flavones, isorhametin, narigenin, and pelargonidin inhibit only NF-κB along with their inhibitory effect on iNOS expression and NO production in activated macrophages. *Mediators Inflamm.* 2007; 2007, 45673(10 pages).

Jeong JC, Kim MS, Kim TH, Kim YK. Kaempferol induces cell death through ERK and Akt-dependent down-regulation of XIAP and surviving in human glioma cells. *Neurochem Res.* 2009;34, 991-1001.

Kim BW, Lee ER, Min HM, Jeong HS, Ahn JY, Kim JY, Choi HY, Choi H, Kim EY, Park SP, Cho SG. Sustained ERK activation is involved in the kaempferol-induced apoptosis of breast cancer cells and is more evident under 3-D culture condition. *Cancer Biol Ther.* 2008;7, 1080-1089.

Kim GN, Kim ES, Kwon YI, Jang HD. Potential mechanism of kaempferol against CU^{2+}-induced oxidative stress through chelating activity and regulation of nuclear factor-erythroid-2-related factor 2 signaling. *Food Sc. Biotechnol.*2012;21, 1469-1475.

Knekt P, Kumpulainen J, Järvinen R, Rissanen H, Heliövaara M, Reunanen, A, Hakulinen T, Aromaa A. Flavonoid intake and risk of chronic diseases. *Am J Clin Nutr.*2002;75, 560-568.

Kohno H, Suzuki R, Curini M, Epifano F, Maltese F, Gonzales SP, Tanaka T. Dietary administration with prenyloxycoumarins, auraptene and collinin, inhibits colitis-related colon carcinogenesis in mice. *Int J Cancer*. 2006;118, 2936-2942.

Kohno K, Suzuki R, Yasui Y, Miyamoto S, Wakabayashi K, Tanaka T. Ursodeoxycholic acid versus sulfasalazine in colitis-related colon carcinogenesis in mice. *Clin Cancer Res.* 2007;13. 2519-2525.

Kwon KH, Murakami A, Tanaka T, Ohigashi H. Dietary rutin but not its aglycone quercetin, ameliorates dextran sulfate sodium-induced experimental colitis in mice: attenuation of pro-inflammatory gene expression. *Biochem Pharmacol.* 2005;69, 395-406.

Kyle JAM, Sharp L, Little J, Duthie GG, McNeill G. Dietary flavonoid intake and colorectal cancer: a case-control study. *Br J Nutr.* 2010;103, 429-436.

Laiyemo AO, Pinsky PF, Marcus PM, Lanza E., Cross AJ, Schatzkin A, Schoen RE. Utilization and yield of surveillance colonoscopy in the Continued Follow-Up Study of the Polyp Prevention Trial. *Clin Gastroenterol Hepatol.* 2009;7, 562-567.

Lanza E, Schatzkin A, Daston C, Corle D, Freedman L, Ballard-Barbash R, Caan B, Lance P, Marshall J, Iber F, Shike M, Weissfeld J, Slattery M, Paskett E, Mateski D, Albert P. Implementation of a 4-y, high-fiber, high-fruit-and-vegetable, low-fat dietary intervention: results of dietary changes in the Polyp Prevention Trial. *Am J Clin Nutr.* 2001;74, 387-401.

Lee KM, Lee KW, Jung SK, Lee EJ, Heo YS, Bode AM, Lubet RA, Lee HY, Dong Z. Kaempferol inhibits UVB-induced COX-2 expression by suppressing Src kinase activity. *Biochem Pharmacol.* 2010;80, 2042-2049.

Lee S, Kim YJ, Kwon S, Lee Y, Choi SY, Park J, Kwon HJ. Inhibitory effects of flavonoids on TNF-α-induced IL-8 gene expression in HEK 293 cells. *BMB Rep.* 2009,42; 265-270.

Leung HWC, Lin CJ, Hour MJ, Yang WH, Wang MY, Lee HZ. Kaempferol induces apoptosis in human lung non-small carcinoma cells accompanied by an induction of antioxidant enzymes. *Food Chem Toxicol.* 2007,45, 2005-2013.

Li W, Du B, Wang T, Wang S, Zhang J. Kaempferol induces apoptosis in human HCT116 colon cancer cells via the ataxia-telangioctasia mutatied-p53 pathway with the involvement of p53 upregulated modulator of apoptosis. *Chem Biol Interact.* 2009;177, 121-127.

Lin J, Zhang SM, Wu K, Willett WC, Fuchs CS, Giovannucci. Flavonoid intake and colorectal cancer risk in men and women. *Am J Epidemiol.* 2006;164, 644-651.

Lin WW and Karin M. A cytokine-mediate link between innate immunity, inflammation, and cancer. *J Clin Invest.* 2007;117, 1175-83.

Leung K, Pinsky P, Laiyemo AO, Lanza E, Schatzkin A, Schoen RE. Ongoing colorectal cancer risk despite surveillance colonoscopy: the Polyp Prevent Trial Continued Follow-up Study. *Gastrointest Endosc.* 2010;71, 111-117.

Nam NH. Naturally occurring NF-κB inhibitors. *Mini Rev Med Chem.* 2006;6, 945-51.

Nieman DC, Henson DA, Gross SJ, Jenkins DP, Davis JM, Murphy EA, Carmichael MD, Dumke CL, Utter AC, McAnulty SR, McAnulty LS, Mayer EP. Quercetin reduces illness but not immune perturbations after intensive exercise. *Med Sci Sports Exerc.* 2007;39, 1561-1569.

Niestroy J, Barbara A, Herbst K, Rode S, van Liempt M, Ross PH. Single and concerted effects of benzo[a]pyrene and flavonoids on the AhR and Nrf2-pathway in the human colon carcinoma cell line Caco-2. *Toxicol In Vitro.* 2011;25, 671-683.

Park S and Choi J. Inhibition of β-catenin/Tcf signaling by flavonoids. *J Cell Biochem.* 2010;110, 1376-1385.

Park MY, Ji GE, Sung MK. Dietary kaempferol suppresses inflammation of dextran sulfate sodium-induced colitis in mice. *Dig Dis Sci.* 2012;57, 355-363.

Reagan-Shaw S, Nihal M, Ahmad N. Dose translation from animal to human studies revisited. *FASEB J.* 2007;22, 659-661.

Reeves P, Nielsen F, Fahey Jr G. AIN-93 purified diets for laboratory rodents: final report of the American Institute of Nutrition ad hoc writing committee on the reformulation of the AIN-76A rodent diet. *J Nutr.* 1993;123, 1939-1951.

SAS. *SAS User's Guide: Statistics, Version 9.2.* SAS Institute Inc., Cary, North Carolina, 2009.

Schatzkin A, Lanza E, Corle D, Lance P, Iber F, Caan B, Shike M, Weissfeld J, Burt R, Cooper MR, Kikendall JW, Cahill J. Lack of effect of a low-fat, high-fiber diet on the recurrence of colorectal adenomas. Polyp Prevention Trial Study Group. *N Engl J Med.* 200;342, 1149-1155.

Sharma V, Joshep C, Ghosh S, Agarwal A, Mishra MK, Sen E. Kaempferol induces apoptosis in glioblastoma cells through oxidative stress. *Mol Cancer Ther.* 2007;6, 2544-2553.

Simons CCJM, Hughes LAE, Arts ICW, Goldbohm RA, van den Brandt PA, Weijenberg MP. Dietary flavonol, flavones, and catechin intake and risk of colorectal cancer in the Netherlands Cohort Study. *Int J Cancer.* 2009;125, 2945-2952.

Tanaka T, Kohno H, Suzuki R, Yamada Y, Sugie S,Mori H. A novel inflammation-related mouse colon carcinogenesis model induced by azoxymethane and dextran sodium sulfate. *Cancer Sci.* 2003;**94**, 965-973.

U.S. Department of Agriculture (USDA), Agricultural Research Service. USDA Database for the Flavonoid Content of Selected Foods, Release 2.1, 2007. Nutrient Data Laboratory web site: http//www.nal.usda.gov/fnic/foodcomp/Data/Flav/Flav02-1.pdf.

Wang BS and Zhang XW. Separation, purification and bioactivity of broccolini leaves

In: Kaempferol
Editors: Garion Villers and Yves Fougere
ISBN: 978-1-62618-515-9

Chapter 6

Multipotent Flavonoid Kaempferol: Molecular Targets and Mechanism of Action and Nanotechnology Applications in Cancer and Human Health

Allen Y. Chen[1] and Yi Charlie Chen[*2]
[1]Department of Pharmaceutical Sciences, West Virginia University, Morgantown, WV, US
[2]Natural Science Division, Alderson-Broaddus College, Philippi, WV, US

Abstract

Kaempferol, a bioactive natural compound widely present in fruits and vegetables, is a polyphenol belonging to the class of flavonoids. Studies for the past several decades have identified the great potential of this flavonoid in the prevention and treatment of human chronic diseases, especially cancer. Kaempferol exerts this wide spectrum of effects by targeting several proteins/enzymes that are implicated in cancer and disease progression. Epidemiological studies have shown an inverse relationship between kaempferol intake and cancer. Kaempferol helps by augmenting the body's antioxidant defence against free radicals, which promote the development of cancer. At the molecular level, kaempferol has been reported to modulate a number of key elements in cellular signal transduction pathways linked to apoptosis, angiogenesis, inflammation, and metastasis. Significantly, kaempferol inhibits cancer cell growth and angiognesis and induces cancer cell apoptosis, while also appearing to preserve normal cell viability, in some cases exerting a protective effect. Because of its outstanding safety profile and its availability in highly purified forms, the clinical implications for kaempferol are growing rapidly. Even though kaempferol is one of the major flavonoids in the human diet, its bioavailability has been debated. Recent studies to improve its bioavailability by

* Corresponding author: Yi Charlie Chen, Alderson-Broaddus College, 101 College Hill Drive, Box 2096, Philippi, WV 26416, Email: chenyc@ab.edu, Tel: 1-304-457-6277, Fax: 1-304-457-6239.

encapsulating it with PEO-PPO-PEO and PLGA nanoparticles are promising. This review focuses mainly on the molecular targets of kaempferol and the mechanisms by which kaempferol exerts its biological effects in the prevention and treatment of human chronic diseases and cancer. A discussion on the application of nanotechnology is also included.

Introduction

Natural products are valuable sources of bioactive compounds and have been considered the single most successful discovery of modern medicine (Nair et al. 2010). Natural dietary agents have received attention from researchers and the general public due to their potential ability to suppress cancers and generally enhance human health (Amin et al. 2009). Flavonoids are polyphenols present in a wide variety of fruits and vegetables (Bosetti et al. 2007) and have been shown to inhibit cancer development (Gonzalez and Riboli 2006; Holick et al. 2007). It has been reported that dietary flavonoids reduce the risk of developing cardiovascular disease (Mink et al. 2007), prostate cancer (Bosetti et al. 2006), colorectal cancer (Theodoratou et al. 2007), and renal cancer (Bosetti et al. 2007) in humans. Flavonoids have also been found to inhibit cell growth and proliferation (Adhami et al. 2007) and induce cell toxicity (Plochmann et al. 2007; Choi et al. 2007) in cancer cells. Kaempferol [3, 5, 7-trihydroxy-2-(4-hydroxyphenyl)-4H-1-benzopyran-4-one] is a multipotent bioactive compound found in numerous plant sources. Studies from the past several decades have shown kaempferol (Fig. 1) to possess antioxidant and antitumor properties and indicate the great potential of this flavonoid in the prevention and treatment of human diseases, especially cancer. Delineated here are the many molecular targets of kaempferol, the mechanisms through which kaempferol acts to fight cancer and disease, and the use of nanotechnology to possibly improve clinical outcomes.

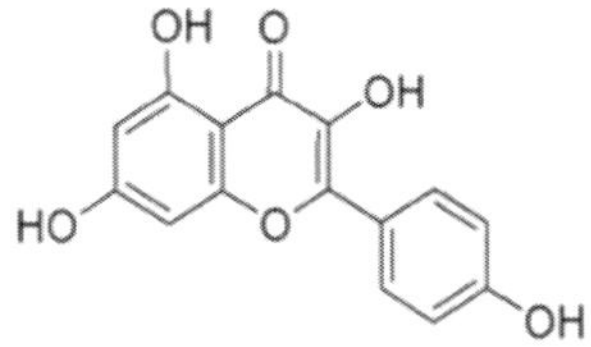

Figure 1. Chemical Structure of Kaempferol.

Natural Sources

As a common plant metabolite, kaempferol is quite the ubiquitous compound. A number of flowering plants, herbs, and spices have been documented to contain kaempferol or some of its numerous glycosides. A kaempferol glucoside is found in the root of Flemingia philippinensis (Fu et al., 2012), and the ethanol extract of Kaempferia galanga L., a common spice used in Asian cuisine, demonstrates antiangiogenic activity, attributed to the presence of kaempferol (He et al., 2012). Fractions of Sonchus arvensis, a traditional Pakistani medicine, exhibit antioxidant properties, likely a function of its innumerable polyphenolic elements including kaempferol (Khan, 2012). And ethyl acetate fractions of the water extract of Geranium nepalense, an anti-inflammatory Chinese herbal medicine, reveal quantities of one of its glycosides (Lu et al., 2012). Even cinnamon trees bear this pervasive flavonol, actually carrying a total of nine of its glycosides (Lin & Chang, 2012), but one can look to many fruit and vegetable sources for a healthy dosing as well. Endive (Mascherpa et al., 2012), broccoli (Somerset & Johannot, 2008), teas, and a wealth of berries and currants (Mikulic-Petkovsek et al., 2012) are thought to represent some of the main sources of kaempferol in the human diet. The list of plants that contain this compound is quite large, and only a small portion is detailed here. Recent evidence suggests that flavonoids, kaempferol and quercetin in particular, play a significant role in maintaining normal plant development (Misyura et al., 2012), so it comes as no surprise that kaempferol is such a universal chemical. Several population studies, which will be discussed further in the chapter, have noted a negative correlation between cancer incidence and kaempferol intake, so being cognizant of the best sources of kaempferol may prove to be valuable in the future.

Attacking the Malignancy

Kaempferol possesses a host of effects that focus on various targets throughout the body. The most studied of kaempferol's effects are its cancer fighting properties, which are quite broad in scope. Discussed here first is the ability of kaempferol to inhibit the growth of cancer cells, in many cases killing them through induction of apoptosis. Even in the narrow context of its cytotoxic effects, however, kaempferol manipulates several cellular mechanisms at multiple levels, including those regulating the cell cycle and those signaling for or against apoptosis (Fig. 2).

Myc represents a major oncogene that helps drive the development and maintenance of cancer. It codes for a transcription factor integral to cell proliferation and therefore is a suitable focus for chemotherapy (Soucek et al., 2008). A consequence of heightened Myc levels appears to be a drop in p21 concentrations (Luo et al., 2010). Since p21 is a cyclin-dependent kinase inhibitor, unbridled Myc production can allow the cell cycle to spiral out of control. Blocking Myc activity can help restore p21 levels and restrict tumor growth. Recent evidence suggests that kaempferol, when administered with cisplatin, antagonizes Myc mRNA levels (Luo et al., 2010). Kaempferol and cisplatin synergistically reduce Myc activity and increase p21 concentrations, thereby inhibiting cellular growth and, in this case, actually inducing apoptosis in OVCAR-3 ovarian cancer cells.

Disrupting Myc represents only one pathway through which kaempferol affects the cell cycle. Kaempferol treatment has been found to diminish the presence of both cyclin A and B, in addition to that of cyclin-dependent kinase 1 (CDK1) (Choi & Ahn, 2008). Since CDK1 is a major regulator of the G2 checkpoint, the introduction of kaempferol has been consistently found to result in cell cycle arrest at the G2/M phase. One study has determined that kaempferol treatment leads to phosphorylation of CDK1 at the Tyr 15 residue (Casagrande & Darbon, 2001), thereby inactivating the protein and its ability to initiate the M phase. Ostensibly, kaempferol has been shown to interfere in a number of steps central to mitosis in cancer cells, but kaempferol's growth inhibiting effects reach further still. Tumor suppressor protein p53 has been demonstrated to be up-regulated by the presence of kaempferol (Choi &

Ahn, 2008). One of the many functions of p53 is to activate p21, which as stated previously is a cyclin-dependent kinase inhibitor. Once p21 is free to bind CDK complexes, the cancer will have to improvise in order to reinitiate the cell cycle. This is one of the reasons why kaempferol has been shown to produce p-53 dependent growth inhibition in colon cancer cells (Li et al., 2009). Overall, kaempferol treatment has led to G2/M phase arrest in leukemia cells (Ren et al., 2010), esophageal squamous cell carcinoma (Zhang et al., 2009), and HeLa human cervix carcinoma cells (Xu et al., 2008). This of course does not represent a complete list; kaempferol's growth inhibitory effect impairs a wide range of cancer cells, highlighting its worth as a potential treatment option.

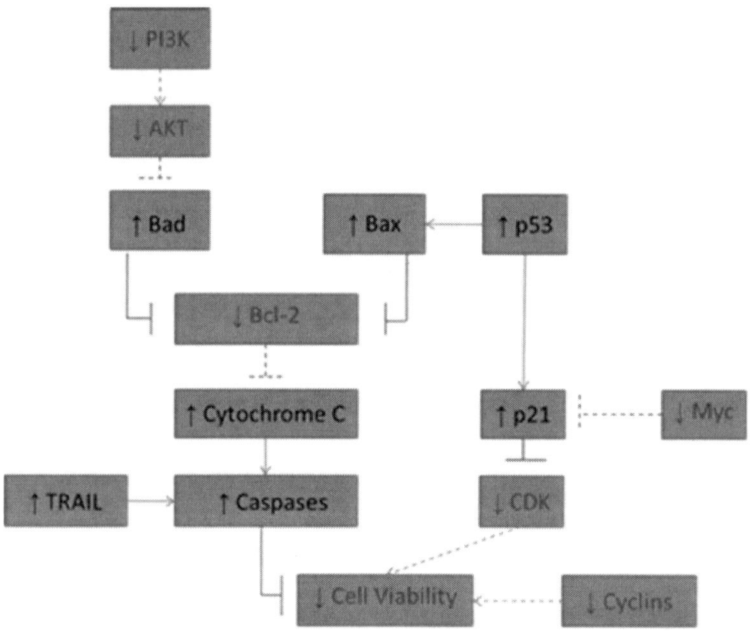

Figure 2. Effects of Kaempferol on Apoptotic and Cell Cycle Signaling Pathways.

While its capacity to halt tumor cell proliferation is valuable, kaempferol's tendency to induce apoptosis in these cells is perhaps a more potent property. Keeping tumor size static marks a step in the right direction, but actually shrinking tumors is infinitely more desirable. Kaempferol's previously cited ability to increase p53 levels applies to more than simple cell cycle arrest. The presence of kaempferol has been found to result in phosphorylation of p53 at the Ser-15 residue, which has been linked to increased apoptosis in cancer cells (Choi & Ahn, 2008). One of the many functions of p53 is to heighten levels of the pro-apoptotic protein Bax, which operates within the intrinsic pathway of apoptosis. Bax inactivates the anti-apoptotic protein Bcl-2, which is subsequently unable to block the flow of cytochrome c from mitochondria. This sequence of events culminates in the activation of the caspase cascade, and the cell begins to tear itself apart. Kaempferol treatment in ovarian cancer cells demonstrated an increase in concentrations of p53, Bax, and Bad, another protein which interferes with Bcl-2 (Luo et al., 2011). Caspase levels were predictably increased as well, while Bcl-2 levels were unsurprisingly reduced. Incidence of apoptosis was lowered by treatment with a caspase-9 inhibitor, confirming that the intrinsic apoptosis pathway is indeed modified by kaempferol. Similar results were obtained from evaluating kaempferol's effects on esophageal squamous carcinoma Eca-109 cells (Li et al., 2011).

Furthermore, kaempferol appears to manipulate the extrinsic apoptosis pathway as well. Kaempferol treatment has been found to potentiate the effects of tumor necrosis factor-related apoptosis-inducing ligand (TRAIL) (Siegelin et al., 2008). One of the functions of this extracellular protein is to bind death receptors which turn on caspase-8, a mediator of the extrinsic apoptosis pathway. When lymphoblastoid cells were exposed to a kaempferol glycoside, the cells initiated their programmed death by exhibiting increased caspase-3 and caspase-8 activity (Bhouri et al., 2011). Similar results were obtained from treatment of human myeloid leukemia cells, along with cytochrome c release, indicating simultaneous effects on both the intrinsic and extrinsic pathways (Benyahia et al., 2004). Further still, kaempferol seems to regulate upstream control of the caspase cascade. The anti-apoptotic protein survivin is known to bind to and inactivate caspase-3 and caspase-7 (Tamm et al., 1998). The introduction of kaempferol has produced apoptosis as well as suppressed survivin activity in both HL-60 leukemia cells (Ren et al., 2010), and human glioma cells (Siegelin et al., 2008; Jeong et al., 2009). At a quick glance, kaempferol does seem to encourage apoptosis from many points of attack. Since cancer cells will have to institute a number of adaptations to survive, its broad spectrum of effects is especially valuable in preventing resistance to treatment.

The reactions of a number of apoptotic proteins to the presence of kaempferol have already been discussed, but kaempferol's actions on their upstream regulators have yet to be investigated here. Cells are of course sensitive to a number of extracellular messengers, many of which are growth factors that bind to receptors on the cell surface. These receptors are responsible for converting the extracellular message into activation of certain genes, transmitting the message throughout the cell by means of a cascade of protein activation. One of these pathways is the PI3K/AKT pathway, which is known to promote growth and cell survival (Vivanco & Sawyers, 2002). This sequence of proteins has been implicated in a number of different cancers, and recent evidence suggests that kaempferol may target this pathway. Kaempferol may actually attach to the PI3K protein at its ATP-binding site (Lee et al., 2010), blocking the protein from activating its downstream AKT target. This is significant in that AKT is a recognized promoter of angiogenesis (Luo et al., 2009), as well as a deactivator of the pro-apoptotic BAD protein (Datta et al., 1997). AKT works to phosphorylate BAD, thereby allowing Bcl-2 to continue disrupting apoptosis. By eliminating PI3K, kaempferol thus facilitates apoptosis by incapacitating the proteins downstream of PI3K. An independent study found that kaempferol inhibited PI3K expression and AKT phosphorylation in leukemia cell lines K562 and U937 (Marfe et al., 2009), which contributed to the observed cell death. Kaempferol also has effects on the MAPK signal transduction pathway, though conflicting results have been obtained. While some suggest that kaempferol induces apoptosis through lowering MAPK/ERK activity (Jeong et al., 2009), others make the claim that treatment prompts cell death through sustaining pathway activation (Kim et al., 2008). Strangely enough, one study found that MAPK initiation is a necessary condition for kaempferol-induced apoptosis to occur in lung cancer A549 cells (Nguyen et al., 2003). It goes on to posit that sustained activation by kaempferol results in apoptosis, while transient activation by growth factors like IGF-1 leads to proliferation. MAPK likely plays different roles among different cell types, which explains why others have linked cancer to increased

MAPK activity (Furukawa et al., 2012). Further investigation into the diversity of kaempferol/MAPK interaction may be warranted.

An exciting new development in kaempferol research is its potential to serve as a pan-inhibitor of human histone deacetylases (HDAC). HDACs modify histone proteins by removing acetyl groups, which amplifies the bonding forces between histone and DNA (Carew et al., 2008). This increased binding discourages gene expression, including a host of regulatory proteins which function in preventing cancer. Preliminary research shows that HDAC inhibitors may be promising anticancer agents, with a tendency to arrest cell growth and induce apoptosis in malignant cells (Wang et al., 2012). Recently, kaempferol has been demonstrated to bind to the zinc ion of HDACs, inhibiting all human HDACs of class I, II, and IV (Berger et al., 2012). This newly identified epigenetic activity of kaempferol may be one of the contributing factors to the observed reduction in cancer cell viability. Additional research is necessary before a final judgment can be made, but this novel property certainly opens up further possible clinical applications.

Much has been discussed concerning kaempferol's cancer fighting properties, but that is not representative of the entirety of its worth. When considering kaempferol's cytotoxic effects, it is essential to keep in mind that they apply strictly to the malignant cells. While many chemotherapy drugs carry serious systemic side effects, kaempferol seems to elicit a dramatically diminished response from normal body cells, in some cases exerting a protective effect. Studies have shown that a range of healthy cells, from embryonic kidney and liver cells (Xu et al., 2008) to retinal ganglion cells (Ondricek et al., 2012), respond favorably to kaempferol treatment, exhibiting signs of toxicity only at much higher levels. Kaempferol has been found to protect neurons from toxin-induced apoptosis (Filomeni et al., 2012) and preserve pancreatic beta cell viability and function (Zhang & Liu, 2011). An extract containing a kaempferol glycoside among other flavonoids was shown to promote healing in human skin and inhibit bacterial growth (Czemplik et al., 2012). Aside from guarding normal cells though, numerous other applications exist in improving general health. Kaempferol can defend cognitive functioning by reducing oxidative stress (Lei et al., 2012) and may even be effective in managing diabetes and obesity (Park et al., 2012). Kaempferol might also have a future in enhancing outcomes in cases of osteoporosis (Kumar et al., 2012). Furthermore, high blood pressure could possibly be managed by kaempferol treatment through ACE inhibition (Guerrero et al., 2012), and a biflavonoid of kaempferol has been documented to exhibit anti-inflammatory and analgesic effects (Ye et al., 2012). In summation, when evaluating the efficacy of kaempferol as a chemotherapy agent, one should consider the relative lack of risk in treatment. In some cases this compound can be likened to a vitamin.

Containing the Malignancy

The previous sections focused on the myriad ways in which kaempferol directly reduces viability and encourages cell death. Of course, there exist other strategies to successfully fighting cancer. Among its available techniques, kaempferol can employ attrition warfare. Capable of more than just apoptosis, kaempferol is able to starve cancer cells of vital sources of energy. It will modify the tumor's environment, walling it off from healthy body cells.

This move can enhance kaempferol's apoptotic effects and augment the response to treatment, in addition to stopping the progression of disease (Fig. 3).

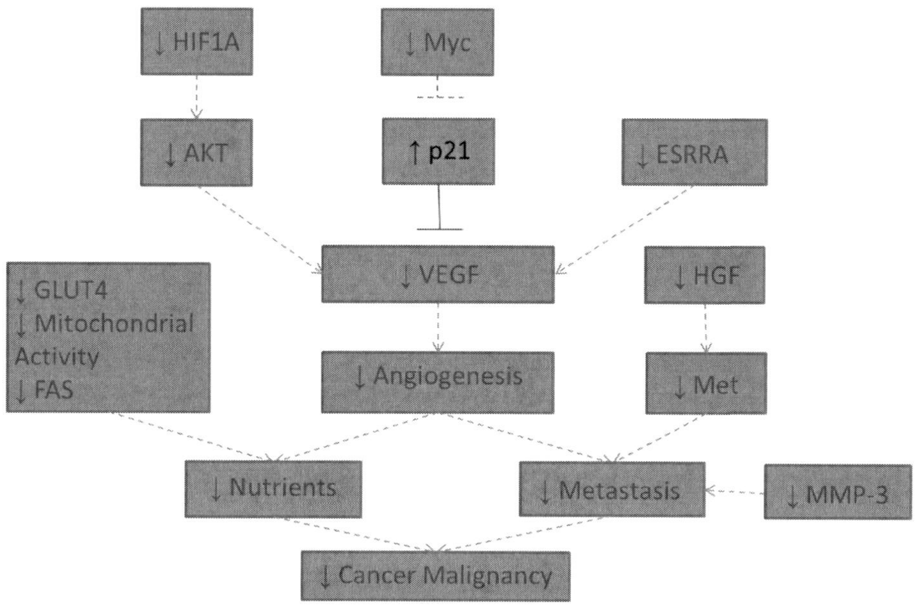

Figure 3. Effects of Kaempferol on Cancer Nutrition, Angiogenesis, and Metastasis.

Growing cells demand a continuous supply of oxygen and nutrients in order to proliferate. Cancer cells are no different. Without a personalized network of blood vessels, the tumor will eventually drown in its own wastes after reaching a certain size. The development of new blood vessels is termed angiogenesis, and the process is regulated by a number of proteins, one being vascular endothelial growth factor (VEGF). By disrupting VEGF activity, kaempferol may actually be able to limit the size to which a tumor can grow. Research has shown kaempferol to interfere in VEGF secretion, revealing lowered protein and mRNA levels (Luo et al., 2009). This can be attributed to the downregulation of a number of VEGF controller proteins. Kaempferol has been demonstrated to target HIF1A as well as ESRRA, both of which are upstream of VEGF. HIF1A typically activates VEGF through the PI3K/AKT pathway, which as mentioned before, kaempferol inhibits. Ostensibly, kaempferol upsets VEGF expression at multiple sites. Further still, another study found kaempferol to lower Myc levels through downregulation of ERK and NFκB (Luo et al., 2012b). Myc, as discussed previously, is an oncogene that tends to encourage cancer development by interfering in p21 expression. This is significant in that p21 is a VEGF antagonist. So in addition to the HIF1A and ESSRA pathways, kaempferol may actually block VEGF production through an ERK-NFκB-Myc-p21 pathway, which likely ensures kaempferol's potency in halting the process of angiogenesis. Preliminary research has shown kaempferol to inhibit in-vitro angiogenesis (Luo et al., 2009), but these studies are scarce. Further investigation is required to better characterize its antiangiogenic effects.

Though stopping blood vessel growth will cut off tumor cells from a constant flow of nutrients, kaempferol can amplify this effect by hindering the utilization of any energy sources already available. In the presence of kaempferol, HeLa cells were unable to bring in glucose through glucose transporter 4, which complexes with the flavonoid (Filomeni et al., 2010). Kaempferol was also found to inhibit the mitochondrial respiratory chain at Complex I. Energetically starved, the HeLa cells initiated the process of autophagy as a survival mechanism. In addition to impairing glucose metabolism, kaempferol interferes with lipogenesis by inhibiting fatty acid synthase (FAS) activity (Brusselmans et al., 2005). FAS is significantly upregulated in cancer cells, where it produces copious amounts of fatty acids used for energy storage and membrane structure. The introduction of kaempferol was found to induce apoptosis in breast and prostate cancer cells. Addition of palmitate, the final product of FAS activity, attenuated the cytotoxic effects, confirming that inhibition of FAS represents an important aspect of kaempferol's chemotherapeutic properties. Based on these results, it's apparent that, in addition to physically removing sources of energy by blocking the formation of new blood vessels, kaempferol can further sabotage the metabolic machinery of cancer cells. This is one of the reasons why kaempferol may be effective in managing obesity (Park et al., 2012).

Cutting off supply lines is one way to contain a tumor, but one of the hallmarks of cancer is its tendency to disseminate to other areas of the body. Separated from a source of nutrients, the cells may be further stressed to move to a new area, and once the cancer has spread, the prognosis generally becomes much darker. The introduction of kaempferol can actually institute measures against metastasis, keeping the tumor glued to one area. Kaempferol's ability to prevent angiogenesis does more than simply creating a hypoxic environment. New blood vessels provide additional opportunities for the cancer to metastasize (Eveno & Pocard, 2012). Though malignant cells can still travel through existing capillaries and lymph vessels, halting angiogenesis can significantly lower the chances of the cancer spreading.

Aside from its antiangiogenic properties, kaempferol alters the functioning of other proteins essential to metastasis. To reach other body areas, cancer cells must break down the surrounding extracellular matrix (ECM) that keeps the cells in one place. Central to this process are the matrix metalloproteinases (MMPs), which not only dissolve a number of ECM components including laminin, elastin, and the collagen subtypes, but also promote tumor growth (Curran & Murray, 1999). The presence of kaempferol has been shown to disrupt activity of MMP-3, actually preventing the in-vitro invasion of MDA-MB-231 breast cancer cells (Phromnoi et al., 2009). In addition to its MMP effects, kaempferol treatment can disrupt hepatocyte growth factor (HGF)/Met signaling. Activation of Met, the receptor for HGF, and subsequent phosphorylation of AKT have been implicated in tumor growth and metastasis (Birchmeier et al., 2003). Kaempferol was observed to inhibit this pathway in medulloblastoma cell line DAOY, consequently interfering with the in-vitro migration of the cells in Boyden chambers (Labbé et al., 2009). Treatment prevented the formation of actin-rich membrane ruffles and thus was able to keep the cells frozen in place. Kaempferol shows remarkable promise as an antimetastatic agent, but these effects demand further elucidation in vivo. Cancers exist in a dynamic environment, in a pool of countless chemical signals and factors. Stopping metastasis in vitro may represent a completely different mechanism from halting it in the context of the human body. More research must be conducted before kaempferol's efficacy can be truly evaluated.

Inflammation

In comparison to its effects on metastasis and angiogenesis, kaempferol's anti-inflammatory properties have been documented and studied much more intensively. Flavonoids as a whole have been characterized as potent antioxidant and anti-inflammatory compounds, and kaempferol is no exception. Chronic inflammation has long been associated with the development of cancer (Hussain & Harris, 2007). Some inflammatory processes culminate in the release of reactive oxygen species (ROS), which may act as antimicrobials—important when the inflammation is triggered by an infection. However, ROS will likely damage a number of healthy body cells as well. An accumulation of ROS can lead to mutations, and if the right proto-oncogenes are affected, the inflammation can bring about the genesis of cancer (Rakoff-Nahoum, 2006). Inflammatory stimuli also generally result in necrosis. To replace lost cells, inflammatory signaling cascades also call for surviving cells in the afflicted area to proliferate. This encouragement to grow can spiral out of control if tumor cells are present, so when combined with ROS, inflammation can create the perfect environment for the development of cancer (Rakoff-Nahoum, 2006). Kaempferol's wealth of cancer fighting effects has already been drawn out here, and it may actually augment those effects by upsetting the inflammatory response.

Inflammation is propagated by a number of chemical messengers referred to as cytokines, which include the interleukins and the tumor necrosis factors. Expression of interleukin-1 beta and tumor necrosis factor-alpha in J774.2 macrophages were both shown to be reduced by kaempferol treatment (Kowalski et al., 2005). These two pro-inflammatory cytokines mediate a number of autoimmune and chronic inflammatory disorders, and the anti-inflammatory properties of flavonoids can be partially attributed to their actions on these messengers. Additionally, kaempferol appears to modulate proteins downstream of TNF-alpha. One of the functions of TNF is to promote upregulation of interleukin-8, another pro-inflammatory cytokine with angiogenic effects (Lee et al., 2009). TNF also seems to activate NF-κB as well as elevate ROS levels. Kaempferol lowers TNF-associated IL-8 activity, NF-κB stimulation, and concentration of ROS in HEK 293 cells (Lee et al., 2009). These processes secondary to TNF stimulation are integral to the inflammatory response, which suggests kaempferol to be an effective treatment option for TNF-mediated inflammatory diseases. IL-4 is another pro-inflammatory cytokine implicated in a number of chronic diseases. To exert its effects, IL-4 activates JAK3, which then phosphorylates STAT6, a transcription factor responsible for turning on inflammatory genes. Kaempferol has been demonstrated to block JAK3 from carrying out its task, effectively negating IL-4 mediated effects (Cortes et al., 2007). This bears implications for IL-2 as well, which employs JAK3 to phosphorylate STAT5. The available evidence suggests kaempferol to be an effective anti-inflammatory agent, but as will be seen in the next section, there exist a number of issues in coaxing kaempferol to function within the human body as it does in vitro.

Nanotechnology

Up to this point, an incredible amount of research has been conducted detailing the in vitro effects of dietary flavonoids including kaempferol. The question still remains, though, as to whether kaempferol is effective in helping real patients suffering from cancer. Encouragingly, a large number of population studies have confirmed that a diet high in flavonoids, namely kaempferol, reduces cancer risk in smokers (Bobe et al., 2008; Cui et al., 2008; Nöthlings et al., 2008). However, results from studies focusing on the nonsmoking population have been more ambivalent. Some prospective studies revealed that over decades, consumption of kaempferol dramatically reduced the risk of cancer (Gates et al., 2007). This finding suggests that kaempferol, as a nontoxic, inexpensive dietary component, is a promising agent for the chemoprevention of ovarian cancers, easily adopted into the lifestyles of most people. On the other hand, some have found extremely limited support for chemoprevention through a diet heavy in flavonoids (Gates et al., 2009), while others maintain no association exists between cancer risk and flavonoid content (Wang et al., 2009). In light of the large catalogue of research documenting the impressive in vitro potential of kaempferol, the best of way of understanding these underwhelming findings is through the bioavailability of kaempferol and flavonoids in general. Like all substances consumed orally, flavonoids are subject to first-pass metabolism by the liver once absorbed through the intestinal wall (Barve et al., 2009). Conjugation reactions typically involve modifying the original compound to a less chemically active one. Before kaempferol can even reach the main circulation, most of it has already been transformed into a handicapped metabolite. Furthermore, since flavonoids are recognized by the body as foreign substances, cells are equipped with a number of pumps designed to escort these alien chemicals outside of cells and membranes (Schinkel & Jonker, 2003). Kaempferol is poorly absorbed into the bloodstream and cannot force its way inside cells, where it can manipulate signaling pathways or inhibit certain protein functions. Because of this efflux, kaempferol's anticancer effects may not be experienced by the body.

Recent advances, however, have imparted hope to overcoming these obstacles in bioavailability. Breast cancer resistance protein (Abcg2) is a transport protein tasked with removing a host of toxic compounds from the cell, including quercetin, another flavonol with promise in treating cancer. Apparently, kaempferol has a higher affinity for Abcg2 than quercetin does. Administration of both flavonols simultaneously revealed that kaempferol blocked the efflux of quercetin, allowing quercetin to remain inside and exert its effects (An et al., 2011). Therefore, kaempferol could possibly be combined with another substance with an even higher affinity for Abcg2, which would leave kaempferol inside to wreak havoc on the malignant cells. In addition, kaempferol has been found to lower mRNA levels of ABCC6, another ATP-binding cassette transporter (Luo et al., 2010). ABCC6 is involved in the transport of cisplatin, a chemotherapy drug, outside of the cell (Zhou et al., 2008). Introduction of kaempferol was found to significantly enhance cisplatin's cytotoxic effects on cancer cells (Luo et al., 2010). Ostensibly, even if kaempferol itself experiences poor bioavailability, at least it seems to improve the bioavailability of other substances meant to fight cancer. These transport proteins are a good potential research focus for increasing the body's access to kaempferol and other drugs. Of course, the oral route doesn't represent the only method of drug administration. Another effective pharmaceutical dosage form is transdermal delivery, allowing the medication to absorb through the skin. This method of administration allows the drug to avoid first-pass metabolism and enter the systemic circulation directly. One study suggests using submicron emulsions as a vehicle for the topical application of kaempferol (Chao et al., 2012). Employing this technique was shown to be significantly more effective than kaempferol dissolved in isopropyl myristate. Transdermal delivery of flavonoids shows great potential, but this area is relatively uninvestigated and calls for more extensive research.

Perhaps the most promising and innovative technique to improving bioavailability, though, is through nanotechnology. The coating of certain chemicals with a layer of nanoparticles increases their permeability and the amount that reaches systemic circulation. The capsule of nanoparticles can help shield kaempferol from efflux transporters and coax cells to transport the nanoparticle complex inwards, in addition to preserving its structural integrity. Research done on EGCG, another flavonoid with potential in cancer treatment, has already demonstrated promise in nanochemoprevention. EGCG was found to be easily coated with an external matrix of chitosan and caseinophosphopeptides, which significantly improves the in vitro transport of EGCG over cell monolayers (Hu et al., 2012). The nanoparticle coating seems to have low cytotoxicity to normal cells as well. Nano research has also been conducted focusing on enhancing the bioavailability of kaempferol specifically. Encapsulation with poly(ethylene oxide)-poly(propylene oxide)-poly(ethylene oxide) (PEO-PPO-PEO) nanoparticles appears to significantly reduce cancer cell viability, as does coating with poly(DL-lactic acid-co-glycolic acid) (PLGA) nanoparticles (Luo et al., 2012a). Both encapsulations are more potent than kaempferol treatment alone. The PEO-PPO-PEO formulation shows a higher effectiveness in lowering cancer viability, but the PLGA covered kaempferol preferentially kills malignant cells. Future research could possibly focus on utilizing a special targeting mechanism sensitive to folate, which cancerous cells tend to overproduce (Sunoqrot et al., 2012). Folate-targeted kaempferol complexes could be incorporated into nanoparticles to achieve high targeting efficacy against folate-overexpressing cancerous cells while limiting effects on normal cells.

Nanochemoprevention represents an exciting field with many new avenues to explore, which means many unanswered questions. Work done in vivo is scarce, and it still remains to be seen whether nanoparticles can really help augment kaempferol's anticancer effects in live patients. Further investigation is a must before we can make sense of kaempferol's true worth.

References

Adhami VM, Malik A, Zaman N, Sarfaraz S, Siddiqui IA, et al. Combined inhibitory effects of green tea polyphenols and selective cyclooxygenase-2 inhibitors on the growth of human prostate cancer cells both in vitro and in vivo. *Clin Cancer Res* 2007;13:1611–1619.

Amin AR, Kucuk O, Khuri FR, Shin DM. 2009. Perspectives for cancer prevention with natural compounds. *J Clin Oncol.* 27(16): 2712–2725.

An G, Gallegos J, Morris ME. The bioflavonoid kaempferol is an Abcg2 substrate and inhibits Abcg2-mediated quercetin efflux. *Drug Metab Dispos.* 2011 Mar;39(3):426-32.

Barve A, Chen C, Hebbar V, Desiderio J, Saw CL, Kong AN. Metabolism, oral bioavailability and pharmacokinetics of chemopreventive kaempferol in rats. *Biopharm Drug Dispos.* 2009 Oct;30(7):356-65.

Benyahia S, Benayache S, Benayache F, Quintana J, López M, León F, Hernández JC, Estévez F, Bermejo J. Isolation from Eucalyptus occidentalis and identification of a new kaempferol derivative that induces apoptosis in human myeloid leukemia cells. *J Nat Prod.* 2004 Apr;67(4):527-31.

Berger A, Venturelli S, Kallnischkies M, Böcker A, Busch C, Weiland T, Noor S, Leischner C, Weiss TS, Lauer UM, Bischoff SC, Bitzer M. Kaempferol, a new nutrition-derived pan-inhibitor of human histone deacetylases. *J Nutr Biochem.* 2012 Nov 14.

Bhouri W, Bouhlel I, Boubaker J, Kilani S, Ghedira K, Ghedira LC. Induction of apoptosis in human lymphoblastoid cells by kaempferol 3-O-β-isorhamninoside and rhamnocitrin 3-O-β-isorhamninoside from Rhamnus alaternus L. (Rhamnaceae). *Cell Prolif.* 2011 Jun;44(3):283-90.

Birchmeier C, Birchmeier W, Gherardi E, Vande Woude GF. Met, metastasis, motility and more. *Nat Rev Mol Cell Biol.* 2003 Dec;4(12):915-25.

Bobe G, Weinstein SJ, Albanes D, Hirvonen T, Ashby J, Taylor PR, Virtamo J, Stolzenberg-Solomon RZ. Flavonoid intake and risk of pancreatic cancer in male smokers (Finland). *Cancer Epidemiol Biomarkers Prev.* 2008 Mar;17(3):553-62.

Bosetti C, Rossi M, McLaughlin JK, Negri E, Talamini R, et al. Flavonoids and the risk of renal cell carcinoma. *Cancer Epidemiol Biomarkers Prev* 2007;16:98–101.

Bosetti C, Bravi F, Talamini R, Parpinel M, Gnagnarella P, et al. Flavonoids and prostate cancer risk: a study in Italy. *Nutr Cancer* 2006;56:123–127.

Brusselmans K, Vrolix R, Verhoeven G, Swinnen JV. Induction of cancer cell apoptosis by flavonoids is associated with their ability to inhibit fatty acid synthase activity. *J Biol Chem.* 2005 Feb 18;280(7):5636-45.

Carew JS, Giles FJ, Nawrocki ST. Histone deacetylase inhibitors: mechanisms of cell death and promise in combination cancer therapy. *Cancer Lett.* 2008 Sep 28;269(1):7-17.

Casagrande F, Darbon JM. Effects of structurally related flavonoids on cell cycle progression of human melanoma cells: regulation of cyclin-dependent kinases CDK2 and CDK1. *Biochem Pharmacol.* 2001 May 15;61(10):1205-15.

Chao Y, Huang CT, Fu LT, Huang YB, Tsai YH, Wu PC. The effect of submicron emulsion systems on transdermal delivery of kaempferol. *Chem Pharm Bull* (Tokyo). 2012;60(9):1171-5.

Choi EJ, Ahn WS. Kaempferol induced the apoptosis via cell cycle arrest in human breast cancer MDA-MB-453 cells. *Nutr Res Pract.* 2008 Winter;2(4):322-5.

Choi EJ, Kim T, Lee MS. Pro-apoptotic effect and cytotoxicity of genistein and genistin in human ovarian cancer SK-OV-3 cells. *Life Sci* 2007;80:1403–1408.

Cortes JR, Perez-G M, Rivas MD, Zamorano J. Kaempferol inhibits IL-4-induced STAT6 activation by specifically targeting JAK3. *J Immunol.* 2007 Sep 15;179(6):3881-7.

Cui Y, Morgenstern H, Greenland S, Tashkin DP, Mao JT, Cai L, Cozen W, Mack TM, Lu QY, Zhang ZF. Dietary flavonoid intake and lung cancer--a population-based case-control study. Cancer. 2008 May 15;112(10):2241-8.

Curran S, Murray GI. Matrix metalloproteinases in tumour invasion and metastasis. *J Pathol.* 1999 Nov;189(3):300-8.

Czemplik M, Kulma A, Bazela K, Szopa J. The biomedical potential of genetically modified flax seeds overexpressing the glucosyltransferase gene. *BMC Complement Altern Med.* 2012 Dec 10;12(1):251.

Datta SR, Dudek H, Tao X, Masters S, Fu H, Gotoh Y, Greenberg ME. Akt phosphorylation of BAD couples survival signals to the cell-intrinsic death machinery. *Cell.* 1997 Oct 17;91(2):231-41.

Eveno C, Pocard M. VEGF levels and the angiogenic potential of the microenvironment can affect surgical strategy for colorectal liver metastasis. *Cell Adh Migr*. 2012 Nov 1;6(6).

Filomeni G, Desideri E, Cardaci S, Graziani I, Piccirillo S, Rotilio G, Ciriolo MR. Carcinoma cells activate AMP-activated protein kinase-dependent autophagy as survival response to kaempferol-mediated energetic impairment. *Autophagy*. 2010 Feb;6(2):202-16.

Filomeni G, Graziani I, De Zio D, Dini L, Centonze D, Rotilio G, Ciriolo MR. Neuroprotection of kaempferol by autophagy in models of rotenone-mediated acute toxicity: possible implications for Parkinson's disease. *Neurobiol Aging*. 2012 Apr;33(4):767-85.

Fu M, Deng D, Huang R, Zhang N, Su Z, Qiu SX. A new flavanocoumarin from the root of Flemingia philippinensis. *Nat Prod Res*. 2012 Sep 13.

Furukawa T, Tanji E, Kuboki Y, Hatori T, Yamamoto M, Shimizu K, Shibata N, Shiratori K. Targeting of MAPK-associated molecules identifies SON as a prime target to attenuate the proliferation and tumorigenicity of pancreatic cancer cells. *Mol Cancer*. 2012 Dec 10;11(1):88.

Gates MA, Tworoger SS, Hecht JL, De Vivo I, Rosner B, Hankinson SE. A prospective study of dietary flavonoid intake and incidence of epithelial ovarian cancer. *Int J Cancer*. 2007 Nov 15;121(10):2225-32.

Gates MA, Vitonis AF, Tworoger SS, Rosner B, Titus-Ernstoff L, Hankinson SE, Cramer DW. Flavonoid intake and ovarian cancer risk in a population-based case-control study. *Int J Cancer*. 2009 Apr 15;124(8):1918-25.

Gonzalez CA, Riboli E. Diet and cancer prevention: where we are, where we are going. *Nutr Cance*r 2006;56:225–231.

Guerrero L, Castillo J, Quiñones M, Garcia-Vallvé S, Arola L, Pujadas G, Muguerza B. Inhibition of Angiotensin-converting enzyme activity by flavonoids: structure-activity relationship studies. *PLoS One*. 2012;7(11):e49493.

He ZH, Yue GG, Lau CB, Ge W, But PP. Antiangiogenic Effects and Mechanisms of trans-Ethyl p-Methoxycinnamate from Kaempferia galangal L. *J Agric Food Chem*. 2012 Nov 14;60(45):11309-17.

Holick CN, Giovannucci EL, Rosner B, Stampfer MJ, Michaud DS. Prospective study of intake of fruit, vegetables, and carotenoids and the risk of adult glioma. *Am J Clin Nutr* 2007;85:877–886.

Hu B, Ting Y, Yang X, Tang W, Zeng X, Huang Q. Nanochemoprevention by encapsulation of (-)-epigallocatechin-3-gallate with bioactive peptides/chitosan nanoparticles for enhancement of its bioavailability. *Chem Commun* (Camb). 2012 Feb 28;48(18):2421-3.

Hussain SP, Harris CC. Inflammation and cancer: an ancient link with novel potentials. *Int J Cancer*. 2007 Dec 1;121(11):2373-80.

Jeong JC, Kim MS, Kim TH, Kim YK. Kaempferol induces cell death through ERK and Akt-dependent down-regulation of XIAP and survivin in human glioma cells. *Neurochem Res*. 2009 May;34(5):991-1001.

Khan RA. Evaluation of flavonoids and diverse antioxidant activities of Sonchus arvensis. *Chem Cent J*. 2012 Oct 29;6(1):126.

Kim BW, Lee ER, Min HM, Jeong HS, Ahn JY, Kim JH, Choi HY, Choi H, Kim EY, Park SP, Cho SG. Sustained ERK activation is involved in the kaempferol-induced apoptosis of breast cancer cells and is more evident under 3-D culture condition. *Cancer Biol Ther*. 2008 Jul;7(7):1080-9.

Kowalski J, Samojedny A, Paul M, Pietsz G, Wilczok T. Effect of apigenin, kaempferol and resveratrol on the expression of interleukin-1beta and tumor necrosis factor-alpha genes in J774.2 macrophages. *Pharmacol Rep.* 2005 May-Jun;57(3):390-4.

Kumar A, Gupta GK, Khedgikar V, Gautam J, Kushwaha P, Changkija B, Nagar GK, Gupta V, Verma A, Dwivedi AK, Chattopadhyay N, Mishra PR, Trivedi R. In vivo efficacy studies of layer-by-layer nano-matrix bearing kaempferol for the conditions of osteoporosis: A study in ovariectomized rat model. *Eur J Pharm Biopharm.* 2012 Nov;82(3):508-17.

Labbé D, Provençal M, Lamy S, Boivin D, Gingras D, Béliveau R. The flavonols quercetin, kaempferol, and myricetin inhibit hepatocyte growth factor-induced medulloblastoma cell migration. *J Nutr.* 2009 Apr;139(4):646-52.

Lee KM, Lee DE, Seo SK, Hwang MK, Heo YS, Lee KW, Lee HJ. Phosphatidylinositol 3-kinase, a novel target molecule for the inhibitory effects of kaempferol on neoplastic cell transformation. *Carcinogenesis.* 2010 Aug;31(8):1338-43.

Lee S, Kim YJ, Kwon S, Lee Y, Choi SY, Park J, Kwon HJ. Inhibitory effects of flavonoids on TNF-alpha-induced IL-8 gene expression in HEK 293 cells. *BMB Rep.* 2009 May 31;42(5):265-70.

Lei Y, Chen J, Zhang W, Fu W, Wu G, Wei H, Wang Q, Ruan J. In vivo investigation on the potential of galangin, kaempferol and myricetin for protection of D-galactose-induced cognitive impairment. *Food Chem.* 2012 Dec 15;135(4):2702-7.

Li RJ, Mei JZ, Liu GJ. *Kaempferol-induced apoptosis of human esophageal squamous carcinoma Eca-109 cells and the mechanism.* Nan Fang Yi Ke Da Xue Xue Bao. 2011 Aug;31(8):1440-2.

Li W, Du B, Wang T, Wang S, Zhang J. Kaempferol induces apoptosis in human HCT116 colon cancer cells via the Ataxia-Telangiectasia Mutated-p53 pathway with the involvement of p53 Upregulated Modulator of Apoptosis. *Chem Biol Interact.* 2009 Jan 27;177(2):121-7.

Lin HY, Chang ST. Kaempferol glycosides from the twigs of Cinnamomum osmophloeum and their nitric oxide production inhibitory activities. *Carbohydr Res.* 2012 Dec 15;364:49-53.

Lu CH, Li YY, Li LJ, Liang LY, Shen YM. Anti-inflammatory activities of fractions from Geranium nepalense and related polyphenols. *Drug Discov Ther.* 2012 Aug;6(4):194-7.

Luo H, Daddysman MK, Rankin GO, Jiang BH, Chen YC. Kaempferol enhances cisplatin's effect on ovarian cancer cells through promoting apoptosis caused by down regulation of cMyc. *Cancer Cell Int.* 2010 May 11;10:16.

Luo H, Jiang B, Li B, Li Z, Jiang BH, Chen YC. Kaempferol nanoparticles achieve strong and selective inhibition of ovarian cancer cell viability. *Int J Nanomedicine.* 2012a;7:3951-9.

Luo H, Rankin GO, Juliano N, Jiang BH, Chen YC. Kaempferol inhibits VEGF expression and in vitro angiogenesis through a novel ERK-NFκB-cMyc-p21 pathway. *Food Chem.* 2012b Jan 15;130(2):321-328.

Luo H, Rankin GO, Li Z, Depriest L, Chen YC. Kaempferol induces apoptosis in ovarian cancer cells through activating p53 in the intrinsic pathway. *Food Chem.* 2011 Sep 15;128(2):513-519.

Luo H, Rankin GO, Liu L, Daddysman MK, Jiang BH, Chen YC. Kaempferol inhibits angiogenesis and VEGF expression through both HIF dependent and independent pathways in human ovarian cancer cells. *Nutr Cancer.* 2009;61(4):554-63.

Marfe G, Tafani M, Indelicato M, Sinibaldi-Salimei P, Reali V, Pucci B, Fini M, Russo MA. Kaempferol induces apoptosis in two different cell lines via Akt inactivation, Bax and SIRT3 activation, and mitochondrial dysfunction. *J Cell Biochem.* 2009 Mar 1;106(4):643-50.

Mascherpa D, Carazzone C, Marrubini G, Gazzani G, Papetti A. Identification of Phenolic Constituents in Cichorium endivia Var. crispum and Var. latifolium Salads by High-Performance Liquid Chromatography with Diode Array Detection and Electrospray Ioniziation Tandem Mass Spectrometry. *J Agric Food Chem.* 2012 Dec 12;60(49):12142-50.

Mikulic-Petkovsek M, Slatnar A, Stampar F, Veberic R. HPLC–MSn identification and quantification of flavonol glycosides in 28 wild and cultivated berry species. *Food Chem.* 2012 Dec 15;135(4):2138-46.

Mink PJ, Scrafford CG, Barraj LM, Harnack L, Hong CP, et al. Flavonoid intake and cardiovascular disease mortality: a prospective study in postmenopausal women. *Am J Clin Nutr* 2007;85:895–909.

Misyura M, Colasanti J, Rothstein SJ. Physiological and genetic analysis of Arabidopsis thaliana anthocyanin biosynthesis mutants under chronic adverse environmental conditions. *J Exp Bot.* 2012 Nov 16.

Nair HB, Sung B, Yadav VR, Kannappan R, Chaturvedi MM, Aggarwal BB. 2010. Delivery of anti-inflammatory nutraceuticals by nanoparticles for the prevention and treatment of cancer. *Biochem. Pharmacol.* 80(12): 1833–1843.

Nguyen TT, Tran E, Ong CK, Lee SK, Do PT, Huynh TT, Nguyen TH, Lee JJ, Tan Y, Ong CS, Huynh H. Kaempferol-induced growth inhibition and apoptosis in A549 lung cancer cells is mediated by activation of MEK-MAPK. *J Cell Physiol.* 2003 Oct;197(1):110-21.

Nöthlings U, Murphy SP, Wilkens LR, Boeing H, Schulze MB, Bueno-de-Mesquita HB, Michaud DS, Roddam A, Rohrmann S, Tjønneland A, Clavel-Chapelon F, Trichopoulou A, Sieri S, Rodriguez L, Ye W, Jenab M, Kolonel LN. A food pattern that is predictive of flavonol intake and risk of pancreatic cancer. *Am J Clin Nutr.* 2008 Dec;88(6):1653-62.

Ondricek AJ, Kashyap AK, Thamake SI, Vishwanatha JK. A comparative study of phytoestrogen action in mitigating apoptosis induced by oxidative stress. *In Vivo.* 2012 Sep-Oct;26(5):765-75.

Park UH, Jeong JC, Jang JS, Sung MR, Youn H, Lee SJ, Kim EJ, Um SJ. Negative regulation of adipogenesis by kaempferol, a component of Rhizoma Polygonati falcatum in 3T3-L1 cells. *Biol Pharm Bull.* 2012;35(9):1525-33.

Phromnoi K, Yodkeeree S, Anuchapreeda S, Limtrakul P. Inhibition of MMP-3 activity and invasion of the MDA-MB-231 human invasive breast carcinoma cell line by bioflavonoids. *Acta Pharmacol Sin.* 2009 Aug;30(8):1169-76.

Plochmann K, Korte G, Koutsilieri E, Richling E, Riederer P, et al. Structure-activity relationships of flavonoid-induced cytotoxicity on human leukemia cells. *Arch Biochem Biophys* 2007;460:1–9.

Rakoff-Nahoum S. Why cancer and inflammation? *Yale J Biol Med.* 2006 Dec;79(3-4):123-30.

Ren HJ, Hao HJ, Shi YJ, Meng XM, Han YQ. *Apoptosis-inducing effect of quercetin and kaempferol on human HL-60 cells and its mechanism.* Zhongguo Shi Yan Xue Ye Xue Za Zhi. 2010 Jun;18(3):629-33.

Schinkel AH, Jonker JW. Mammalian drug efflux transporters of the ATP binding cassette (ABC) family: an overview. *Adv Drug Deliv Rev*. 2003 Jan 21;55(1):3-29.

Siegelin MD, Reuss DE, Habel A, Herold-Mende C, von Deimling A. The flavonoid kaempferol sensitizes human glioma cells to TRAIL-mediated apoptosis by proteasomal degradation of survivin. *Mol Cancer Ther*. 2008 Nov;7(11):3566-74.

Somerset, SM, Johannot L. Dietary flavonoid sources in Australian adults. *Nutr Cancer.* 2008;60(4):442-9.

Soucek L, Whitfield J, Martins CP, Finch AJ, Murphy DJ, Sodir NM, Karnezis AN, Swigart LB, Nasi S, Evan GI. Modelling Myc inhibition as a cancer therapy. *Nature*. 2008 Oct 2;455(7213):679-83.

Sunoqrot S, Bae JW, Pearson RM, Shyu K, Liu Y, Kim DH, Hong S. Temporal control over cellular targeting through hybridization of folate-targeted dendrimers and PEG-PLA nanoparticles. *Biomacromolecules*. 2012 Apr 9;13(4):1223-30.

Tamm I, Wang Y, Sausville E, Scudiero DA, Vigna N, Oltersdorf T, Reed JC. IAP-family protein survivin inhibits caspase activity and apoptosis induced by Fas (CD95), Bax, caspases, and anticancer drugs. *Cancer Res*. 1998 Dec 1;58(23):5315-20.

Theodoratou E, Kyle J, Cetnarskyj R, Farrington SM, Tenesa A, et al. Dietary flavonoids and the risk of colorectal cancer. *Cancer Epidemiol Biomarkers Prev* 2007;16:684–693.

Vivanco I, Sawyers CL. The phosphatidylinositol 3-Kinase AKT pathway in human cancer. *Nat Rev Cancer*. 2002 Jul;2(7):489-501.

Wang G, He J, Zhao J, Yun W, Xie C, Taub JW, Azmi A, Mohammad RM, Dong Y, Kong W, Guo Y, Ge Y. Class I and Class II Histone Deacetylases Are Potential Therapeutic Targets for Treating Pancreatic Cancer. *PLoS One*. 2012;7(12):e52095.

Wang L, Lee IM, Zhang SM, Blumberg JB, Buring JE, Sesso HD. Dietary intake of selected flavonols, flavones, and flavonoid-rich foods and risk of cancer in middle-aged and older women. *Am J Clin Nutr*. 2009 Mar;89(3):905-12.

Xu W, Liu J, Li C, Wu HZ, Liu YW. Kaempferol-7-O-beta-D-glucoside (KG) isolated from Smilax china L. rhizome induces G2/M phase arrest and apoptosis on HeLa cells in a p53-independent manner. *Cancer Lett*. 2008 Jun 18;264(2):229-40.

Ye Y, Guo Y, Luo YT. Anti-Inflammatory and Analgesic Activities of a Novel Biflavonoid from Shells of Camellia oleifera. *Int J Mol Sci*. 2012 Sep 27;13(10):12401-11.

Zhang Q, Zhao XH, Wang ZJ. Cytotoxicity of flavones and flavonols to a human esophageal squamous cell carcinoma cell line (KYSE-510) by induction of G2/M arrest and apoptosis. *Toxicol In Vitro*. 2009 Aug;23(5):797-807.

Zhang Y, Liu D. Flavonol kaempferol improves chronic hyperglycemia-impaired pancreatic beta-cell viability and insulin secretory function. *Eur J Pharmacol.* 2011 Nov 16;670(1):325-32.

Zhou SF, Wang LL, Di YM, Xue CC, Duan W, Li CG, Li Y. Substrates and inhibitors of human multidrug resistance associated proteins and the implications in drug development. *Curr Med Chem*. 2008;15(20):1981-2039.

In: Kaempferol
Editors: Garion Villers and Yves Fougere
ISBN: 978-1-62618-515-9

Chapter 7

Metal Complexes of Kaempferol and their Speciation in Human Plasma

Andrija Ciric*[*1], *Ivan Jakovljevic*[1], *Milica Cvijovic*[2], *Milena Jelikic-Stankov*[3] and *Predrag Djurdjevic*[1]
[1]Faculty of Science, University of Kragujevac, Kragujevac, Serbia
[2]Faculty of Agronomy, University of Kragujevac, Čačak, Serbia
[3]Faculty of Pharmacy, University of Belgrade, Belgrade, Serbia

Flavonoids are a large class of phenolic compounds which are sub classified as flavones, flavonols, isoflavones, flavanones and catechins, chalcones and anthocyanidins depending on phenyl substituent in the C_2 or C_3 position in *benzo-γ-pyrone* nucleus. Interest in the bioflavonoids is related to their diversity, biological significance as secondary plant metabolites and ecological role [1], use as chemotaxonomic markers [2], impact on fruit quality [3], physiological effects [4–6] and industrial applications [7].

The flavonoids are potent antioxidants, free radical scavengers [8] and metal chelators; they inhibit lipid peroxidation [9] and exhibit various physiological activities [10–15], including anti-inflammatory [16], anti-allergic, anti-carcinogenic, antihypertensive and anti-arthritic activities [17].

It is well known that blood serum contains essential amino acids, 12 essential metal ions at least another 100 ligands as well as numerous low molecular weight complexes. In addition, there are α/β macro globulins such as albumin and transferin weighing up to 100000 Daltons. It has been well demonstrated that metal ions play a large number of important roles in biological systems [18–20]. Metal ions which are considered essential to human life included calcium, magnesium, manganese, iron, cobalt, copper and zinc [21]. Furthermore man is constantly being challenged by pollutants included the toxic metal ions mercury, cadmium and lead.

In each flavonoid molecule, there are three domains that can likely interact with metal ions, *i.e.*, the 3',4'-dihydroxy group located on the B ring, the 3-hydroxy or 5-hydroxy and the

[*] corresponding author: andrija@kg.ac.rs, tel: +38134300261, fax: +38134335040.

4-carbonyl groups in the C ring. Generally, the chelating properties of flavonoids toward metal ions have been attributed to the presence of the 3- or 5-hydroxypyran-4-one, rather than the *ortho*-hydroxyl groups in the B ring. [22]

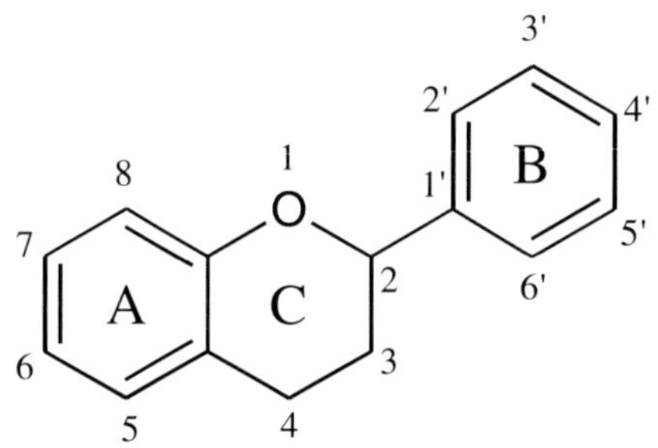

Figure 1. General formula of flavones.

Flavonoids exhibit antioxidant properties through chelating with transition metals, primarily Fe(II), Fe(III) and Cu(II), which participate in reactions generating free radicals [23]. Metal–flavonoid chelates are considerably more potent free radical scavengers than the parent flavonoids and play a prominent role in protecting from oxidative stress. Flavonoids may sequester metal ions by chelating and preventing metal-mediated generation of free radicals and, accordingly, may protect the potential biological targets from oxidative stress. Thus, the overall anti-oxidant action of flavonoids appears to be a combination of a direct reaction with free radicals and chelating the metal ions responsible for the production of reactive oxygen species. Moreover, experimental data have shown that the chelates are considerably more effective free radical scavengers than the free flavonoids. Kostyuk *et al.* [24] found that complexes of rutin, dihydroquercetin or epicatechin with Fe(II), Fe(III), Cu(II) or Zn(II) are more effective radical scavengers than the free flavonoids, due to the acquisition of additional superoxide dismutating centers. These complexes show elevated efficiency in protecting red blood cells against asbestos-induced oxidative injury *in vitro*. According to the same authors, the Cu–rutin complex was found to be the most effective anti-oxidant against asbestos-induced lipid peroxidation in pulmonary tissue *in vivo*. Moridani *et al.* [25] found that the Fe(III) complexes of flavonoids were much more effective than the free flavonoids in protecting isolated rat hepatocytes against hypoxia–reoxygenation injury. By using the 1,1-diphenyl-2-picrylhydrazyl radical scavenging method, de Souza and de Giovani [26] found antioxidant activities of the quercetin, rutin, galangin, and catechin complexes more effecttive then free flavonoids. Afanas'ev *et al.* [27] found that Fe(II)– and Cu(II)–rutin complexes were more efficient free radical scavengers *in vitro* and *ex vivo*. These complexes decreased the production of oxygen radicals by xanthine oxidase, rat liver microsomes and the rat peritoneal macrophages, as well as the generation of oxygen radicals by bronchioalveolar cell from bleomycin-treated rats by 2–30 times compared to the parent rutin. The anti-oxidative activities of morin and its Pd(II)- and Pt(II)-complexes were also investigated [28]. The anti-oxidative effects (scavenging superoxide radicals) of the complexes were greater than that of morin itself, while the Pt(II)-complex exhibited stronger scavenging efficacy than the Pd(II)-complex. Both the Pd(II)- and Pt(II)-complexes showed an inhibitory effect on lipid peroxides which was greater than that of free morin. Due to antioxidative mechanisms, morin complexes with La(III), Gd(III) and Lu(III) ions against three bacterial strains, *i.e.*, *Escherichia coli, Klebsiella pneumoniae* and *Staphylococcus aureus* showed inhibitory action [29]. Complexes of flavonoids play an important role in limiting metal bioavailability and suppressing metal toxicity. For example, aluminum has been implicated in neurological and bone disorder. The complexation of Al(III) by quercetin reduces aluminum overload in the diet [30]. Forming complexes, flavonoids appear to be a suitable antidote for heavy metal poisoning *in vivo* [31]. Quercetin, as an active biological ligand, might be an appropriate Mo(VI) chelator in the case of molybdenum deficiency caused by irradiation, since the use of molybdenum salts is undesirable because of their high toxicity [32].

Depending on the experimental conditions, as well as with an excess of the metal ion, Torreggiani *et al.* [33] found that two chelating processes occurred consecutively, implicating two binding sites in the Cu(II)–quercetin complex.

In human blood plasma the metal ions present may be classified into four distinct fractions: those which are incorporated rigidly into the metaloproteins and are non-exchangeable; those which are relatively loosely bound by other types of protein and are in labile equilibrium with similar ions in solutions; those which are complexed by numerous low molecular weight ligands present and the free metal ions [34].

The low molecular weight complexes are believed to be involved as intermediates with metal ions which are inserted into or removed from certain metalloenzyimes or carrier proteins; in the transfer of certain metal ions across membranes; in keeping essential metal ions solution and in altering the potential of certain redox couples.

Human serum albumin (HSA), the most prominent protein in the plasma, binds different classes of ligands at multiple sites. HSA provides a depot for many compounds, affects the pharmacokinetics between plasma proteins (serum albumin) and small molecule-drugs have been an interesting field of research in life science, chemistry and clinical medicine. The binding of metal ions such as Cu^{2+}, Ni^{2+}, Zn^{2+}, and Co^{2+} to serum albumins has been widely reported [35-37]. Many metal ions can form complexes with drug molecules, thereby affecting some of the characteristics of the drug.

Chemical Speciation in Blood Plasma

Heavy metals in blood plasma are mainly bound by albumin (HSA) and transferin. Harris has shown that labile zinc in blood serum is almost completed (98%) bound to serum albumin and transferin [38]; this is the case for many other heavy metals. A small fraction is bound to low molecular weight complexes. The particular interest in low molecular complexes comes from the necessity of prior knowledge of chemical reaction in blood plasma, which determines intestinal absorption, deposition into bones or tissues, passage through the blood-brain barrier, excretion via the kidney or liver.

The studies on the interaction between plasma proteins (serum albumin) and small molecule drugs have been an interesting filed of research in life science, chemistry and clinical medicine. In the work of [39] the shift of the maximum λ_{em} of HSA is indicative of changes in the immediate environment of Trp (Triptophan) residues. The hydrophobic groups are in the interior of the tertiary structure and the polar groups are on the surface of native proteins. The emission of HSA may be blue-shifted if the indole group of Trp is **insuline** the native protein, and its emission may be red-shifted when the protein is unfolded. The result suggests that a greater change in the immediate environment of the Trp residues occurred when flavonoid was in close proximity to the Trp residues. Thus, the molecular conformation of the protein was affected. These results also indicated that the quenching effect of flavonoids on HSA fluorescence depended on the structure of flavonoid.

The metal complexes of flavonoids may be also bond to HSA. The flavonoid molecule may be tied up not only to free metal ions in blood plasma but also it may disturb the labile equilibria between metal ions and LMW ligands in plasma as well as between metal ions and HSA. In this was biodistribution of metal ions may be altered.

Computer simulation and speciation details are especially important in predicting biodistribution of metal ions. This is where computer modeling can be extremely helpful because mathematically there is almost no limit to the concentration that can be employed; whereas, current technique such as atomic absorption or ion-selective electrodes have detection limits in the range of parts per billion.

Perrin and Agarwal [40] were the first to attempt a computer simulation of the complex formation equilibria between low molecular weight ligands and selected metal ions found in human blood serum. They concentrated on copper and zinc ions with essential amino acids. Their limiting factor depended on the available computer programs that handled only restricted numbers of metals and ligands at one time. To overcome this problem May et al. [41] wrote a new program to evaluate constituent concentration in large equilibrium, called ECCLES. Numerous publications, in human blood plasma were produced. ECCLES however is not Windows oriented possesses no graphical output and cannot easily handle data input. Therefore, we used in this study the program HySS from HyperQuad suite family of programs [42].

Computer Simulation with the HySS Program

Hyperquad Simulation and Speciation (HySS) [42] is a computer program from a group of computer programs, developed in several versions (HySS1, HySS2, HySS2003, HySS2006, HySS2009) for Windows OS (95, 98, NT, 2000, XP) on personal computers. The program has two main functions:

- Potentiometric titration curves simulation and
- Determination of the concentrations of species in solution with a constant volume system to display speciation diagrams.

Structure of HySS software is based to solving the system of mass balance equations:

$$T_A = [A] + \sum_i a_i \beta_i [A]^{a_i} [B]^{b_i} \ldots + \sum_j p_j C_j$$

and

$$T_B = [B] + \sum_i a_i \beta_i [A]^{a_i} [B]^{b_i} \ldots + \sum_j q_j C_j ,$$

where A, B and so on are reagents, a, b, ..., p, q, and so on are stoichiometrices indexes, β is the (cumulative) equilibrium constant, and C is the molar concentration of insoluble species (the amount of solid components divided by the volume of solution) and is used if there are insoluble species. With each insoluble species solubility products: is associated.

Association and Dissociation Constants in HySS

In organic chemistry and biochemistry it is customary to use pKa values acid *dissociation* equilibria. For a monobasic acid with formula AH, the acid *dissociation* constant, $\mathbf{K_a}$, is defined by

$$HA \leftrightarrows H^+ + A^-; K_a = \frac{[H^+][A^-]}{[HA]}; pK_a = -\log_{10} K_a = \log_{10}(1/K_a) \quad (1)$$

On the other hand stability constants for metal complexes, and binding constants for host-guest complexes are generally expressed as *association* constants. It is therefore necessary to use *association* constants for all equilibria.

The stepwise constant, K, for the formation of the same complex from M and L in two steps is given by equations:

$$M + L \leftrightarrows ML;\ K_1 = \frac{[ML]}{[M][L]}$$

$$ML + L \leftrightarrows ML_2\ ;\ K_2 = \frac{[ML_2]}{[ML][L]}$$

Therefore $\beta_2 = K_1K_2$ and $\log\beta_2 = \log K_1 + \log K_2$. A cumulative constant can always be expressed as the product of stepwise constants. Each stability constant should be defined with reference to an equilibrium expression, as in the examples above.

Hydrolysis Constants

Metal ion hydrolysis can be expressed, omitting electrical charges and water molecules coordinated to the metal, as

$$M + OH \leftrightarrows M(OH)\colon \beta = \frac{[M(OH)]}{[M][OH]}$$

$$H_2O \leftrightarrows H^+ + OH^-;\ K_w = [H^+][OH^-]$$

Combining these two equilibria one obtains

$$M + H_2O \leftrightarrows M(OH) + H^+$$

for which hydrolysis constant β* may be expressed as β* = β× K_W. In HySS these calculation β* is used.

Speciation of Flavonoids in Blood Plasma

The absorption and subsequent distribution, metabolism and excretion of flavonoids in humans is little studied. Absorption of flavonoids from the diet was long considered to be negligible, as they are bound to sugars as glycosides (with the exception of catechins). Only free flavonoids without a sugar molecule, the so-called aglycones, were considered to be able to pass the gut wall, and no enzymes that can split these predominantly R-glycosidic bonds are secreted into the gut or present in the intestinal wall [43].

In 1995 Hollman et al. [44], based on indirect evidence, proposed that flavonoid glycosides actually could be absorbed intact in the small intestine, using the sodium dependent glucose transporter 1 (SGLT1). It was demonstrated that the efficiency of such absorption was dramatically suppressed by efflux of at least some flavonoid glycosides by the apical transporter multidrug resistance-associated protein 2 (MRP2) [45]. It was also demonstrated that if the flavonoid glycosides could gain entry to the intestinal epithelial cells (enterocytes), which may include shedded cells, they might be hydrolyzed by a broad-specific β-glucosidase enzyme (BShG) [46, 47]. Of even greater interest was the finding that lactase phloridzin hydrolase (LPH), located in the brush border of the mammalian small intestine could perform this hydrolysis, at least for some flavonoid glycosides [48]. After the high vegetable diet the detectable values for hesperetin ranged between 40 and 2958 nmol/dm^3. The corresponding range for naringenin was 90 – 2120 nmol/dm^3 and for quercetin it was 7 – 214 nmol/dm^3 [49].

So far, in literature maximum number of flavonoid molecules in a metal complex never exceeded two. The stoichiometric composition of flavonoid complexes is typically determined by the spectroscopic and potenciometric methods; mass spectrometry also can be used. The most reliable results were obtained by the spectrophotometric method of continual variation of equimolar solutions [50], method of continual variation of non-equimolar solutions is not recommended for metal–flavonoids complexes because of low reproducibility. It was found that the Bent–French method [51] is not sufficiently accurate for complexes with a 1:2 stoichiometric ratio, but is acceptable for complexes with a 1:1 composition. Complexes with the highest coordination number are typically formed in slightly acidic or neutral pH, rarely in basic media. The optimal pH for complex formation, although strongly dependent on the features of the metal ion, is around pH 6. Complex formation at pH values lower than 3.0 is difficult because the flavonoids are predominantly present in their undissociated form. Although high pH values favor deprotonation of flavonoids and, consequently, more complex species, at higher pH values metal ions are often involved in side reaction (hydrolysis) and hydroxo-complexes are formed.

Complexation with flavonoids as unidentate or bidentate ligands leads to the formation of complexes that contain protons in addition to the metal ion and ligand (so-called protonated complexes) which tend to dissociate at higher pH values. Thus, the bathochromic shift that can be observed in the absorption spectra of metal–flavonoid complexes at higher pH values can be attributed to the dissociation of the protonated complexes, rather than the formation of complexes with different stoichiometric compositions. In Table 1 are given stability constants of some metal-flavonoid complexes.

Table 1. Stability constants of metal-flavonoid complexes

Flavonoid	Metal ion	Metal-flavonoid ratio	Stability constants logβ	Ref
Rutin	Cu(II)	1:2	10.76	52
	Zn(II)	1:1	4.68	53
	Pb(II)	1:2	13.81	54
	Ni(II)	1:2	8.95	55
	Co(II)	1:1	6.04	56
	MoO_4^{2-}	1:1	8.01	57
	WO_4^{2-}	1:2	13.44	58
	Eu(III)	1:2	10.59	59
	UO_2(II)	1:1	6.57	60
	Pd(II)	1:2	10.15	61
	$TiO(C_2O_4)_2^{2-}$	1:2	10.80	62
Quercetin	Ni(II)	1:1	5.57	63
	Co(II)	1:1	4.87	64
	Pd(II)	1:1	6.05	65
	$TiO(C_2O_4)_2^{2-}$	1:2	11.84	66,67
Morin	Cu(II)	1:2	4.94	68
	Zn(II)	1:2	6.74	69
	WO_4^{2-}	1:2	11.6	70
	Pd(II)	1:1	4.55	71
	$TiO(C_2O_4)_2^{2-}$	1:2	7.35	72
	Ba(II)	1:1	4.55	73
3-Hydroxyflavone	Zn(II)	1:1	8.51	74
	Pd(II)	1:1	7.74	75
	Ni(II)	1:1	7.63	76
	Co(II)	1:1	10.87	77
	MoO_4^{2-}	1:2	15.13	78
	WO_4^{2-}	1:2	16.45	79
	Eu(II)	1:2	13.47	80
	UO_2(II)	1:1	8.68	81
	$TiO(C_2O_4)_2^{2-}$	1:2	16.65	82
	Mn(II)	1:1	5.43	83
Hesperidin	Cu(II)	1:2	5.78	84
	UO_2(II)	1:2	7.00	85
	Al(III)	1:1	4.54	86

Complex Formation and Speciation of Kaempferol

Chemicaly kaempferol belong to the class of weak acids due to ionization of its phenolic groups. Dissociation of kaempferol is schematically depicted in figure 2.

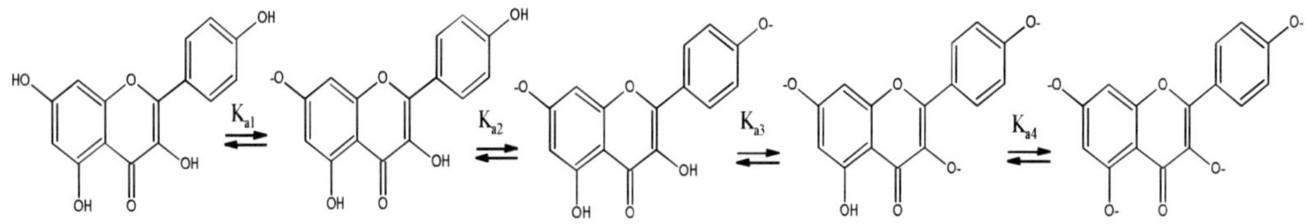

Figure 2. Stepwise dissociation of kaempferol.

From literature [87] it can be seen that in kaempferol most highly acidic properties are possessed by hydroxyl group in position 7. The 4'-OH group will undergo dissociation next. It is possible that the two constants in kaempferol are due to the 3-OH and 5-OH groups and their values of ionization constants are given in Table 2.

Table 2. Ionization constant (pKa ± Δσ) of kaempferol [87]

Compound	pK_{a1}	pK_{a2}	pK_{a3}	pK_{a4}
	7-OH	4'-OH	3-OH	5-OH
Kaempferol	8.2 ± 0.2	9.5 ± 0.1	10.5	12.5

Since, the capacity for ionization of the 3- and 5-hydroxy groups is considerable and changes differently under the influence of substituents in position 7 of 4', it cannot definitely affirm the assignment of the two latter constants that have made. Acidic properties of kaempferol are determined by its hydroxyl groups present in different position in the molecule and their capacity for ionization. In the second place, the capacity for ionization of each hydroxyl group is not constant in all types of flavonoid compound but can vary considerably according to the nature of the substitution of the flavones skeleton.

The distribution diagram of various kaempferol species in aqueous solution is presented in Figure 3. As could been seen from the Figure 3 kaempferol exists as neutral acid species at pH values lower than 8.0 (tetraprotic acid), and as protonated ion: triprotonated between pH 8 and 9.5, diprotonated ion between pH 9.5 and 10.5 and monoprotonated ion between pH 10.5 and 12.5 and as fully deprotonated specie at pH values higher than 13. All its forms are capable to bind metal ions by displacing proton form OH^- group(s) and/or by shearing electron pairs with negative oxygen donors. Thus variety of complexes in various state of protonation may be formed in solution. Most effective forms for metal binding are those with highest negative charge, but under physiological conditions the tetra and triprotic acid forms dominate. These kaempferol species should be good chelators for metal ions which have high affinity for negatively charge oxygen donor and which could easily displace hydrogen from OH- group. Such metal ions are usually classified as hard acids according to Pearson [88]. To this group belong - $Fe^{3+,}$ Al^{3+}, Mg^{2+}, Ca^{2+}. On the other hand essential metal ions Cu^{2+} and Zn^{2+}, are at the border line. There affinity to oxygen is less pronounced than for the hard metal acids so one may expect that their complexes with kaempferol would have less stability than those with iron and aluminium.

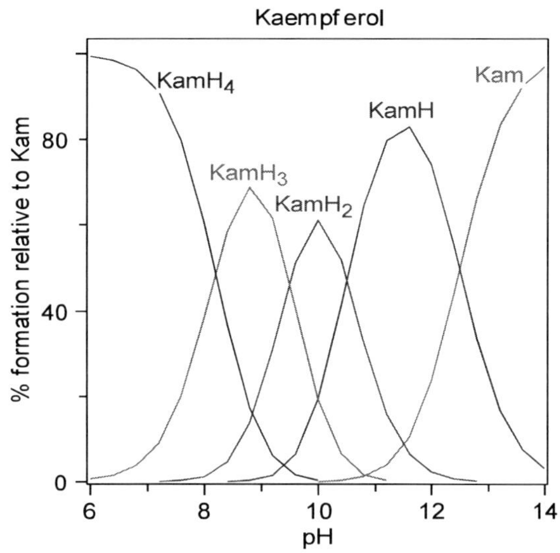

Figure 3. The distribution diagram of the different species as a function of pH for kaempferol

So far in the literature a very limited number of complexes of kaempferol with metals could be found. Its binding constants have been determined for serum albumin and iron and its binding to aluminium have been studied. The stability constants of known kaempferol complexes are presented in Table 3.

Table 3. Stability constants of kaempferol complexes

Complex	logβ	Ref
Kaempferol-HAS	6.79	89
Kaempferol-Fe	2.56	90

To the purpose of the modeling kaempferol speciation in blood plasma its binding to essential metal ion copper(II) must be known. In this paper we have studied the complexation between copper(II) and kaempferol by UV/Vis spectroscopy using continuous variation method. In **Figure 4** the UV/Vis spectra of water solution of kaempferol ($C_{kampferol} = 4\times10^{-5}$ mol/dm^3) and copper(II) ($C_{Cu} = 4\times10^{-5}$ mol/dm^3) + kaempferol ($C_{kaempferol} = 4\times10^{-5}$ mol/dm^3) in water have shown. The solutions were buffered with HEPES (2-[4-(2-hydroxyethyl)piperazin-1-yl]ethanesulfonic acid) (C = 0.05 mol/dm^3, pH = 7.4). Kaempferol gives simple pattern in spectral region 300 – 450 nm with a single asymmetric band with maximum absorption at 365 nm, while in the presence of copper this band splits in to two bands one with hypsochromic and one with bathochromic shift. Bathochromic band (**λ_{max}** = 413 nm) is characteristic for complex and was used to determine the identity and stability of the copper kamepferol complex. Series of solutions have been prepared and their absorbance was read at 413 nm.

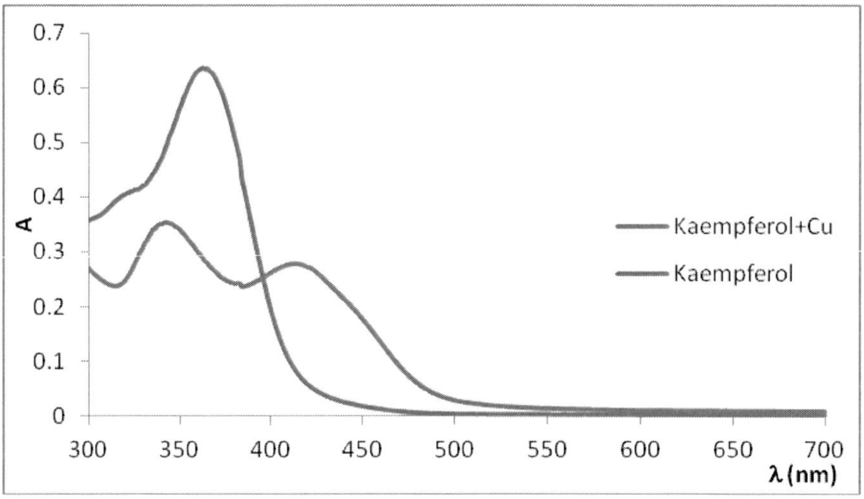

Figure 4. UV/Vis spectra of kaempferol and kaempferol copper complex.

By applying continuous variation method and mole ratio method to spectral data it was concluded that the complex between kaempferol and copper(II) ion possesses stiochometry 1 : 1. (Figure 5.)

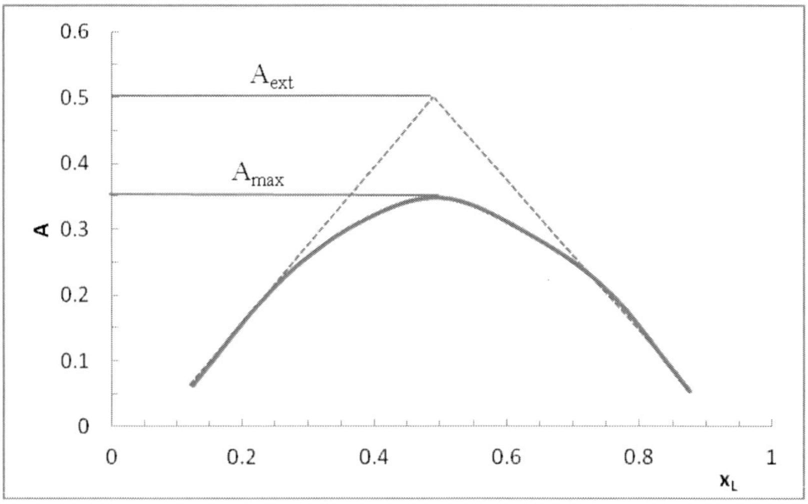

Figure 5. Job's curve for kaempferol copper complex.

The stability of the complex was calculated through the formula:

$$\beta_1 = \frac{[ML]}{[M][L]} = \frac{\frac{A_{max}}{A_{ext}} C_x}{\left[C_M - \frac{A_{max}}{A_{ext}} C_x\right]\left[C_L - \frac{A_{max}}{A_{ext}} C_x\right]}$$

where A_{ext} stands for the extrapolated absorbance, A_{max} for the actual absorbance, C_M concentration of metal ion, C_L concentration of ligand and C_x the concentration at maximum absorbance.

The stability constant of the complex was found to be $\log\beta = 5.15 \pm 0.04$.

The above presented data were used in computer simulation of kaempferol blood plasma speciation. Computer speciation was performed using the database of the blood composition and computer program HySS. The model have included 8 metal ions (Ca^{2+}, Mg^{2+}, Zn^{2+}, Pb^{2+}, Mn^{2+}, Fe^{2+}, Fe^{3+}, Cu^{2+}) and 43 ligands (26 amino acids, 7 inorganic ligands, 8 carboxyl acids and 2 proteins) with total number of 5941 complexes. To the existing data base the kaempferol species and complexes were added. The stability constants of kaempferol with Zn^{2+} and Ca^{2+} have been estimated using LFER method.

The result of calculation indicate that concentration of HSA-kaempferol is 99.9% , Fe-Kam is 0.0001% and Cu-Kam is 0.0109% . Bearing in mind small plasma concentration of kaempferol (7×10^{-8} mol/dm^3) it is unlikely that its binding to Fe^{3+} can change the biodistribution of Fe^{3+}.

May and Williams [91] suggested that the efficacy of a chelating agent for mobilizing a metal ion from the labile metal protein complex in blood plasma could be expressed in terms of a PLASMA MOBILIZING INDEX (PMI). Formulated in this way PMI provides a good measure of the relative thermodynamic ability of any chelating agent to compete for a given metal ion in blood plasma, provided only that the metal ion concentration in the low molecular weight fraction remains small compared with that bound to plasma proteins such as albumin and transferin.

The PMI factor is useful particularly because it is independent of the exact extent of metal protein binding. The PMI factor does not take into account any kinetic contributions to reach equilibrium. PMI is defined as:

$$PMI = \frac{[LMW\ complex]\ in\ chelator\ or\ drug}{[LMW\ complexes]\ in\ normal\ blood\ plasma}$$

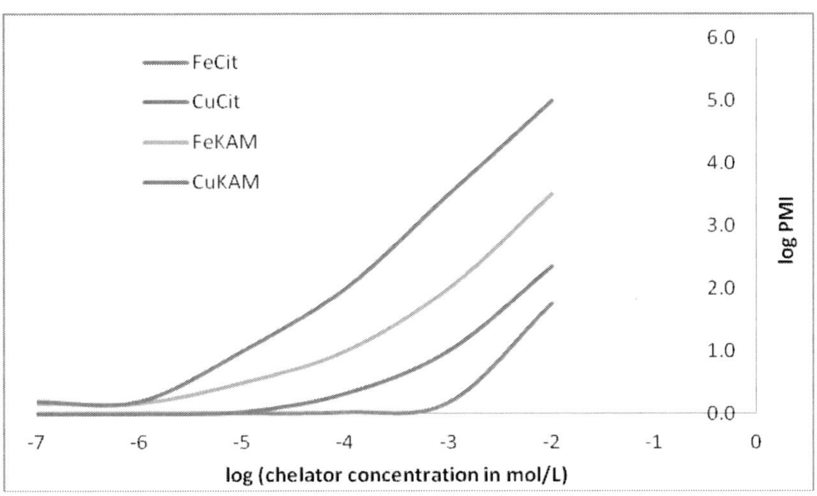

Figure 6. PMI curves for Kaempferol PMI vs. log chelator concentration.

In Figure 6. a PMI graph for kaempferol complexation with Fe^{3+} and Cu^{2+} ions is presented. It indicates that citrate possesses lower capacity of Fe^{3+} and Cu^{2+} mobilization than kaempferol. On the other hand kaempferol cannot change the biodistribution of Fe^{3+} and Cu^{2+} in plasma.

References

[1] Harborne J.B., Do natural plant phenols play a role in ecology?, *Acta Hort.*, 1994, 381, 36 – 43.

[2] Kamiya S., Esaki S., Konishi G., Flavonoids in Citrus hybrids., *Agric. Biol. Chem.* 1979, 43, 1529 – 1536.

[3] Rouseff R.L., Flavonoids and citrus quality, *ACS Symp. Ser.*, 1980, 143, 83 – 863.

[4] Forkmann G., *The 16th International Conference of Groupe Polyphenols*, Lisbon, 1992, 16, 19 – 27.

[5] Herrmann K., *Review on nonessential constituents of vegetables*. III. Carrots, celery, parsnips, beets, spinach, lettuce, endives, chicory, rhubarb, and artichokes. Z. Lebtwsm. Unters. Forsch., 1978, 167, 262 – 273.

[6] Cody V., Middleton E,, Harborne J.B., Beretz A., *Plant Flavonoids in Biology and Medicine II: Biochemical, Cellular and Medicinal Properties*, Alan R. Liss, New York, 1988, 107 – 1214.

[7] Ortuno A., Garcia-Puig D., Fuster M.D., Perez M.L., Sabater F., Porras I., Garcia-Lidon A., Del Rio J.A., Flavanone and Nootkatone Levels in Different Varieties of Grapefruit and Pummelo, *J. Agric. Food Chem.,* 1995, 43,1 – 5.

[8] Sato M., Ramarathnam N., Suzuki Y., Ohkubo T., Takeuchi M., Ochi H., Varietal differences in the phenolic content and superoxide radical scavenging potential of wines from different sources *J. Agric. Food Chem.*, 1996, 44, 37 – 41.

[9] Cook N.C., Samman S., Flavonoids chemistry, metabolism, cardioprotective effects, and dietary sources. Nutr. Biochem. 1996 7: 66 – 76.

[10] Middleton E., *Chem. Abstr.*, 1976, 84, 426 – 435.

[11] Liebovitz B, Mueller J. Bioflavonoids and polyphenols: medical applications. *J Opt. Nutr.*, 1993, 2, 17 – 35.

[12] Dakora F.D., Plant flavonoids: biological molecules for useful exploitation., Aust. *J. Plant. Physiol.*, 1995, 22, 87 – 99.

[13] Raghavan K., Buolamwini J.K., Fesen M.R., Pommier Y., Kohn K.W., Weinstein J.N., Three-dimensional quantitative structure-activity relationship (QSAR) of HIV integrase inhibitors: a comparative molecular field analysis (CoMFA) study. *J Med Chem.* 1995, 38, 890 – 897.

[14] Das A, Wang J.H., Lien E.J., Carcinogenicity, mutagenicity and cancer preventing activities of flavonoids: a structure-system-activity relationship (SSAR) analysis. *Prog Drug Res.*, 1994, 42, 133 – 166.

[15] Thompson L.U., Potential health benefits and problems associated with antinutrients in foods, *Food Res. Int.,* 1993, 26, 131 – 149.

[16] Read M.A., Flavonoids: Naturally occurring anti-inflammatory agents. *Am. J. Pathol.* 1995,147, 235 – 237.

[17] Ficarra R., Ficarra P., Tommasini S., Calabro M.L., Ragusa S., Barbera R., Rapisarda A., *Leaf extracts of some Cordia species: analgesic and anti-inflammatory activities as well as their chromatographic analysis, Farmaco.*, 1995, 50, 245 – 256.
[18] Williams D.R., *The Metals of Life*, Van Nostrand Reinhold, London, 1971
[19] Williams D.R., Charles C. Thomas, *An Introduction to Bioinorganic Chemistry,* Springfield, Illinois, 1976.
[20] Eichhorn G., *Inorganic Biochemistry*, Elsevier, Amsterdam, 1973.
[21] Fraga C. G., Relevance, essentiality and toxicity of trace elements in human health. *Molec. Asp. Med.*, 2005, 26, 235 – 244.
[22] Perin D.D., Agarwal R.P., *Metal Ions in Biological Systems*, H. Sigel, Marcel Dekker, New York, 1973.
[23] Pietta P.G., Flavonoids as antioxidants. *J. Nat. Prod.*, 2000, 63 1035 – 1042.
[24] Fenton H.J.H., Oxidation of tartaric acid in the presence of iron. *J. Chem. Soc. Trans.,* 1894, 65, 889 – 910.
[25] Kostyuk V.A., Potapovich A.I., Vladykovskaya E.N., Korkina L.G., Afanas'ev I.B. Influence of metal ions on flavonoid protection against asbestos-induced cell injury. *Arch Biochem Biophys.*, 2001, 385, 129 – 137.
[26] Moridani M.Y., Pourahmad J., Bui H., Siraki A., O'Brien P.J. Dietary flavonoid iron complexes as cytoprotective superoxideradical scavengers. *Free Radic. Biol. Med.*, 2003, 34, 243 – 253.
[27] De Souza R.F., De Giovani W.F.Antioxidant properties of complexes of flavonoids with metal ions. *Redox. Rep.,* 2004, 9, 97 – 104.
[28] Anafas'ev I.B., Ostrakhovitch E.A., Mikhal E.V., Ibragimova G.A., Korkina L.G., Enhancement of antioxidant and anti-inflammatory activities of bioflavonoid rutin by complexation with transition metals. *Biochem. Pharmacol.*, 2001, 61, 677 – 684.
[29] Tang H., Wang X., Yang S., Wang L. Synthesis, characterization, and biological activities of Pt(II) and Pd(II) complexes with 2′,3′,4′,5,7-pentahydroxy flavone. *Rare Metals,* 2004, 23, 38 – 42.
[30] Kopacz M., Woznicka E., Gruszecka J. Antibacterial activity of morin and its complexes La(III), Gd(III) and Lu(III) ions. *Acta Pol. Pharm.*, 2005, 62, 65 – 67.
[31] Deng Z., Coudray C., Gouzoux L., Mazur A., Rayssiguier Y., Pépin D. Effects of acute and chronic coingestion of AlCl3 with citrate or polyphenolic acids on tissue retention and distribution of aluminum in rats. *Biol. Trace. Elem. Res.,* 2000, 76, 245 – 256.
[32] Wang S.X., Zhang F.J., Feng Q.P., Li Y.L., Synthesis, characterization, and antibacterial activity of transition metal complexes with 5-hydroxy-7,4'-dimethoxyflavone. *J. Inorg. Biochem.*, 1992, 46, 251- 257 .
[33] Yuldashev Kh R., Makhkamov Kh. M., Sharipov Kh. T., Aliev Kh. U. Synthesis and study by IR and UV methods of spectral analysis of a complex of Mo(VI) with quercetin. *Chem. Nat. Comp.*, 1999, 35, 420-421.
[34] Torreggiani A., Tamba M., Trinchero A., Bonora S. Copper(II)–Quercetin complexes in aqueous solutions: spectroscopic and kinetic properties. *J. Mol. Struct.*, 744–747 (2005) 759 – 766.
[35] May P.M., Linder P.W., Williams D.R. Computer Simulation of Metal-ion Equilibria in Biofluids: Models for the Low-molecular-weight Complex Distribution of Calcium(II), Magnesium(II), Manganese(II), Iron(III), Copper(II), Zinc(II), and Lead(II) Ions in Human Blood Plasma. *J. Chem. Soc. Dalton Trans.* 1977, 588 – 595.

[36] Liang H., Ouyang D., Hu X.Y., Tai J.Z., He J.T., Zhou Y.Q., Structural studies on metal serum albumin - III. Slow conformational transition of HSA and BSA induced by Ni^{2+} ion. *Acta Chim. Sinica.*, 1998, 56, 662 – 667.

[37] Liang H., Huang J., Tu C.Q., Zhang M., Zhou Y.Q., Shen P.W., The subsequent effect of interaction between Co^{2+} and human serum albumin or bovine serum albumin. *J. Inorg. Biochem.*, 2001, 85, 167 – 171.

[38] Xiao J., Cao H., Wang Y., Zhao J., Wei X., Glycosylation of dietary flavonoids decreases the affinities for plasma protein. *J. Agric. Food Chem.*, 2009, 57, 6642 – 6648.

[39] Harris W.R., Keen C.K., *Calculations of the Distribution of Zinc in a Computer Model of Human Serum, Nutrition,* 1989, 119, 1677 – 1682.

[40] Yang F., Wang J., Liu C., Xu X., Li W., Xia Z., Xiao J., Pb^{2+}, Cu^{2+}, Zn^{2+}, Mg^{2+} and Mn^{2+} reduce the affinities of flavones, genistein and kaempferol for human serum albumin in vitro. Arch. Biol. Sci. 2011, 63, 623 – 634.

[41] Perin P.P., Agarwal R.P., *Metal ions in Biological systems.*, H. Sigel, Marcel Dekker, New York, 1973, 2, 168.

[42] May P.M., Linder P.W., Williams D.R., Ambivalent effect of protein binding on computed distributions of metal ions complexed by ligands in blood plasma., *Experientia*, 1976, 32, 1492 – 1494.

[43] Alderighi L., Gans P., Ienco A., Peters D., Sabatini A, Vacca A. Hyperquad simulation and speciation (HySS): a utility program for the investigation of equilibria involving soluble and partially soluble species. *Coord. Chem. Rev.* 1999, 184, 311 – 318.

[44] Griffiths L.A. Mammalian metabolism of flavonoids. In: Harborne J, Mabry T, eds. *The Flavonoids; Advances in Research*. London: Chapman and Hall, 1982, 68, 1 – 71.

[45] Hollman P.C.H., de Vries J.H.M., van Leeuwen S.D., Mengelers M.J.B., Katan M.B. Absorption of dietary quercetin glycosides and quercetin in healthy ileostomy volunteers. *Am. J. Clin. Nutr.*, 1995, 62, 1276 – 1282.

[46] Walgren R.A., Karnaky K.J., Lindenmayer G.E., Walle T. Efflux of dietary flavonoid quercetin 4V-h-glucoside across human intestinal Caco-2 cell monolayers by apical multidrug resistanceassociated protein-2. *J. Pharmacol. Exp. Ther.*, 2000, 294, 830 – 836.

[47] Day A.J., DuPont M.S., Ridley S., Rhodes M., Rhodes M.J.C., Morgan M.R.A., Williamson G. Deglycosylation of flavonoid and isoflavonoid glycosides by human small intestine and liver h-glucosidase activity. *FEBS Lett.,* 1998, 436, 71 – 75.

[48] Ioku K., Pongpiriyadacha Y., Konishi Y., Takei Y., Nakatani N., Terao J., h-Glucosidase activity in the rat small intestine toward quercetin monoglucosides. *Biosci. Biotechnol. Biochem.*, 1998, 62, 1428 – 1431.

[49] Day A.J., Canada F.J., Diaz J.C., Kroon P.A., Mclauchlan R., Faulds C.B., Plumb G.W., Morgan M.R.A., Williamson G. Dietary flavonoid and isoflavone glycosides are hydrolysed by the lactase site of lactase phlorizin hydrolase. *FEBS Lett.,* 2000, 468, 166 – 170.

[50] Erlund I., Silaste M.L., Alfthan G., Rantala M., Kesaniemi Y.A., Aro A. Plasma concentrations of the flavonoids hesperetin, naringenin and quercetin in human subjects following their habitual diets, and diets high or low in fruit and vegetables. *Europ. J. Clin. Nutr.*, 2002, 56, 891 – 898.

[51] Job P. Recherches sur la formation de complexes mineraux en solution, et sur leur stabilite (in Paris). *Ann. Chim.* 1928, 9, 113 – 203.

[52] Bent H., French C. The Structure of Ferric Thiocyanate and its Dissociation in Aqueous Solution. *J. Am. Chem. Soc.* 1941, 63, 568 - 572.

[53] Radovic Z., Malešev D. Copper(II)-rutin complex compound. *Pharmazie*, 1984, 39, 870.

[54] Malešev D., Radovic Z., *Pharmazie*, 1985, 40, 133.

[55] Radovic Z., Malešev D., Spectrophotometric investigations of the complex of Lead^{2+} and rutin. *Mikrochim. Acta* 1985, 86, 247-252.

[56] Radovic Z., Malešev D., *Acta Polon. Pharm. XLIV*, 1987, 5, 433.

[57] Malešev D., Radovic Z., *Pharmazie*, 1987, 42, 57.

[58] Malešev D., Radovic Z., Jelikic–Stankov M., Bogavac M. Investigation of molybdate(II) rutin complex and the determination of the dissociation-constant of rutin in water ethanol mixture. *Anal. Lett.,* 1991, 24, 1159-2719.

[59] Malešev D., Radovic Z., Jelikic–Stankov M., *Pharmazie*, 1991, 46, 534.

[60] Malešev D., Radovic Z., Jelikic–Stankov M., Investigation of Europium (III)-Rutin Complex in Water-Ethanolic Solution. *Spectrosc. Lett.* 26 (1993)1985-1995.

[61] Kuntic V., Malešev D., Radovic Z., Kosani M., Mio U., Vukojevic V. Spectrophotometric investigation of uranil(II)-rutin complex in 70% ethanol. *J. Agric. Food Chem.*, 1998, 46, 5139-5142.

[62] Kuntic V., Kosanic M., Malešev D., Radovic Z., Spectrophotometric investigation of Pd(II)-rutin complexes and its application to rutin determination in tablets. *Pharmazie*, 1998, 53, 724-726.

[63] Kuntic V., Malešev D., Radovic Z., Vukojevic V. Spectrophotometric investigation of complexing reaction between rutin and titanyloxalate anion in 50% ethanol. *Monaths. Chem.*, 2000, 131, 769-778.

[64] Radovic Z., Malešev D. Investigation of the Complex of Ni(II) and Quercetin. *Arch. Pharm.*, 1987, 320, 188-189.

[65] Malešev D., Radovic Z., Cobalt-Quercetine-complexes. *Pharmazie*, 1987, 42, 59.

[66] Kuntic V., Blagojevic S., Malešev D., Radovic Z., Bogavac M. Spectrophotometric investigation of the Pd(II)-quercetin complex in 50% ethanol. *Monatsh. Chem.*, 1998, 129, 41-48.

[67] Pejic N., Kuntic V., Malešev D. Potassium titanyloxalate as analytical reagent for micro-quantitative determination of quercetin, *Pharmazie*, 2002, 57, 216-

[68] Kuntic V., Pejic N., Micic S., Vukojevic V., Vujic Z, Malešev D. Određivanje kvercetina u farmaceutskim preparatima preko reakcije sa kalijum-titaniloksalatom - određivanje konstanti stabilnosti kvercetin-titaniloksalatnog kompleksa. *J. Serb. Chem Soc.*, 2005, 70, 753-763.

[69] Radovic Z., Malešev D. Ispitivanje kompleksnog jedinjenja bakar(II)-morin. *Arh. Farm.* 1983, 33, 221-226.

[70] Malešev D., Radovic Z., Spektrofotometrijsko ispitivanje kompleksa cink(II)-morin u vodenom-alkoholnim rastvorima. *Bull. Soc. Chim.*, 1983, 48, 745-750.

[71] Malešev D., Radovic Z., Jelikic-Stankov M. Investigation of tungstate-morin complex in 50% ethanol. *J. Serb.Chem. Soc.,* 1993, 58, 557-562.

[72] Blagojevic S., Malešev D., Radovic Z., Spectrophotometric pH-metric investigation of Pd(II)-morin complex in 70% ethanolic solution. *Arh. Farm.*, 1999, 5-6, 387-396.

[73] Blagojevic S., Aleksic M., Malešev D., Radovic Z. Spektroskopsko ispitivanje kompleksa titanil oksalata-morin u 50% v/v etanolu. *Arh. Farm.*, 2003, 53, 17-26.
[74] Radovic Z., Malešev D., Uber barium-morin complex. *Pharmazie*, 1988, 43, 135-
[75] Radovic Z., Malešev D., *Arh. Farm.*, 1983, 4, 147.
[76] Malešev D., Radovic Z., Jelikic–Stankov M. Spectrophotometric and pH-metric inestigation of Pb(II)-3-hydroxyflavone complex in 70% ethanol. *Arh. Farm.*, 1991, 46, 223-227.
[77] Radovic Z., Malešev D. Study of the nickel(II)-3-hydroxyflavone complex in water-ethanol mixture. *Arh. Farm.* 42 (1992) 27-33.
[78] Radovic Z., Malešev D. Study of Cobalt(II)-3-hydroxyflavone complex. *Pharmazie*, 1989, 44, 496.
[79] Malešev D., Radovic Z., Jelikic–Stankov M. Investigation of the molybdate(II)-3-hydroxyflavone complex. *Monatsh. Chem.*, 1991, 122, 429-434.
[80] Malešev D., Radovic Z., Jelikic–Stankov M. Investigation of WO_4^{2-}-3-Hydroxyflavone Complex in Water - Ethanolic Solution. *Spectrosc. Lett.*, 1992, 25, 1089-
[81] Malešev D., Radovic Z., Jelikic–Stankov M., Ispitivanje kompleksnog jedinjenja Europijum(III)-3-Hidroksiflavon u 70% etanolu. *Arh. Farm.*, 1992, 42 163-167.
[82] Malešev D., Radovic Z., Jelikic–Stankov M. Investigation of the uranil(II)-3-Hydroxyflavone complex. *Monatsh. Chem.*, 1990, 121, 455-
[83] Aleksic M., Blagojevic S., Malešev D., Radovic Z. Spektrofotometrijsko ispitivanje kompleksa između titanil oksalata i 3-hidroksiflavona u vodeno-etanolnoj smeši. *J. Serb. Chem. Soc.*, 2000, 65, 631-638.
[84] Malešev D., Radovic Z., Jelikic–Stankov M. Investigation of manganese(II)-3-hydroxyflavone complex. *Pharmazie*, 1989, 44, 863.
[85] Kuntic V., Blagojevic S., Malešev D., Radovic Z. Spectrophotometric investigation of Cu(II)-hesperidin complex in 50% ethanol. *Pharmazie*, 1999, 54, 548-549.
[86] Kuntic V., Kosanic M., Malešev D., Radovic Z., Mio U. Spectrophotometric investigation of the uranyl(II)-hesperidin complex in 70% methanol. *J. Serb. Chem. Soc.*, 1998, 36, 565-572.
[87] Malešev D., Radovic Z., Kuntic V., Kosanic M. Spectrophotometric determination of hesperidin by Al(III)-hesperidin complex in water-methanol solution. *Anal. Lett.*, 1997, 30, 917-926.
[88] Tyukavkina N.A., Pogodaeva N.N., Ultraviolet absorption of flavonoids. VIII. Ionization constants of kaempferol and quercetin, *Chemistry of natural compound*, 1975, 11, 741 – 743.
[89] Pearson R.G. Hard and Soft Acids and Bases. J. Am. Chem. Soc. 1963, 85, 3533-3539.
[90] Fan Yang, Jie Wang, Chunxi Liu, Xiaochen Xu, Wenjun Li, Zhihua Xia, Jianbo Xiao, Pb^{2+}, Cu^{2+}, Zn^{2+}, Mg^{2+} and Mn^{2+} reduce the affinities of flavones, genistein and kaempferol for human serum albumin in vitro, *Arch. Biol. Sci., Belgrade*, 2011, 63, 623 – 634.
[91] Li Hua, Yu Yan-qiu, Yang Bin, *Study on reaction kinetic between flavonoid and Fe(II)*, Zhongguo Shiyan Fangjixue Zazhi, 2010, 16, 99 – 102.
[92] May P.M., Williams D.R., Computer simulation of chelation therapy Plasma mobilizing index as a replacement for effective stability constant., *FEBS Letters* 1977, 78, 134 – 138.

In: Kaempferol
Editors: Garion Villers and Yves Fougere
ISBN: 978-1-62618-515-9

Chapter 8

The Action of Kaempferol on Oral Cavity Cancer and Airway Secretion

Kyung-Su Kim*
Department of Otorhinolaryngology and Human Barrier Research Institute, Yonsei University College of Medicine, Seoul, Korea

Abstract

Ginkgo biloba extract has long been used as an anticancer drug and expectorant in Oriental medicine. Because *Ginkgo biloba* extract is a mixed compound comprising kaempferol and other components, it may be preferable to use only kaempferol for treatments due to possible adverse reactions with the other components. Furthermore, the action mechanism and potency of kaempferol compared to *Ginkgo biloba* extract have not been revealed. Therefore, this chapter gives a brief survey of the current knowledge on kaempferol and its potential use as an anti-oral cancer agent and also as an anti-hypersecretory agent.

In previous studies, *Ginkgo biloba* extract has been shown to induce apoptosis in oral cavity cancer cells. In the case of kaempferol, the results of cell proliferation assays on oral cavity cancer cell lines such as SCC-1483, SCC-25, and SCC-QLL1 showed that kaempferol significantly inhibited cellular proliferation. Flow cytometry showed that the anti-proliferative effect was due to increased apoptosis. Kaempferol showed cleavage of poly (ADP-ribose) polymerase (PARP). Therefore, kaempferol effectively induces caspase-3-dependent apoptosis of oral cavity cancer cells and can be considered a possible anti-oral cancer agent.

In addition to its anticancer effects, *Ginkgo biloba* extract suppresses IL-1β-induced *MUC5AC* gene expression in NCI-H292 cells, and therefore, kaempferol may be used as a suppressor of mucin production. The RT-PCR and real-time PCR results showed that kaempferol suppressed *MUC5AC* mRNA expression in a dose-dependent manner, with significant inhibition starting at 40 μM (equal concentration to about a 1/12or 1/13 dose of *Ginkgo biloba* extract).

* Corresponding author: Professor Kyung-Su Kim, Department of Otorhinolaryngology, Gangnam Severance Hospital, Yonsei University College of Medicine, 211 Eonju-ro, Gangnam-gu, 135-720 Seoul, Korea. Tel: +82 2 2019 3463, Fax: +82 2 3463 4750; E-mail address: ydrhinol@yuhs.ac.

Kaempferol suppressed phosphorylation of ERK and p38 kinases. Therefore, kaempferol may overcome the dose problem of *Ginkgo biloba* extract to play a clinically valuable role in controlling mucin hypersecretion in airway inflammation.

Introduction

The flavonoid kaempferol (3,5,7-trihydroxy-2-(4-hydroxyphenyl)-4H-1-benzopyran-4-one) has been commonly identified in edible plants and traditional medicine.

It is found in a wide variety of edible plants such as onion, kale, endive, tea, broccoli, cabbage, leak, strawberries, grapes, propolis and *Ginkgo biloba* extracts [1].

It has been associated with a reduced risk of developing cancer and cardiovascular diseases in some epidemiologic studies [2-6]. Kaempferol has been shown to have a wide range of pharmacological activities including cardio-protective, neuro-protective, anti-diabetic, anti-osteoporotic, analgesic, anti-allergic, antioxidant, antimicrobial, anti-inflammatory, and anticancer activities [1]. Its anticancer effect has been shown in hepatocellular carcinoma, breast cancer, lung cancer and prostate cancer [7]. However, its anticancer effect has not been shown in oral cavity cancer cells. Additionally, kaempferol has shown an anti-inflammatory effect in some cell types and clinical studies, but there have been no reports about its effect on suppressing mucin production [8, 9]. Therefore, in this chapter, the effects of kaempferol on oral cavity cancer and airway secretion are mentioned and discussed.

Kaempferol as a Chemopreventive Agent Against Oral Cavity Cancer

The idea to assess kaempferol's potential anticancer effects on oral cavity cancer was inspired by a study on the *Ginkgo biloba* extract, EGb 761. EGb 761 has been known to have neuro-protective, antioxidant, anti-inflammatory, and anticancer effects on various cells and tissues [10]. Among these effects, the anticancer effect on oral cavity cancer cells was recently revealed. EGb 761 may be considered a possible chemo-preventive agent against oral cavity cancer because it induced the apoptosis of SCC-1483 oral cavity cancer cells and activated caspase-3 [10]. The mitogen activated protein kinase (MAPK) pathway in cells treated with EGb 761 was investigated to better understand the signal transduction involved in EGb 761-mediated apoptosis. In SCC-1483, 250 μg/mL EGb 761 activated extra cellular signal-regulated kinase (ERK), and apoptosis was noted. In the study using a MAPK/ERK kinase 1 (MEK1)-dominant negative construct, phosphorylation of ERK and p90 ribosomal S 6 protein kinase (RSK1) were inhibited, and this inhibition led to the inhibition of apoptosis. In the study using an RSK1-dominant negative construct and a cAMP response element binding protein (CREB)-dominant negative construct, the inhibition of apoptosis was also noted. Therefore, the apoptosis of SCC-1483 cells by EGb 761 is linked to the activation of ERK and occurs via the ERK MAPK/RSK1/CREB signal transduction pathway [11]. Even though the induction of apoptosis by EGb 761 in oral cavity cancer cells is manifest, the limitation of its commercial use is the dose required (250 μg/mL). The acute toxicity of EGb

761 was reported to be 1.1-7.73 g/kg, and the chronic toxicity was not noted in a dose of 500 mg/kg/day in mice [12, 13]. However, the use of EGb 761 should be carefully considered because it can accumulate in the nervous system, glands, and eyes [13]. Exploration of the EGb 761 components included some studies using IPS 2000, ginkgolide B, quercetin, and kaempferol, but a study on the anticancer effect of these components on oral cavity cancer cells has not been conducted [8, 9].

EGb 761 is composed of about 250 compounds, and the most important ingredients are flavonoids (kaempferol, quercetin and isorhamnetin), constituting 24% of the total; terpenoids (ginkgolides and bilobalide), making up 6% of the total; and ginkgolic acid (less than 5 ppm) [14-16]. In our study, bilobalide, ginkgolides A and B, kaempferol, and quercetin were individually studied to elucidate their effects on oral cavity cancer cells.

The results showed that kaempferol and quercetin inhibited oral cavity cancer cell proliferation through apoptosis. Various oral cavity cancer cell lines such as SCC-1483, SCC-25, and SCC-QLL1 were used and showed effects of apoptosis and cleavage of poly (ADP-ribose) polymerase with the involvement of caspase-3 in response to treatment [17]. Therefore, kaempferol and quercetin can be considered anti-oral cavity cancer agents. Of note, 40 μM of kaempferol (11.5 μg/mL) showed the same effects as those elicited by EGb 761 at 250 μg/mL. Not only is the dose relatively low, but the use of kaempferol is safer and more effective in inducing apoptosis. Caspase-dependent apoptosis by kaempferol is also noted in prostate cancer cells (LNCaP) and breast cancer cells (MDA-MB-231), and this study corresponded with the above results [18].

According to the above experimental results, kaempferol can be used as an anticancer agent against oral cavity cancer. The possible anticancer activities can be explained by numerous activities of kaempferol. For instance, low concentrations of kaempferol may protect DNA from damage induced by carcinogens [19-22] and can induce apoptosis [18, 23-29]. Kaempferol is a potent antioxidant and can prevent the accumulation or reduce excessive cellular levels of reactive oxygen species [30-33]. It has an anti-inflammatory activity and may play an important role in cancer preventive activity [34]. Kaempferol may inhibit P450 and prevent the activation of carcinogenic agents [35, 36].

As to cancer prevention, the activation of RSK2 is known to be involved in carcinogenesis. Kaempferol was identified as the specific inhibitor of this kinase and suppresses breast cancer cells and prostate cancer cells [37, 38]. It has been found to induce the activation of caspase-3, 7, and 9 [24-26, 39, 40].

The data on oral cavity cancer cells are similar to this mechanism in the aspect of caspase-3. Several other mechanisms may participate in kaempferol-induced anticancer activity, such as the activation of apoptosis-inducing factor, the decrease of Bcl-2 and Bcl-X1, the increase of Bax and Bad, and the activation of PARP [25-27, 29, 41]. Even though several mechanisms including PARP activation and caspase-3 activation may be involved in the kaempferol-induced anticancer effect on oral cavity cancer cells, the exact mechanism has yet to be revealed.

The pharmacokinetics of kaempferol is a problem needing to be solved before its commercial use. Kaempferol is mainly absorbed in the small intestine by passive diffusion, facilitated diffusion or active transport [42]. Kaempferol can be metabolized in the small intestine and colon. After absorption, it is extensively metabolized in the liver, and kaempferol and some kaempferol glycosides enter into systemic circulation [7, 43-48]. The percentage of kaempferol excreted in urine has been found to be 1.9-2.5 % [43, 46].

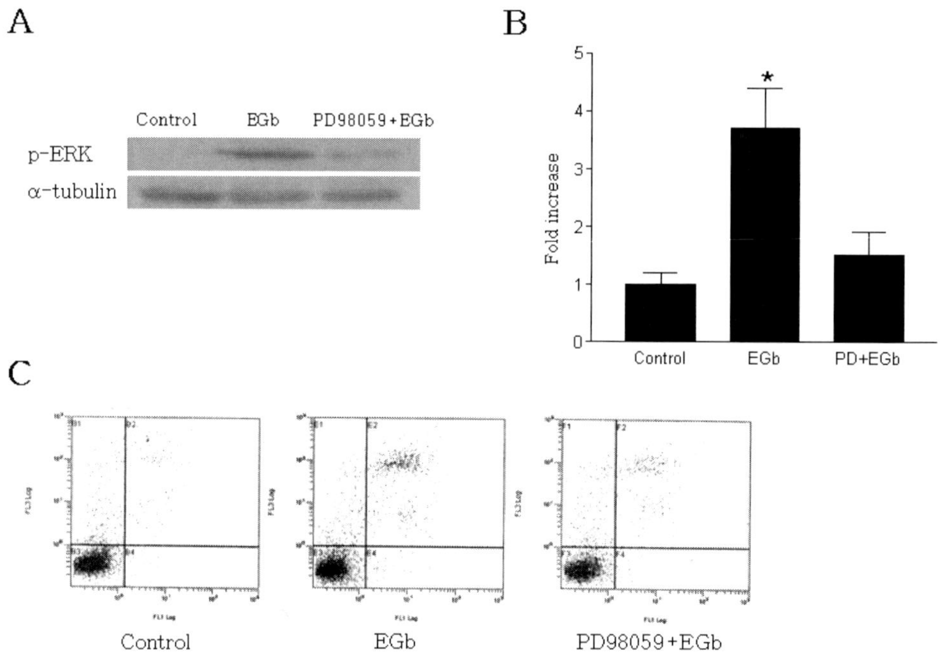

Figure 1. Inhibition of EGb 761-induced apoptosis by ERK inhibitor. SCC-1483 cells were separately treated with ethanol (control), 250 μg/mLEGb 761, or pre-treated with 10 μM PD98059 and 250 μg/mLEGb 761 for 24 hours. Western blot analysis (a) and flow cytometry (b, c) were performed. A. EGb 761 induces the activation of p-ERK, but this activation is inhibited by PD98059 (PD98059+EGb). B. EGb 761 induces apoptosis in SCC 1483 cells, but pre-treatment with PD98059 abrogated the effect of EGb 761 on the levels in the controls. C. Dot plots of apoptosis data. The figures shown are representative of three independent experiments. *p<0.05 versus the control. *(Permitted quotation from Korean J. Otolaryngol 2006;49:1183).*

The bioavailability of kaempferol is known to be 2%, indicating low to moderate absorption. In contrast, other polyphenols show 2-20% bioavailability [7]. After oral intake, kaempferol is detected in plasma at nanomolar concentrations. Ingestion of 246 mg kaempferol/kg endive led to a mean maximum plasma concentration of 100 nM 5.8 hours after ingestion [43]. The mean intake of kaempferol was 4.7 mg/day, and the mean plasma concentration was 10.7 nM [43].

The intake of 14.97 mg kaempferol/day led to a plasma concentration of 57.86 nM [49]. The intake of 27 mg kaempferol from tea led to a plasma concentration of 52 nM [46]. Thus, a plasma concentration 40 μM of kaempferol is considered a very high concentration. However, the hopeful experimental results exist. Even when the IC_{50} of kaempferol on human osteosarcoma was 150 μM, the oral administration of 25 or 50 mg/kg of kaempferol significantly reduced tumor weights and volume.

This indicates that tumor suppression observed in mice after the oral intake of kaempferol may be caused by its metabolites or by an unknown indirect effect [23]. Therefore, an *in vivo* study on the oral intake of kaempferol is necessary to solve this low bioavailability problem. After this is resolved, its commercial use can be made possible in the future.

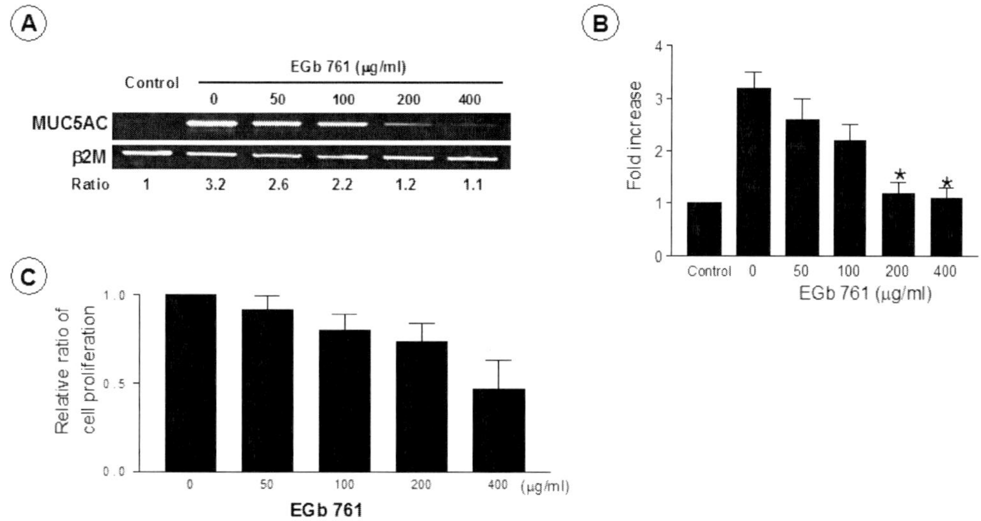

Figure 2. Suppression of *MUC5AC* gene expression by EGb 761. A. NCI-H292 cells were pretreated with 0, 50, 100, 200, and 400 μg/mLEGb 761 for 1 hour and then treated with 10 ng/ml IL-1β for 24 hours. RT-PCR for *MUC5AC* mRNA expression was performed on each group. Compared with the EGb 761-untreated cells, the expression of *MUC5AC* mRNA is significantly suppressed from the treatment with 200 μg/mLEGb 761. B. Fold-increase of *MUC5AC* mRNA treated with the various concentrations of EGb 761. C. The percentage of living cells through cell proliferation assay shows no differences compared to the EGb 761-untreated cells. The figures shown are representative of three independent experiments. *$p<0.05$ compared with EGb 761-untreated cells. *(Permitted quotation from J. Rhinol 2007;14(1):52).*

Kaempferol as an Anti-hypersecretory Agent in Airway Inflammation

The idea of studying the anti-hypersecretory action of kaempferol also came from a study on EGb 761. In addition to the anticancer effects of EGb 761, it has been shown to have a beneficial effect on pulmonary diseases. One of the components of EGb 761, ginkgolide, showed a clinical effect on asthma. The possible mechanisms for this action may be the inhibition of platelet-activating factor, its receptor, and T lymphocytes [13, 50].

Meanwhile, in inflammatory airway diseases such as rhinitis, sinusitis, and bronchitis, mucin-hypersecretion results in suppression of ciliary movement, induction of a bacterial infection and respiratory problems.

To date, 20 human mucin genes have been revealed, and among them, *MUC5AC* plays a major role in hypersecretory airway diseases [51-56]. Thus, EGb 761 was studied in this context to reveal whether it produces anti-hypersecretory effects on IL-1β-induced *MUC5AC* secretion.

According to the results in NCI-H292 cells, EGb 761 suppressed the IL-1β-induced *MUC5AC* gene and protein expression via the ERK and p38 MAPK pathways [57]. With these results, EGb 761 was verified as an anti-hypersecretory agent on *MUC5AC* secretion in human airway epithelial cells, but the potent dose started at 200 μg/mL. This dose is below the acute and chronic toxic doses shown in the study of EGb 761's anticancer effects.

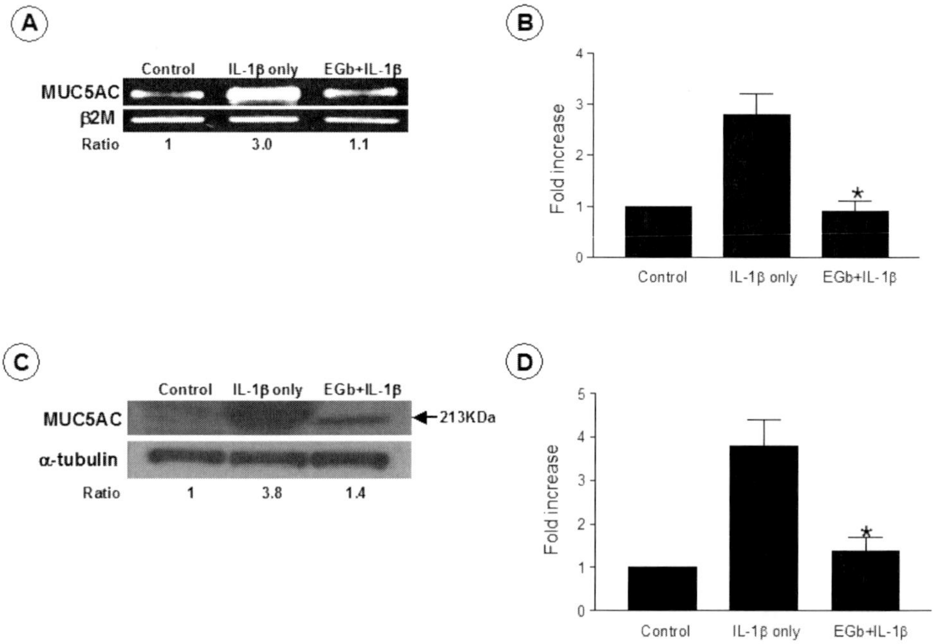

Figure 3. Suppression of *MUC5AC* gene and protein expression by EGb 761. A. NCI-H292 cells were pretreated with 200 μg/mLEGb 761 for 1 hour and treated with 10 ng/ml IL-1β (EGb+IL-1β) for 24 hours. The expression of *MUC5AC* mRNA exhibited a 3.0-fold increase with IL-1β treatment, but this increase was significantly inhibited to the level of the control group by pretreatment with EGb 761. B. Fold-increase of *MUC5AC* mRNA by the treatment with IL-1β only and EGb 761+IL-1β. C. Western blot analysis for MUC5AC protein expression was performed on each group, and α-tubulin was employed as the control. The expression of MUC5AC protein (213 kDa) induced by IL-1β is suppressed by pretreatment with EGb 761. D. Fold-increase of MUC5AC protein by treatment with IL-1β only and EGb 761+IL-1β. The figures shown are representative of three independent experiments. *$p<0.05$ compared with IL-1β-treated cells. *(Permitted quotation from J. Rhinol 2007;14(1):52).*

As in the anticancer findings, the use of individual components of EGb 761 was recommended for safety and effectiveness. Among the components of EGb 761 we studied were kaempferol, quercetin, ginkgolide A and B and bilobalide. Suppression of IL-1β-induced *MUC5AC* gene expression and the ERK and p38 MAPK pathways were related to treatment with 40 μMkaempferol and quercetin [58].

Similar to the anticancer findings, 40 μM of kaempferol (11.5 μg/mL) showed the same effects as 200 μg/mL of EGb 761. This dose is sufficiently low, and the use of kaempferol is safer and more effective in suppressing *MUC5AC* gene expression than EGb 761. Therefore, kaempferol can be regarded as an anti-hypersecretory agent against respiratory hypersecretory diseases. The effect of kaempferol on ciliary movement of the respiratory mucosa is the next problem to be solved for its clinical use. However, data on this effect were not found in the literature review.

In an experiment using human nasal mucosa, 40 μM of quercetin, did not inhibit the ciliary movement for 24 hours of treatment. Thus, quercetin may be safely used in clinical trials, but kaempferol should first be tested [59].

Conclusion

Through this text, it is emphasized that kaempferol effectively induces apoptosis of oral cavity cancer cells and can be considered as a possible anticancer agent against oral cancer. Kaempferol may also be regarded as an anti-hypersecretory agent against respiratory hypersecretory diseases. In vivo study and clinical trials are necessary next steps to exploring the use of kaempferol as a pharmaceutical agent.

References

[1] Calderon-Montano JM, Burgos-Moron E, Perez-Guerrero C, Lopez-Lazaro M. A review on the dietary flavonoid kaempferol. *Mini reviews in medicinal chemistry.* 2011;11(4):298-344.

[2] Garcia-Closas R, Gonzalez CA, Agudo A, Riboli E. Intake of specific carotenoids and flavonoids and the risk of gastric cancer in Spain. Cancer causes and control: *CCC.* 1999;10(1):71-5.

[3] Hertog MG, Feskens EJ, Hollman PC, Katan MB, Kromhout D. Dietary antioxidant flavonoids and risk of coronary heart disease: the Zutphen Elderly Study. *Lancet.* 1993;342(8878):1007-11.

[4] Le Marchand L. Cancer preventive effects of flavonoids--a review. Biomedicine and pharmacotherapy = Biomedecine and pharmacotherapie. 2002;56(6):296-301.

[5] Maron DJ. Flavonoids for reduction of atherosclerotic risk. Current atherosclerosis reports. 2004;6(1):73-8.

[6] Neuhouser ML. Dietary flavonoids and cancer risk: evidence from human population studies. *Nutrition and cancer. 2004*;50(1):1-7.

[7] Barve A, Chen C, Hebbar V, Desiderio J, Saw CL, Kong AN. Metabolism, oral bioavailability and pharmacokinetics of chemopreventive kaempferol in rats. *Biopharmaceutics and drug disposition*. 2009;30(7):356-65.

[8] Mutoh M, Takahashi M, Fukuda K, Matsushima-Hibiya Y, Mutoh H, Sugimura T, et al. Suppression of cyclooxygenase-2 promoter-dependent transcriptional activity in colon cancer cells by chemopreventive agents with a resorcin-type structure. *Carcinogenesis.* 2000;21(5):959-63.

[9] Papadopoulos V, Kapsis A, Li H, Amri H, Hardwick M, Culty M, et al. Drug-induced inhibition of the peripheral-type benzodiazepine receptor expression and cell proliferation in human breast cancer cells. *Anticancer research.* 2000;20(5A):2835-47.

[10] Kim KS, Rhee KH, Yoon JH, Lee JG, Lee JH, Yoo JB. Ginkgo biloba extract (EGb 761) induces apoptosis by the activation of caspase-3 in oral cavity cancer cells. *Oral oncology*. 2005;41(4):383-9.

[11] Chang JH, Yoon J-H, Choi EC, Lee KW, Cho CI, Kim K-S. Mechanism of Apoptosis induced by *Ginkgo Biloba* Extract (Egb 761) in Oral Cavity Cnacer Cell Lines. *Korean J. Otolaryngol.* 2006;49:1181-7.

[12] Drieu K. *Rokan*, *Ginkgo biloba*: Recent results in pharmacology and clinic. Funfgeld E, editor. Berlin: Springer-Verlag; 1988. 32-6 p.

[13] DeFeudis F. *Ginkgo biloba extract* (Egb 761): Pharmacological activities and clinical applications. Paris: *Elsevier*; 1991. 143-6 p.
[14] DeFeudis FV, Papadopoulos V, Drieu K. Ginkgo biloba extracts and cancer: a research area in its infancy. *Fundamental and clinical pharmacology*. 2003;17(4):405-17.
[15] Kleijnen J, Knipschild P. Ginkgo biloba. *Lancet*. 1992;340(8828):1136-9.
[16] Jacobs BP, Browner WS. Ginkgo biloba: a living fossil. *The American journal of medicine*. 2000;108(4):341-2.
[17] Kang JW, Kim JH, Song K, Kim SH, Yoon JH, Kim KS. Kaempferol and quercetin, components of Ginkgo biloba extract (EGb 761), induce caspase-3-dependent apoptosis in oral cavity cancer cells. Phytotherapy research: *PTR*. 2010;24 Suppl 1:S77-82.
[18] Brusselmans K, Vrolix R, Verhoeven G, Swinnen JV. Induction of cancer cell apoptosis by flavonoids is associated with their ability to inhibit fatty acid synthase activity. *The Journal of biological chemistry*. 2005;280(7):5636-45.
[19] Cemeli E, Schmid TE, Anderson D. Modulation by flavonoids of DNA damage induced by estrogen-like compounds. *Environmental and molecular mutagenesis*. 2004;44(5):420-6.
[20] Francis AR, Shetty TK, Bhattacharya RK. Modulating effect of plant flavonoids on the mutagenicity of N-methyl-N'-nitro-N-nitrosoguanidine. *Carcinogenesis*. 1989;10(10):1953-5.
[21] Francis AR, Shetty TK, Bhattacharya RK. Modifying role of dietary factors on the mutagenicity of aflatoxin B1: in vitro effect of plant flavonoids. *Mutation research*. 1989;222(4):393-401.
[22] Anderson D, Basaran N, Dobrzynska MM, Basaran AA, Yu TW. Modulating effects of flavonoids on food mutagens in human blood and sperm samples in the comet assay. *Teratogenesis, carcinogenesis, and mutagenesis*. 1997;17(2):45-58.
[23] Huang WW, Chiu YJ, Fan MJ, Lu HF, Yeh HF, Li KH, et al. Kaempferol induced apoptosis via endoplasmic reticulum stress and mitochondria-dependent pathway in human osteosarcoma U-2 OS cells. *Molecular nutrition and food research*. 2010;54(11):1585-95.
[24] Kang GY, Lee ER, Kim JH, Jung JW, Lim J, Kim SK, et al. Downregulation of PLK-1 expression in kaempferol-induced apoptosis of MCF-7 cells. *European journal of pharmacology*. 2009;611(1-3):17-21.
[25] Marfe G, Tafani M, Indelicato M, Sinibaldi-Salimei P, Reali V, Pucci B, et al. Kaempferol induces apoptosis in two different cell lines via Akt inactivation, Bax and SIRT3 activation, and mitochondrial dysfunction. *Journal of cellular biochemistry*. 2009;106(4):643-50.
[26] Nguyen TT, Tran E, Ong CK, Lee SK, Do PT, Huynh TT, et al. Kaempferol-induced growth inhibition and apoptosis in A549 lung cancer cells is mediated by activation of MEK-MAPK. *Journal of cellular physiology*. 2003;197(1):110-21.
[27] Sharma V, Joseph C, Ghosh S, Agarwal A, Mishra MK, Sen E. Kaempferol induces apoptosis in glioblastoma cells through oxidative stress. *Molecular cancer therapeutics*. 2007;6(9):2544-53.
[28] Zhang Y, Chen AY, Li M, Chen C, Yao Q. Ginkgo biloba extract kaempferol inhibits cell proliferation and induces apoptosis in pancreatic cancer cells. *The Journal of surgical research*. 2008;148(1):17-23.

[29] Kim BW, Lee ER, Min HM, Jeong HS, Ahn JY, Kim JH, et al. Sustained ERK activation is involved in the kaempferol-induced apoptosis of breast cancer cells and is more evident under 3-D culture condition. *Cancer biology and therapy.* 2008;7(7):1080-9.
[30] Szatrowski TP, Nathan CF. Production of large amounts of hydrogen peroxide by human tumor cells. *Cancer research.* 1991;51(3):794-8.
[31] Suh YA, Arnold RS, Lassegue B, Shi J, Xu X, Sorescu D, et al. Cell transformation by the superoxide-generating oxidase Mox1. *Nature.* 1999;401(6748):79-82.
[32] Arnold RS, Shi J, Murad E, Whalen AM, Sun CQ, Polavarapu R, et al. Hydrogen peroxide mediates the cell growth and transformation caused by the mitogenic oxidase Nox1. *Proceedings of the National Academy of Sciences of the United States of America.* 2001;98(10):5550-5.
[33] Lopez-Lazaro M. A new view of carcinogenesis and an alternative approach to cancer therapy. *Mol. Med.* 2010;16(3-4):144-53.
[34] Mantovani A, Allavena P, Sica A, Balkwill F. Cancer-related inflammation. *Nature.* 2008;454(7203):436-44.
[35] Chaudhary A, Willett KL. Inhibition of human cytochrome CYP 1 enzymes by flavonoids of St. John's wort. *Toxicology.* 2006;217(2-3):194-205.
[36] Chang TK, Chen J, Yeung EY. Effect of Ginkgo biloba extract on procarcinogen-bioactivating human CYP1 enzymes: identification of isorhamnetin, kaempferol, and quercetin as potent inhibitors of CYP1B1. *Toxicology and applied pharmacology.* 2006;213(1):18-26.
[37] Smith JA, Poteet-Smith CE, Xu Y, Errington TM, Hecht SM, Lannigan DA. Identification of the first specific inhibitor of p90 ribosomal S6 kinase (RSK) reveals an unexpected role for RSK in cancer cell proliferation. *Cancer research.* 2005;65(3):1027-34.
[38] Clark DE, Errington TM, Smith JA, Frierson HF, Jr., Weber MJ, Lannigan DA. The serine/threonine protein kinase, p90 ribosomal S6 kinase, is an important regulator of prostate cancer cell proliferation. *Cancer research.* 2005;65(8):3108-16.
[39] Zhang Q, Zhao XH, Wang ZJ. Flavones and flavonols exert cytotoxic effects on a human oesophageal adenocarcinoma cell line (OE33) by causing G2/M arrest and inducing apoptosis. Food and chemical toxicology: *an international journal published for the British Industrial Biological Research Association.* 2008;46(6):2042-53.
[40] Zhang Q, Zhao XH, Wang ZJ. Cytotoxicity of flavones and flavonols to a human esophageal squamous cell carcinoma cell line (KYSE-510) by induction of G2/M arrest and apoptosis. *Toxicology in vitro: an international journal published in association with BIBRA.* 2009;23(5):797-807.
[41] Leung HW, Lin CJ, Hour MJ, Yang WH, Wang MY, Lee HZ. Kaempferol induces apoptosis in human lung non-small carcinoma cells accompanied by an induction of antioxidant enzymes. Food and chemical toxicology: *an international journal published for the British Industrial Biological Research Association.* 2007;45(10):2005-13.
[42] Crespy V, Morand C, Besson C, Cotelle N, Vezin H, Demigne C, et al. The splanchnic metabolism of flavonoids highly differed according to the nature of the compound. *American journal of physiology Gastrointestinal and liver physiology.* 2003;284(6):G980-8.

[43] DuPont MS, Day AJ, Bennett RN, Mellon FA, Kroon PA. Absorption of kaempferol from endive, a source of kaempferol-3-glucuronide, in humans. *European journal of clinical nutrition*. 2004;58(6):947-54.

[44] Wang FM, Yao TW, Zeng S. Disposition of quercetin and kaempferol in human following an oral administration of Ginkgo biloba extract tablets. *European journal of drug metabolism and pharmacokinetics*. 2003;28(3):173-7.

[45] Bonetti A, Marotti I, Dinelli G. Urinary excretion of kaempferol from common beans (Phaseolus vulgaris L.) in humans. *International journal of food sciences and nutrition*. 2007;58(4):261-9.

[46] de Vries JH, Hollman PC, Meyboom S, Buysman MN, Zock PL, van Staveren WA, et al. Plasma concentrations and urinary excretion of the antioxidant flavonols quercetin and kaempferol as biomarkers for dietary intake. *The American journal of clinical nutrition*. 1998;68(1):60-5.

[47] Radtke J, Linseisen J, Wolfram G. Fasting plasma concentrations of selected flavonoids as markers of their ordinary dietary intake. *European journal of nutrition*. 2002;41(5):203-9.

[48] Wang FM, Yao TW, Zeng S. Determination of quercetin and kaempferol in human urine after orally administrated tablet of ginkgo biloba extract by HPLC. *Journal of pharmaceutical and biomedical analysis*. 2003;33(2):317-21.

[49] Cao J, Zhang Y, Chen W, Zhao X. The relationship between fasting plasma concentrations of selected flavonoids and their ordinary dietary intake. *The British journal of nutrition*. 2010;103(2):249-55.

[50] Meade CJ, Heuer H. PAF antagonism as an approach to the treatment of airway hyperreactivity. *The American review of respiratory disease*. 1991;143(3 Pt 2):S79-82.

[51] Voynow JA, Young LR, Wang Y, Horger T, Rose MC, Fischer BM. Neutrophil elastase increases MUC5AC mRNA and protein expression in respiratory epithelial cells. *The American journal of physiology*. 1999;276(5 Pt 1):L835-43.

[52] Song KS, Lee WJ, Chung KC, Koo JS, Yang EJ, Choi JY, et al. Interleukin-1 beta and tumor necrosis factor-alpha induce MUC5AC overexpression through a mechanism involving ERK/p38 mitogen-activated protein kinases-MSK1-CREB activation in human airway epithelial cells. *The Journal of biological chemistry*. 2003;278(26):23243-50.

[53] Gray T, Nettesheim P, Loftin C, Koo JS, Bonner J, Peddada S, et al. Interleukin-1beta-induced mucin production in human airway epithelium is mediated by cyclooxygenase-2, prostaglandin E2 receptors, and cyclic AMP-protein kinase A signaling. *Molecular pharmacology*. 2004;66(2):337-46.

[54] Hovenberg HW, Davies JR, Herrmann A, Linden CJ, Carlstedt I. MUC5AC, but not MUC2, is a prominent mucin in respiratory secretions. *Glycoconjugate journal*. 1996;13(5):839-47.

[55] Thornton DJ, Howard M, Khan N, Sheehan JK. Identification of two glycoforms of the MUC5B mucin in human respiratory mucus. Evidence for a cysteine-rich sequence repeated within the molecule. *The Journal of biological chemistry*. 1997;272(14):9561-6.

[56] Wickstrom C, Davies JR, Eriksen GV, Veerman EC, Carlstedt I. MUC5B is a major gel-forming, oligomeric mucin from human salivary gland, respiratory tract and

endocervix: identification of glycoforms and C-terminal cleavage. *The Biochemical journal*. 1998;334 (Pt 3):685-93.

[57] Chang JH, Kim JH, Lee KW, Cho CI, Chun JH, Kim KS. Suppression of IL-1b-induced MUC5AC Gene Expression by Ginkgo biloba extract (Egb 761) in human airway epithelial cells. *Journal of Rhinology*. 2007;14(1):49-55.

[58] Kwon SH, Nam JI, Kim SH, Kim JH, Yoon JH, Kim KS. Kaempferol and quercetin, essential ingredients in Ginkgo biloba extract, inhibit interleukin-1beta-induced MUC5AC gene expression in human airway epithelial cells. Phytotherapy research: *PTR*. 2009;23(12):1708-12.

[59] Chang JH, Song KJ, Kim HJ, Kim JH, Kim NH, Kim KS. Dietary polyphenols affect MUC5AC expression and ciliary movement in respiratory cells and nasal mucosa. *American journal of rhinology and allergy*. 2010;24(2):e59-62.

In: Kaempferol
Editors: Garion Villers and Yves Fougere
ISBN: 978-1-62618-515-9

Chapter 9

Kaempferol Regulates Intestinal Tight Junction Permeability

Takuya Suzuki*

Department of Biofunctional Science and Technology,
Graduate School of Biosphere Science, Hiroshima University, Japan

Abstract

A primary function of gastrointestinal epithelial cells is to provide a physical barrier to the diffusion of pathogens, toxins, and allergens from the external environment into the tissues. Defects in the barrier function are involved in the pathogenesis of various intestinal diseases, such as inflammatory bowel diseases, irritable bowel disease, and food allergies. This intestinal barrier is organized through interactions among several barrier components, including the adhesive mucous gel layer, the mucosal immune system, and intercellular tight junctions (TJs). The TJ is a multi-protein complex, which locates around the apical end of the lateral membrane of polarized epithelial cells, and selectively regulates the paracellular passage of ions, molecules and water. The TJ complex is composed of transmembrane proteins that interact with the perijunctional actin cytoskeleton through intracellular plaque proteins. TJ permeability and assembly are dynamically regulated by various intracellular signaling molecules, such as protein kinase C, mitogen-activated protein kinases, and protein phosphatases. Kaempferol is one of the most abundant flavonoids in human diets and exhibits various biological effects, such as anti-inflammatory and anti-cancer activities. A recent study has demonstrated that kaempferol participates in the regulation of intestinal TJ barrier function. This chapter presents the recent advances in our understanding of the kaempferol-mediated suppressive effect on intestinal TJ permeability with a particular focus on intracellular molecular mechanisms.

* Corresponding author: Takuya Suzuki, Ph. D., 1-4-4, Kagamiyama, Higashi-Hiroshima, 739-8528, Japan. TEL: 81-82-424-7984. Fax: 81-82-424-7916; E-mail: takuya@hiroshima-u.ac.jp.

Intestinal Barrier and Diseases

A multifunctional epithelium lines the luminal surface of the gastrointestinal mucosa. As a primary function, the epithelium provides a physical barrier to luminal inflammatory substances, including bacteria, bacterial products, and dietary antigens (Figure 1) [1]. Impairment of the barrier function causes diffusion of these inflammatory substances into mucosal tissues and circulation system, resulting in chronic activation of the immune system in tissues. Clinical studies have demonstrated the presence of intestinal barrier defects in several diseases, including inflammatory bowel diseases (IBDs) [2], celiac disease [3], and irritable bowel disease [4], indicating the involvement of defects in the pathogenesis of these diseases. Some findings suggest that intestinal barrier impairment could be an initial event prior to the onset or recurrence of such diseases. For example, intestinal permeability to macromolecules is increased, not only in patients with Crohn's disease and ulcerative colitis, but also in their apparently healthy first-degree relatives [5]. Non-involved segments of the intestines from Crohn's disease patients also exhibit hyperpermeability [6]. In addition, barrier impairment may also contribute to the progression or development of diseases through the production of inflammatory cytokines, such as TNF-α and IFN-γ [7]. According to these findings, the protection and/or promotion of barrier integrity might afford both preventive and therapeutic tools for diseases associated with barrier impairment.

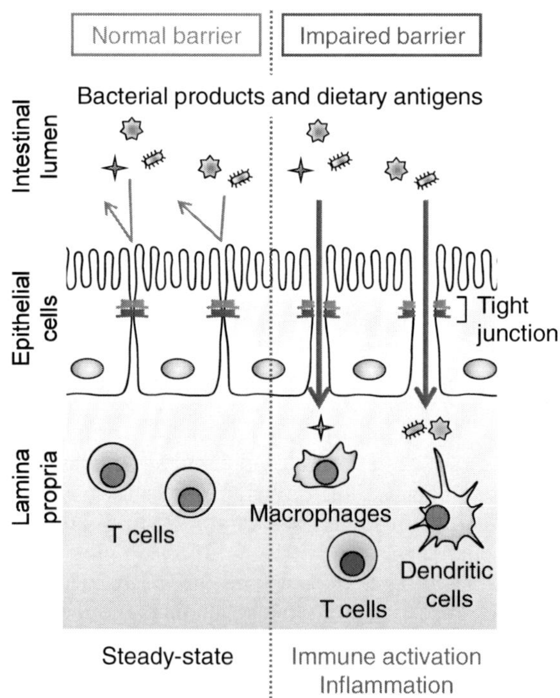

Figure 1. Barrier function of intestinal tight junctions (TJs).
The intestinal epithelium provides a physical barrier to luminal inflammatory substances. Impaired barrier function allows the diffusion of these inflammatory substances into mucosal tissues and circulation system, resulting in chronic activation of the immune system in tissues.

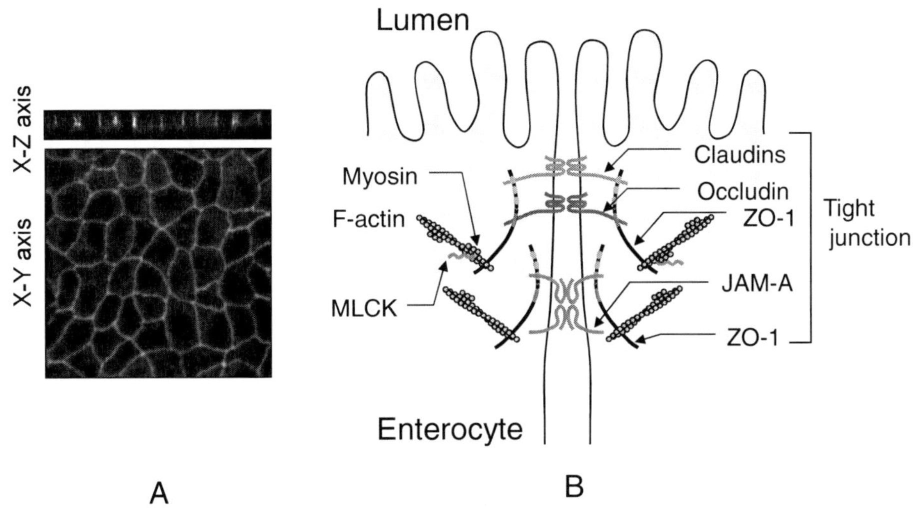

Figure 2. Structure of the intestinal tight junction (TJ) barrier. (A) Confocal images of an immunolabeled TJ protein, zonula occludens (ZO)-1, collected from human intestinal Caco-2 cells. Immunofluorescence of ZO-1, indicating the localization of the TJ protein complex, is observed at the apical portion of the lateral membrane of epithelial cells. (B) The TJ is a multiple protein complex consisting of transmembrane and cytosolic proteins and prevents the paracellular diffusion of noxious substances into tissues. The TJ complex is directly associated with the perijunctional actin cytoskeletal rings.

Intestinal Tight Junction (TJ) Structure

The intestinal barrier is organized through interactions among a number of different components, including the mucus gel layer, anti-bacterial peptides, and intercellular TJs. The TJ is a multiple protein complex consisting of transmembrane and cytosolic proteins [8]. The TJ protein complex locates at the apical portion of the lateral membrane of epithelial cells and prevents the paracellular diffusion of noxious substances into tissues (Figure 2). To date, 4 transmembrane TJ proteins, claudins [9], occludin [10], tricellulin [11] and junctional adhesion molecule (JAM) [12], have been identified. The intracellular domains of transmembrane proteins are directly or indirectly associated with cytosolic plaque proteins, such as zonula occludens (ZO) proteins [13], which in turn anchor the transmembrane proteins to the perijunctional actin cytoskeleton ring. This actin cytoskeletal interaction has a pivotal role in maintaining the localization of the TJ complex and barrier integrity [14].

Claudins, occludin and tricellulin are tetraspanin membrane proteins and contain 2 extracellular loops directly regulating paracellular permeability. Claudins are a multigene family with at least 24 members in mice and humans [15]. Several reports demonstrate that claudins play an essential role in determining the size- and charge selectivity of the paracellular pathways [16]. Each claudin member shows a unique expression pattern in tissues and cell lines. In mouse intestines, claudin-1, -2, -3, -5, -7, -8, -10, -12, -13, -14, -15, -17 and -18 are detected at the gene level [15]. The claudin members are roughly divided into two types: those forming the barrier and those forming pores. To our knowledge, claudin-1, -3, -4, -5, -8, -9, -10, -11, -14 and -18 can be categorized as barrier-forming claudins, while claudin-2, -7, -12, -15 and -17 are pore-forming claudins. Occludin is the first integral membrane TJ protein to be identified [10]. The function of occludin is not yet fully understood, but many studies using animals and cell cultures have shown that occludin has a crucial role in TJ structure and permeability in the intestinal epithelium. Al-Sadi *et al.* have demonstrated that occludin regulates macromolecule permeability in the intestines [17]. Tricellulin uniquely and preferably localizes at tricellular junctions, at which more than 3 adjacent cells make contact, although it is also observed at bicellular junctions along with claudins and occludin. Krug *et al.* have extensively examined the roles of tricellulin in the intestinal barrier at both tricellular and bicellular junctions [18]. The JAM family belongs to the immunoglobulin superfamily and its members exhibit multiple functions including barrier formation, virus recognition, and platelet activation [19]. In the intestines, at least, JAM-A [12], JAM-4 [20] and CAR (Coxsackievirus and adenovirus receptor) [21] are expressed and involved in TJ barrier function. Several intracellular plaque proteins, including ZO proteins and cinglin, have been identified in the TJ complex [8, 13]. Although the roles of these intracellular TJ proteins are still under investigation, many of them are considered to play a fundamental role as the molecular scaffold in the TJ structure through their multiple domains; e.g., their PDZ and SH3 domains. In addition, a recent study has demonstrated that ZO-1 contributes to the assembly of the TJ structure [22].

Evidence of alteration in intestinal TJ structures in human diseases is still accumulating, but some observations have been reported from patient biopsies. For example, in colon biopsies from ulcerative colitis patients, decreased occludin, claudin-1, claudin-4, tricellulin and JAM-A expression and redistribution, and increased claudin-2 expression were observed [23, 24]. Structurally, the TJ strand count and meshwork depth are reduced in ulcerative colitis patients [25]. Similar observations have been reported in patients with Crohn's disease [2, 24].

Intestinal TJ Regulation by Kaempferol

Information on the role of kaempferol in intestinal TJ regulation is quite limited; however, one study has demonstrated that kaempferol enhances intestinal TJ barrier integrity in human intestinal Caco-2 cells. Kaempferol administration dose-dependently increases transepithelial electrical resistance (TER), an indicator of TJ permeability, in cells, indicating the enhancement of TJ barrier integrity (Figure 3) [26].

The TJ complex directly interacts with the perijunctional actin cytoskeleton and this interaction is essential for maintaining the TJ barrier integrity [14]. Kaempferol administration promotes the assembly of TJ proteins, ZO-1 and -2, occludin, and claudin-1, 3, and 4, at the intercellular junctional complex in cells [26]. Further, the kaempferol-mediated ZO-2 and claudin-4 assembly is, in part, regulated at the total expression level. These findings indicate that kaempferol enhances intestinal TJ barrier integrity through the assembly and expression of TJ proteins.

The molecular mechanisms for the regulation of TJ protein localization and function remain unclear. Several studies have shown that occludin is highly phosphorylated and that its phosphorylation status has a role in occludin localization and TJ integrity in cells [27]. The kaempferol-mediated assembly of occludin at the TJ occurs simultaneously with the promotion of occludin phosphorylation, as indicated by increased multiple occludin bands in immunoblot analysis (Figure 3) [26]. Previous studies have demonstrated that the phosphorylation of occludin at its serine and threonine residues is essential for occludin localization at the TJs [28-30], suggesting that kaempferol promotes occludin assembly at the TJs as well as TJ barrier integrity through occludin phosphorylation. Occludin phosphorylation is regulated by the balance between protein kinases and phosphatase. Some kinases, including PKCζ and η [28, 29] and casein kinase I and II [31, 32], have been identified to directly phosphorylate occludin, leading to the regulation of TJ integrity, whereas protein phosphatase 1 and 2A have been reported to dephosphorylate occludin [33, 34]. Based on these findings, it is possible that kaempferol influences the activities of these kinases and/or phosphatase, resulting in increased occludin phosphorylation and assembly at the TJs.

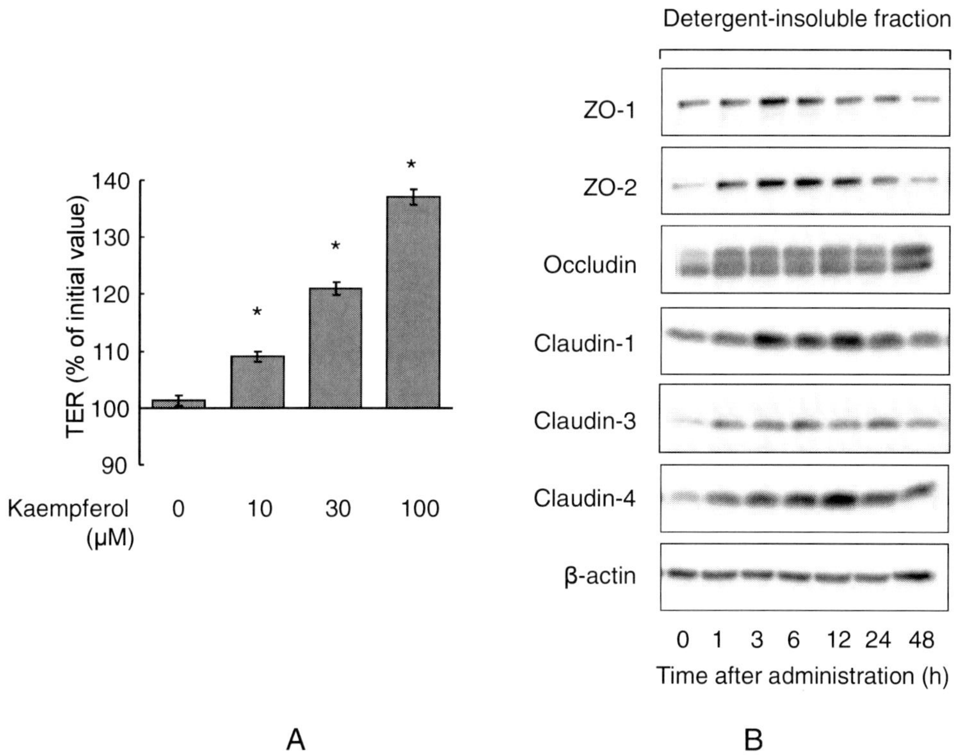

Figure 3. Regulation of the intestinal tight junction (TJ) barrier by kaempferol. (A) Kaempferol administration dose-dependently increases transepithelial electrical resistance (TER), an indicator of TJ barrier integrity, in human intestinal Caco-2 cells. Asterisks indicate significant differences from the control value ($P<0.05$). (B) Kaempferol administration increases the distribution of TJ proteins, ZO-1 and -2, occludin, and claudin-1, 3, and 4, in the detergent-insoluble fractions of Caco-2 cells, indicating increases in these TJ proteins at the intercellular junctional complex of cells.

Extensive studies have addressed the existence of asymmetric membrane lipid microdomains, such as raft and caveola [35]. The lipid microdomains are formed by a considerable amount of membrane cholesterol being clustered with sphingomyelin and glycosphingolipids. Because of their unique physical characteristics, the lipid microdomains attract and sequester a variety of membrane proteins and, therefore, act as platforms for various signaling pathways and are involved in various cellular functions [36]. Recently, the lipid microdomains have been shown to interact with the TJ protein complex [37]. Cell membrane cholesterol depletion using methyl-β-cyclodextrin, which disturbs the structure and function of the lipid microdomains, suppresses the kaempferol-mediated increase in TER, indicating that microdomain structure has a role in kaempferol-mediated barrier modulation [26]. In addition, kaempferol administration induces the redistribution of flottilin-1 and caveolin-1, microdomain-associated proteins, in cells, as evaluated by isopycnic sucrose density gradients. These findings suggest that kaempferol interacts with the lipid microdomain and/or the associated proteins, thereby affecting the physical or physiological characteristics of the microdomain, resulting in the subsequent promotion of TJ integrity. This interaction may also affect the activity of signaling molecules with a role in TJ regulation.

No *in vivo* studies to investigate the effects of kaempferol on intestinal barrier function have been undertaken to date. Some reports, however, have demonstrated that supplemental feeding of kaempferol prevents or ameliorates diseases associated with intestinal barrier defects, although barrier integrity was not examined. Dextran sulfate sodium (DSS)-induced colitis in rodents exhibits many morphological and pathological features, such as mucosal damage, ulceration, impaired barrier function, and inflammatory cytokine production, similar to those observed in human IBDs. Mi-Young *et al.* demonstrated that the supplemental feeding of kaempferol ameliorates DSS-induced colitis in mice, as indicated by the restoration of myeloperoxidase activity and inflammatory mediator expression [38]. The modulation of the intestinal barrier may be one of the possible mechanisms underlying the kaempferol-mediated anti-colitic effects.

Conclusion

Kaempferol has a promotive effect on intestinal TJ barrier integrity. Although the underlying mechanisms remain to be determined, dietary supplementation with kaempferol may afford therapeutic and preventive tools for use in diseases associated with intestinal barrier defects.

References

[1] Turner, J. R. (2009). Intestinal mucosal barrier function in health and disease. *Nat Rev Immunol*, *9*, 799-809.

[2] Zeissig, S., Burgel, N., Gunzel, D., Richter, J., *et al.*, (2007). Changes in expression and distribution of claudin 2, 5 and 8 lead to discontinuous tight junctions and barrier dysfunction in active Crohn's disease. *Gut*, *56*, 61-72.

[3] Smecuol, E., Bai, J. C., Vazquez, H., Kogan, Z., *et al.*, (1997). Gastrointestinal permeability in celiac disease. *Gastroenterology*, *112*, 1129-1136.

[4] Camilleri, M. & Gorman, H. (2007). Intestinal permeability and irritable bowel syndrome. *Neurogastroenterol Motil*, *19*, 545-552.

[5] Hollander, D., Vadheim, C. M., Brettholz, E., Petersen, G. M., et al. (1986). Increased intestinal permeability in patients with Crohn's disease and their relatives. A possible etiologic factor. *Ann Intern Med, 105*, 883-885.
[6] Soderholm, J. D., Peterson, K. H., Olaison, G., Franzen, L. E., *et al.*, (1999). Epithelial permeability to proteins in the noninflamed ileum of Crohn's disease? *Gastroenterology*, *117*, 65-72.
[7] Wang, F., Graham, W. V., Wang, Y., Witkowski, E. D., *et al.*, (2005). Interferon-gamma and tumor necrosis factor-alpha synergize to induce intestinal epithelial barrier dysfunction by up-regulating myosin light chain kinase expression. *Am J Pathol*, *166*, 409-419.
[8] Gonzalez-Mariscal, L., Betanzos, A., Nava, P. & Jaramillo, B. E. (2003). Tight junction proteins. *Prog Biophys Mol Biol*, *81*, 1-44.
[9] Furuse, M., Fujita, K., Hiiragi, T., Fujimoto, K. & Tsukita, S. (1998). Claudin-1 and -2: novel integral membrane proteins localizing at tight junctions with no sequence similarity to occludin. *J Cell Biol*, *141*, 1539-1550.
[10] Furuse, M., Hirase, T., Itoh, M., Nagafuchi, A., *et al.*, (1993). Occludin: a novel integral membrane protein localizing at tight junctions. *J Cell Biol*, *123*, 1777-1788.
[11] Ikenouchi, J., Furuse, M., Furuse, K., Sasaki, H. & Tsukita, S. (2005). Tricellulin constitutes a novel barrier at tricellular contacts of epithelial cells. *J Cell Biol*, *171*, 939-945.
[12] Martin-Padura, I., Lostaglio, S., Schneemann, M., Williams, L., *et al.*, (1998). Junctional adhesion molecule, a novel member of the immunoglobulin superfamily that distributes at intercellular junctions and modulates monocyte transmigration. *J Cell Biol*, *142*, 117-127.
[13] Willott, E., Balda, M. S., Fanning, A. S., Jameson, B., *et al.*, (1993). The tight junction protein ZO-1 is homologous to the Drosophila discs-large tumor suppressor protein of septate junctions. *Proc Natl Acad Sci U S A*, *90*, 7834-7838.
[14] Rao, R. K., Basuroy, S., Rao, V. U., Karnaky Jr, K. J. & Gupta, A. (2002). Tyrosine phosphorylation and dissociation of occludin-ZO-1 and E-cadherin-beta-catenin complexes from the cytoskeleton by oxidative stress. *Biochem J*, *368*, 471-481.
[15] Holmes, J. L., Van Itallie, C. M., Rasmussen, J. E. & Anderson, J. M. (2006). Claudin profiling in the mouse during postnatal intestinal development and along the gastrointestinal tract reveals complex expression patterns. *Gene Expr Patterns*, *6*, 581-588.
[16] Van Itallie, C. M. & Anderson, J. M. (2006). Claudins and epithelial paracellular transport. *Annu Rev Physiol*, *68*, 403-429.
[17] Al-Sadi, R., Khatib, K., Guo, S., Ye, D., *et al.*, (2011). Occludin regulates macromolecule flux across the intestinal epithelial tight junction barrier. *Am J Physiol Gastrointest Liver Physiol*, *300*, G1054-1064.
[18] Krug, S. M., Amasheh, S., Richter, J. F., Milatz, S., *et al.*, (2009). Tricellulin forms a barrier to macromolecules in tricellular tight junctions without affecting ion permeability. *Mol Biol Cell*, *20*, 3713-3724.
[19] Bazzoni, G. (2003). The JAM family of junctional adhesion molecules. *Curr Opin Cell Biol*, *15*, 525-530.

[20] Hirabayashi, S., Tajima, M., Yao, I., Nishimura, W., *et al.*, (2003). JAM4, a junctional cell adhesion molecule interacting with a tight junction protein, MAGI-1. *Mol Cell Biol*, *23*, 4267-4282.

[21] Cohen, C. J., Shieh, J. T., Pickles, R. J., Okegawa, T., *et al.*, (2001). The coxsackievirus and adenovirus receptor is a transmembrane component of the tight junction. *Proc Natl Acad Sci U S A*, *98*, 15191-15196.

[22] Tsukita, S., Katsuno, T., Yamazaki, Y., Umeda, K. & Tamura, A. (2009). Roles of ZO-1 and ZO-2 in establishment of the belt-like adherens and tight junctions with paracellular permselective barrier function. *Ann N Y Acad Sci*, *1165*, 44-52.

[23] Heller, F., Florian, P., Bojarski, C., Richter, J., *et al.*, (2005). Interleukin-13 is the key effector Th2 cytokine in ulcerative colitis that affects epithelial tight junctions, apoptosis, and cell restitution. *Gastroenterology*, *129*, 550-564.

[24] Vetrano, S., Rescigno, M., Cera, M. R., Correale, C., *et al.*, (2008). Unique role of junctional adhesion molecule-a in maintaining mucosal homeostasis in inflammatory bowel disease. *Gastroenterology*, *135*, 173-184.

[25] Schmitz, H., Barmeyer, C., Fromm, M., Runkel, N., *et al.*, (1999). Altered tight junction structure contributes to the impaired epithelial barrier. *Gastroenterology*, *116*, 301-309.

[26] Suzuki, T., Tanabe, S. & Hara, H. (2011). Kaempferol enhances intestinal barrier function through the cytoskeletal association and expression of tight junction proteins in Caco-2 cells. *J Nutr*, *141*, 87-94.

[27] Sakakibara, A., Furuse, M., Saitou, M., Ando-Akatsuka, Y. & Tsukita, S. (1997). Possible involvement of phosphorylation of occludin in tight junction formation. *J Cell Biol*, *137*, 1393-1401.

[28] Suzuki, T., Elias, B. C., Seth, A., Shen, L., *et al.*, (2009). PKC eta regulates occludin phosphorylation and epithelial tight junction integrity. *Proc Natl Acad Sci U S A*, *106*, 61-66.

[29] Jain, S., Suzuki, T., Seth, A., Samak, G. & Rao, R. (2011). Protein kinase Czeta phosphorylates occludin and promotes assembly of epithelial tight junctions. *Biochem J*, *437*, 289-299.

[30] Raleigh, D. R., Boe, D. M., Yu, D., Weber, C. R., *et al.*, (2011). Occludin S408 phosphorylation regulates tight junction protein interactions and. *J Cell Biol*, *193*, 565-582.

[31] McKenzie, J. A., Riento, K. & Ridley, A. J. (2006). Casein kinase I epsilon associates with and phosphorylates the tight junction protein occludin. *FEBS Lett*, *580*, 2388-2394.

[32] Smales, C., Ellis, M., Baumber, R., Hussain, N., *et al.*, (2003). Occludin phosphorylation: identification of an occludin kinase in brain and cell extracts as CK2. *FEBS Lett*, *545*, 161-166.

[33] Nunbhakdi-Craig, V., Machleidt, T., Ogris, E., Bellotto, D., *et al.*, (2002). Protein phosphatase 2A associates with and regulates atypical PKC and the epithelial tight junction complex. *J Cell Biol*, *158*, 967-978.

[34] Seth, A., Sheth, P., Elias, B. C. & Rao, R. (2007). Protein phosphatases 2A and 1 interact with occludin and negatively regulate the assembly of tight junctions in the CACO-2 cell monolayer. *J Biol Chem*, *282*, 11487-11498.

[35] Hooper, N. M. (1999). Detergent-insoluble glycosphingolipid/ cholesterol-rich membrane domains, lipid rafts and caveolae (review). *Mol Membr Biol*, *16*, 145-156.

[36] Simons, K. & Toomre, D. (2000). Lipid rafts and signal transduction. *Nat Rev Mol Cell Biol*, *1*, 31-39.

[37] Nusrat, A., Parkos, C. A., Verkade, P., Foley, C. S., *et al.*, (2000). Tight junctions are membrane microdomains. *J Cell Sci*, *113 (Pt 10)*, 1771-1781.

[38] Park, M. Y., Ji, G. E. & Sung, M. K. (2012). Dietary kaempferol suppresses inflammation of dextran sulfate sodium-induced colitis in mice. *Dig Dis Sci*, *57*, 355-363.

In: Kaempferol
Editors: Garion Villers and Yves Fougere

ISBN: 978-1-62618-515-9

Chapter 10

Identification of Kaempferol in Historical Textiles: A Case Study on Textiles of Byzantine Origin

***Dimitrios Mantzouris*[1] *and Ioannis Karapanagiotis*[2,*]**
[1]Ormylia Foundation, Art Diagnosis Center, Ormylia, Greece
[2]University Ecclesiastical Academy of Thessaloniki,
Department of Management and Conservation of Ecclesiastical
Cultural Heritage Objects, Thessaloniki, Greece

Abstract

Our goal is to present, in brief, the historical use of kaempferol as a dyestuff material for textiles. For this reason we first review previously published reports that describe the identification of kaempferol in textiles of the cultural heritage. Then, a case study is presented where the identification of kaempferol in samples removed from three religious textiles, preserved in monasteries of Mount Athos, is described. Kaempferol is detected in mixture with several other colouring components using high performance liquid chromatography (HPLC) with diode-array detection (DAD). Finally, a brief overview of the various colourants found in historical textiles from Mount Athos is provided.

1. Introduction: Identification of Kaempferol in Historical Textiles

Kaempferol is contained in various plants which have been used as dyes for textiles, since antiquity. Typical examples of natural/plant sources of kaempferol are summarized in Table 1 [1-15]. Moreover, kaempferol is a flavonoid and, as the Latin word *flavus* means, it is a material that yields the yellow colour. Our goal is to discuss the historical use of kaempferol

[*] Corresponding author. E-mail: y.karapanagiotis@aeath.gr.

as a dyestuff material for textiles. For this reason we first summarize previously published reports that describe the identification of kaempferol in historical textiles of the cultural heritage. Then, a case study is presented where the identification of kaempferol in three post Byzantine textiles, preserved in monasteries of Mount Athos, is described in detail (chapter 3). Prior to this, a brief overview of the various colourants found in historical textiles from Mount Athos is provided in chapter 2.

Although kaempferol is easily found in nature and it is contained in several natural organic colourants (e.g. Table 1), its identification in works of the cultural heritage has been rarely reported. To evaluate this comment appropriately, one has to consider that kaempferol is usually contained in dyestuff sources in small amounts and therefore its identification is not easy. This is demonstrated in chapter 3. Furthermore, historical objects are subjected to degradation mechanisms, developed because of their extensive use and/or ageing effects. Consequently, materials contained in historical objects can be decomposed under the long influence of the conditions to which have been exposed. A brief review of the relatively limited number of previously published reports that describe the identification of kaempferol in historical textiles is provided next.

Samples removed from textiles belonging to Pre-Columbian civilizations in Peru were investigated several years ago by Wouters and Rosario-Chirinos using high performance liquid chromatography (HPLC) with diode-array detection (DAD) [11] and more recently by Zhang et al. using HPLC-DAD coupled to mass spectrometry (MS) [16]. Among the various colouring compounds reported in the HPLC results were kaempferol and its derivatives [11, 16]. Trojanowicz and his co-workers identified kaempferol in several microsamples removed from Coptic textiles (4th -12th c. AD) of the National Museum of Warsaw using HPLC with UV-Vis detection and coupled to fluorescence detection [6, 7] or mass spectrometry [17].

Table 1. Examples of natural dyes that contain kaempferol

Dye	Biological source	Reference
Buckthorn/Persian berries	*Rhamnus* trees	[1, 2, 3, 4, 5]
Weld	*Reseda luteola* L.	[6, 7, 8]
Dyer's broom	*Genista tinctoria* L.	[1]
Young fustic	*Cotinus coggygria* Scop.	[1]
Old Fustic	*Chlorophora tinctoria* L. (Gaud.)	[1, 4, 9, 10]
Quercitron bark	*Quercus veluntina* Lam.	[4]
Blue Gum	*Eucalyptus globulus* Labill.	[11]
Bog myrtle	*Myrica gale* L.	[4]
Sawwort	*Serratula tinctoria* L./ *Serratula coronate* L.	[9]
Canadian goldenrod/Yellow wood	*Solidago canadensis* L./ *Solidago virgaurea* L.	[1, 12, 13]
Cotton rose/Gulzuba	*Hibiscus mutabilis*	[14]
Dyer's chamomile	*Anthemis tinctoria* L.	[1]
Marigold	*Tagetes* spp.	[1]
Daphne	*Daphne gnidium* L.	[1]
Onion skins	*Allium cepa* L.	[1]
Kayu kuning	*Maclura cochinchinensis* (Lour.) Corner	[1]
Bastard hemp	*Datisca cannabina* L.	[1, 15]

Chromatography was also employed for the identification of dyes used in the Scottish tartan trade of the late 18th/early 19th c., resulting in the identification of kaempferol [18]. The compound was furthermore detected in three ecclesiastical-religious embroideries worked in the Byzantine tradition and preserved in the National Museum of Art of Romania [15]. Finally, kaempferol was included in the HPLC results reported for ecclesiastical textiles, which are preserved in monasteries from Mount Athos, Greece [19, 20]. An overview of the various colourants found in historical textiles from Mount Athos is provided in the next chapter.

2. Dyes Identified in Historical Textiles from Mount Athos

Mount Athos is the eastern peninsula of the area of Halkidiki in Macedonia, Greece, and it is an UNESCO World Heritage Site. Today there are twenty (20) active monasteries (and several sketes) which preserve the orthodox eastern Byzantine tradition. Ecclesiastical textiles of inestimable religious, spiritual and historical significance can be found in these monasteries. These objects have been rarely studied through modern physicochemical techniques. Our group has recently carried out some systematic studies, aiming at elucidating the origin of the organic colourants contained in a relatively large number of textiles from Mount Athos. Textile collections investigated by our research team are grouped in Table 2 [19-21].

According to Table 2, one hundred twenty six (126) microsamples, removed from forty six (46) textiles from Mount Athos, were analysed in total using HPLC-DAD-MS. The studied objects belong to various monasteries of Mount Athos but the majority of them are preserved in the monastery of Simonos Petra, followed by the Xeropotamou monastery. Natural dyes identified in the historical samples are summarized in Figure 1. The HPLC results (identifications) are presented in a plot: the y-axis corresponds to the number of objects and x-axis describes the identified natural dyes. Consequently, Figure 1 provides an indication about the frequency of use of each natural product by the dyer's of the ecclesiastical objects.

Table 2. Textile collections from monasteries of Mount Athos, which were investigated using HPLC-DAD-MS to identify the organic colourants

Provenance of the objects	Type of the objects	Date	Number of objects	Number of samples	Reference
Monastery of Simonos Petra (Mount Athos)	various ecclesiastical textiles	16th – early 20th c.	30	88	[19]
Monastery of Xeropotamou (Mount Athos)	various ecclesiastical textiles	16th – 19th c.	9	22	[20]
Various monasteries of Mount Athos	ecclesiastical garments (*sakkoi*)	15th – 19th c.	7	16	[21]

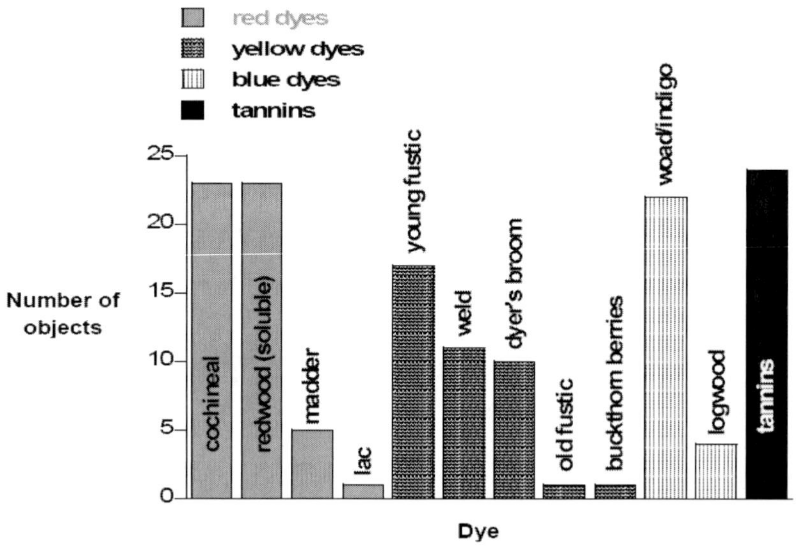

Figure 1. Dyes identified in textile objects from Mount Athos (Table 2), as described in previously published reports of our group [19-21].

The following natural colouring materials were identified in the textiles of Byzantine origin: cochineal, soluble redwood, madder, lac (red dyes); young fustic, weld, dyer's broom, old fustic, buckthorn berries (yellow dyes); indigo/woad, logwood (blue dyes); moreover, tannins were detected in several objects. Tannins have been used both as dyes to produce black dyed fibres and as adhesives to promote the stabilization of colourants on textile substrates. It must be noted that only natural colourants are included in the plot of Figure 1. Except for the natural materials, early, semi-synthetic dyes, such as indigo carmine, fuchsine and rhodamine B were identified in some textiles dated from late 19th to early 20th century [19, 20]; these are not included in the results of Figure 1.

Dyes summarized in Figure 1 were identified by detecting various colouring compounds in the HPLC chromatograms of the tested samples. These compounds were then used as markers for dyestuff identification. Among the several detected colouring compounds, kaempferol, the molecule of our interest in the present article, was identified in three (3) objects, as described in detail next.

3. Identification of Kaempferol in Historical Textiles from Mount Athos

3.1. Experimental

3.1.1. Historical Samples

Samples were removed from three ecclesiastical textiles as follows. Two samples of silk and cotton nature were extracted from an *Epimanikion* (16th c.) and a *Phelonion* (late 19th c.), respectively. The two objects belong to the Monastery of Simonos Petra, Mount Athos.

Another sample of silk nature was removed from an *Orarion* which is dated in 1862 AD and is preserved in the Monastery of Xeropotamou, Mount Athos.

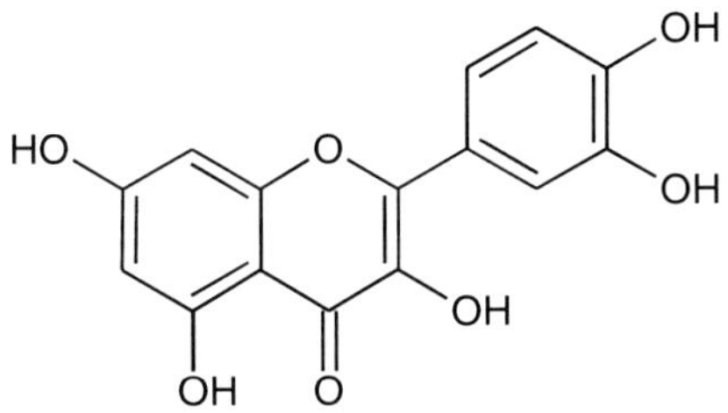

quercetin (**1**)

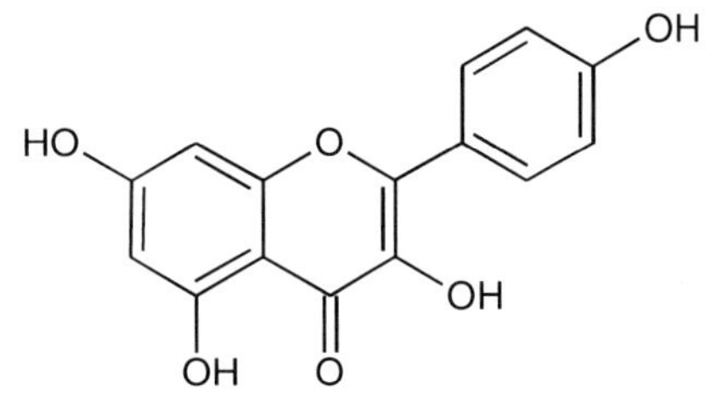

kaempferol (**2**)

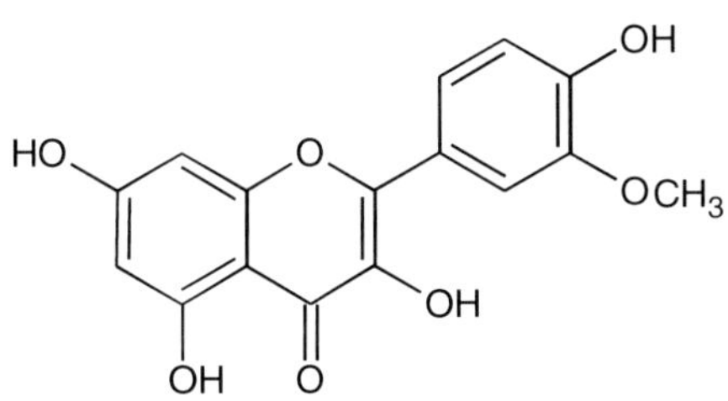

isorhamnetin (**3**)

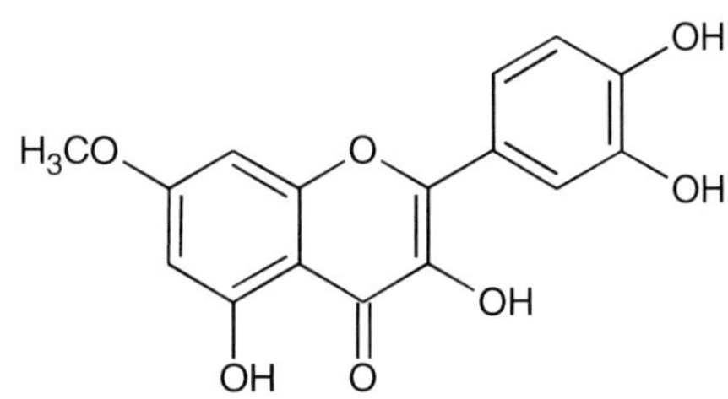

rhamnetin (**4**)

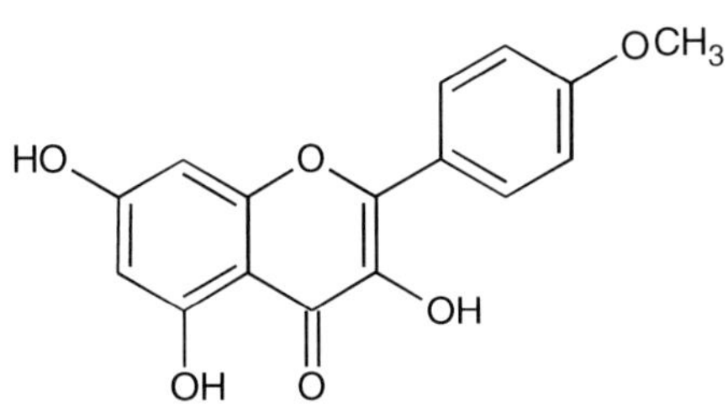

kaempferide (**5**)

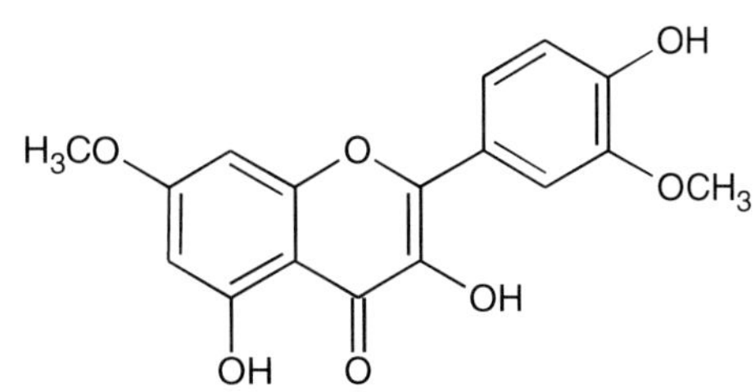

rhamnazin (**6**)

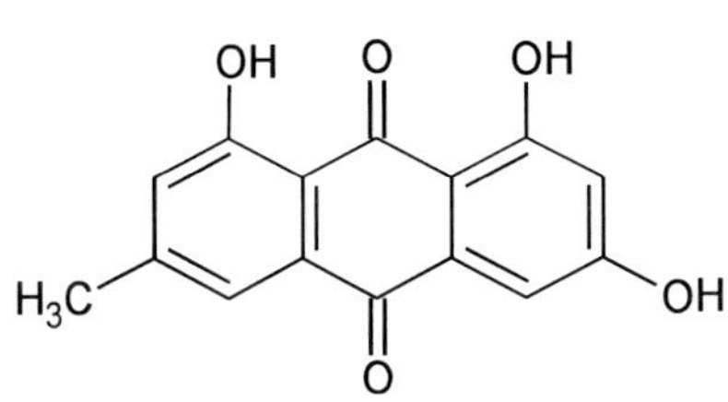

emodin (**7**)

Figure 2. Molecular structures of compounds contained in buckthorn berries.

3.1.2. Materials

Standards used for HPLC identification purposes were kaempferol, rhamnetin, fisetin, apigenin, indigo carmine (Fluka), morin, isorhamnetin, luteolin (Sigma), rhamnazin, kaempferide, chrysoeriol, luteolin-3′, 7-di-O-glucoside, luteolin-7-O-glucoside, apigenin-7-O-glucoside (Extrasynthese), emodin (Acros Organics), sulfuretin (Wako) and fustin, butein, naringenin and quercetin, which were previously produced in pure form [22]. Some of these compounds are contained in buckthorn berries, a dye which is later discussed in detail. The molecular structures of compounds contained in buckthorn berries are shown in Figure 2.

Three dyestuff sources of buckthorn berries, known also as Persian berries (*Rhamnus* trees) were analysed. Ripe (Rhamni maturi) and unripe (Rhamni immaturi) berries, were purchased from Kremer Pigmente. A sample of *Rhamnus petiolaris* L. was a generous gift from Professor Recep Karadag (Marmara University, Turkey).

Solvents and chemicals used for HPLC and sample treatment were of high purity, usually HPLC grade.

3.1.3. Dyestuff Extraction Methods

Prior to HPLC analysis, textile samples have to be treated to extract and solubilise the contained dyestuffs. Acid treatment is usually necessary as most of the dyes are attached on the textile substrates through metal ions (mordants) [23]. Therefore, the extracting solution must be able to disrupt the dye-metal complex.

The widely adopted hydrochloric acid (HCl) extraction method [24] was used to treat the historical samples from the Monastery of Simonos Petra: samples were added to methanolic acidified solution, which was kept at 100 °C for several minutes. HCl was then removed by evaporation (65 °C) under gentle nitrogen flow. The dry residues were dissolved in dimethyl sulfoxide (DMSO). The sample extracted from the *Orarion* (Monastery of Xeropotamou) was treated with the trifluoroacetic acid (TFA) extraction method [25]: sample was immersed in 400 μL of H_2O:MeOH:2M TFA (1:1:2, v/v/v) and kept at 100 °C for several minutes. The liquid phase was then evaporated (65 °C) under gentle nitrogen flow and the dry residue was dissolved in DMSO.

The use of the milder TFA, instead of the harsh HCl, results in very good extraction yields, preserving simultaneously the glycosides, which quite often are contained in yellow flavonoid dyes [25]. It is known that sugar derivatives (glycosides) are decomposed to their parent aglycones, under harsh acidic conditions e.g. using HCl in the extracting solution [16, 19, 23, 25].

3.1.4. Chromatography

Samples were analysed using high performance liquid chromatography (HPLC) coupled to a diode-array-detector (DAD) (Thermoquest) which is consisted of a 4000 quaternary HPLC pump, a SCM 3000 vacuum degasser, an AS3000 auto sampler with column oven and a Rheodyne 7725i injector with a 20 μL sample loop. Analyses were carried out with an Altima C18 (Alltech) column (5 μm particle size, i.d. 250 mm x 3.0 mm) at stable temperature of 35 °C. Gradient elution program utilized two solvents, H_2O-0.1 % TFA and CH_3CN-0.1 % TFA, as described in detail elsewhere [21].

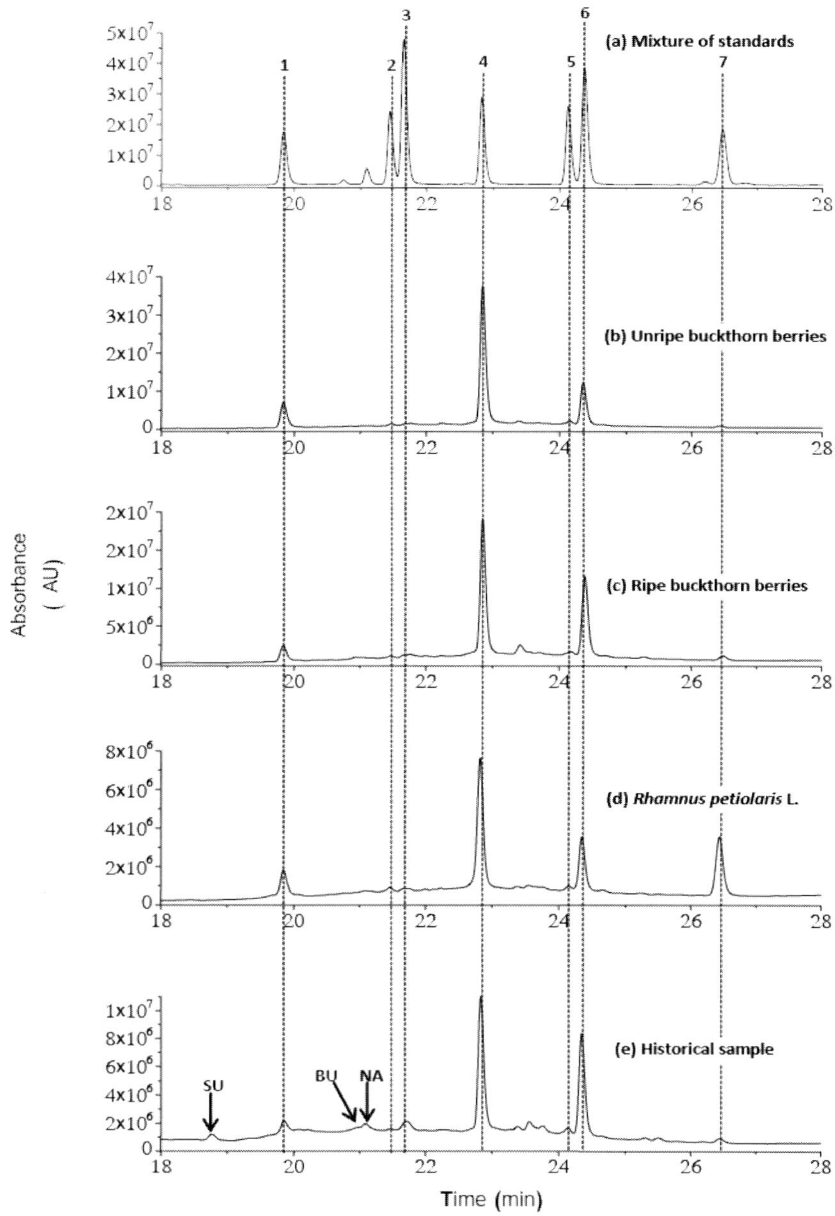

Figure 3. Chromatograms collected at 254 nm for a (a) mixture of standards of quercetin (1), kaempferol (2), isorhamnetin (3), rhamnetin (4), kaempferide (5), rhamnazin (6) and emodin (7), (b) sample of unripe buckthorn berries (Kremer Pigmente), (c) sample of ripe buckthorn berries (Kremer Pigmente), (d) sample of *Rhamnus petiolaris* L. and (e) historical sample, removed from *Epimanikion*. Some components of young fustic, which apparently are not related to buckthorn berries, are detected in the historical sample; these are: sulfuretin (SU), butein (BU), and naringenin (NA). Other components of young fustic, for example fustin and fisetin, were also detected in the historical sample, as indicated in Table 3. However, they corresponded to retention times <18 min and are therefore not included in (e).

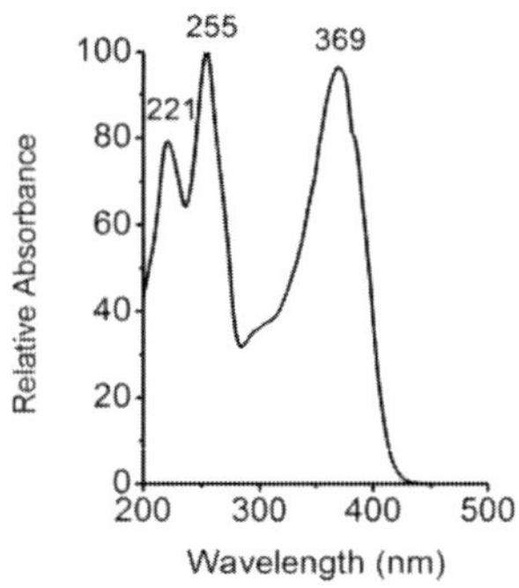

quercetin **(1)**

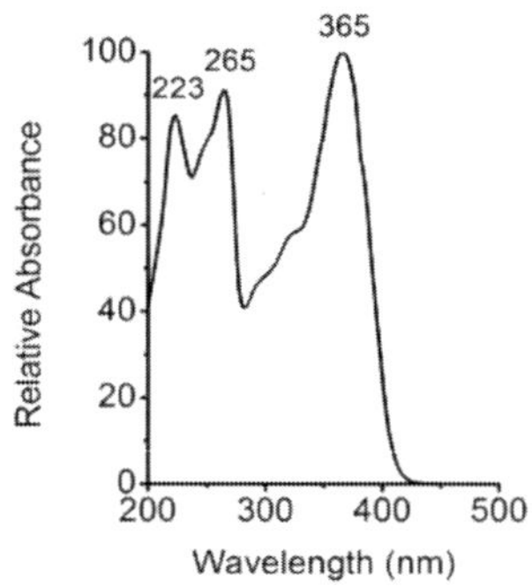

kaempferol **(2)**

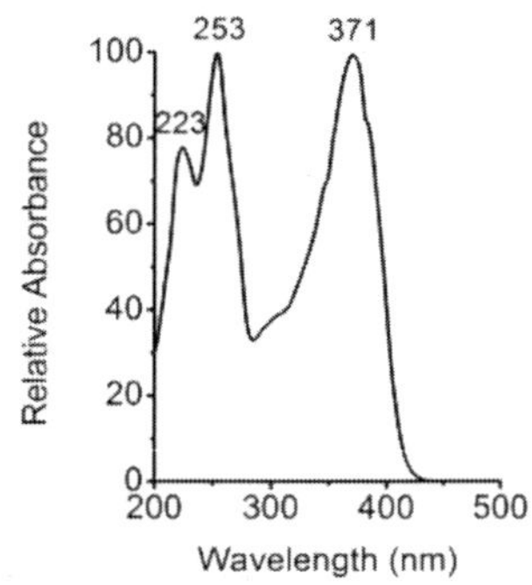

isorhamnetin **(3)**

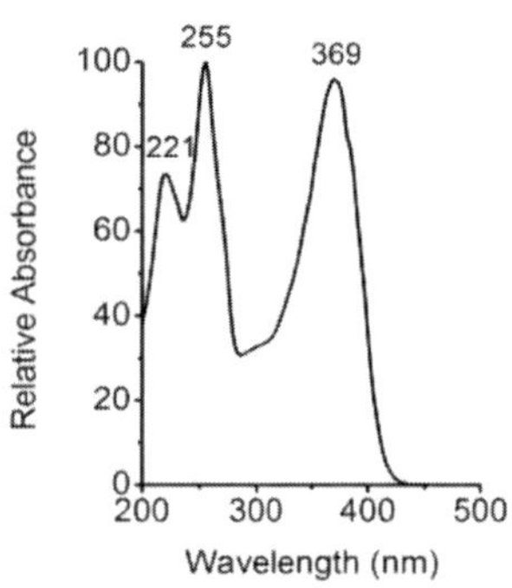

rhamnetin **(4)**

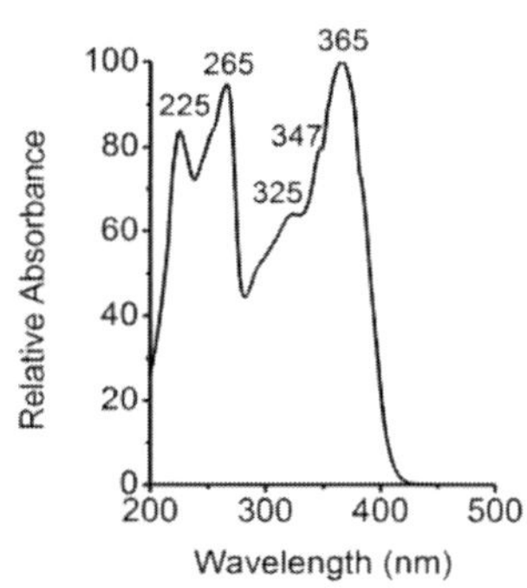

kaempferide **(5)**

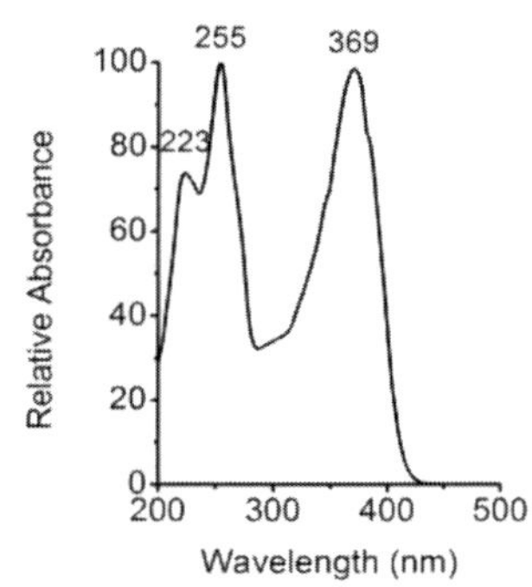

rhamnazin **(6)**

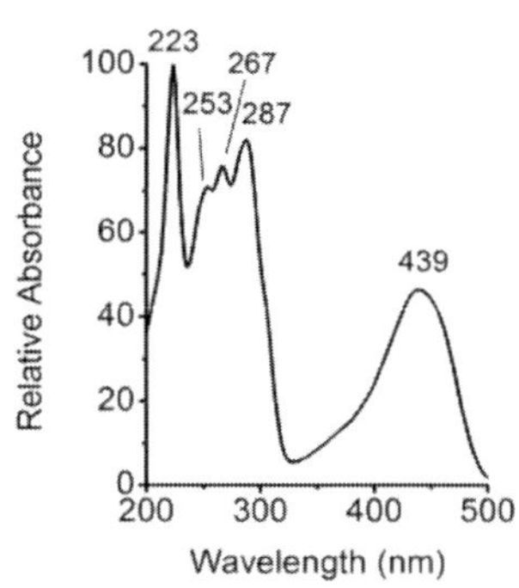

emodin **(7)**

Figure 4. UV-Vis spectra of buckthorn berries components detected in Figure 3.

3.2. Results and Discussion

Figure 3 shows the HPLC analyses of the three *Rhamnus* dyestuff sources included in the study and the identification of buckthorn berries in the silk sample extracted from the *Epimanikion*. In particular, Figure 3a shows the chromatogram obtained from a standard mixture that contained quercetin (1), kaempferol (2), isorhamnetin (3), rhamnetin (4), kaempferide (5), rhamnazin (6), and emodin (7). The structures of these compounds are shown in Figure 2. The corresponding spectra collected with the DAD are shown in Figure 4, which can be used as a database-library for other scientists who are interested in the HPLC analysis of these molecules.

Using the data of Figure 3a and 4 the identification of the aforementioned compounds in the samples of buckthorn berries were obtained as shown in Figures 3b, 3c and 3d. Finally, the chromatogram recorded for the historical sample is shown in Figure 3e. Dyestuff (Figures 3b, 3c and 3d) and historical (Figure 3e) samples were treated with the same HCl method, prior to HPLC analyses, to make the direct comparison of the HPLC graphs obtained for all these materials possible. As shown in Figure 3e, sulfuretin (SU), butein (BU) and naringenin (NA) were also detected in the historical sample. These compounds are contained in young fustic (*Cotinus coggygria* Scop.) and are not related to buckthorn berries [22]. Other components of young fustic such as fustin and fisetin were detected in the chromatogram of the historical sample. However, these two compounds were eluted in retention times <18 min and are therefore not shown in Figure 3e.

Table 3. Identification of kaempferol and other colouring materials, in samples removed from textiles of Mount Athos, using HPLC-DAD

Object		Sample		HPLC results	
Type (monastery)	Date	Fibre/ colour	Treatment method	Identified colouring compounds	Dye
Epimanikion (Monastery of Simonos Petra)	16th c.	silk/yellow	HCl	kaempferol, quercetin, isorhamnetin, rhamnetin, kaempferide, rhamnazin, emodin, fustin, fisetin, sulfuretin, butein, naringenin	buckthorn berries, young fustic
Phelonion (Monastery of Simonos Petra)	Late 19th c.	cotton/yellow	HCl	kaempferol, quercetin	unknown
Orarion (Monastery of Xeropotamou)	1862 AD	silk/green	TFA	kaempferol, morin, indigo carmine, luteolin-3′,7-di-O-glucoside, luteolin-7-O-glucoside, apigenin-7-O-glucoside, luteolin, apigenin, chrysoeriol	old fustic, indigo carmine, weld,

The marker compound for the identification of buckthorn berries is rhamnetin, which is clearly detected, in large amounts in the samples of Figure 3, both in dyestuff sources and historical sample. However, other compounds that are detected in buckthorn berries and are included in Figure 3, including kaempferol, are rarely reported [1]. As noted in the Introduction, this is probably because kaempferol is usually detected in small amounts (trace) in the extracts of the plant dye sources, as it is evidenced by the data of Figure 3.

Kaempferol and several other colouring components were detected in the other two historical samples, which were extracted from the *Phelonion* and *Orarion*. Overall, the HPLC results collected for the historical samples are summarized in Table 3. In the extract of the *Phelonion* sample, kaempferol was found in mixture with quercetin. Both compounds are contained in several plant species and therefore the identification of the dyestuff/biological source used to treat the yellow cotton sample was not possible. As noted in Table 3, the extract was received and submitted to HPLC using HCl for sample treatment.

The sample removed from the *Orarion* was treated with the milder TFA method, which appeared to be efficient to extract kaempferol. In this sample, kaempferol was detected as a minor component of old fustic (*Chlorophora tinctoria* L.). The identification of old fustic was based on the detection of morin, which is the marker compound of this dye. As indicated in Table 1, old fustic contains also kaempferol. Interestingly, maclurin that is usually present in textiles treated with old fustic [23], was not detected in the HPLC chromatogram of the sample. Furthermore, several other colouring compounds were identified in the sample removed from the *Orarion*, as summarized in Table 3. In particular, indigo carmine, which is a blue semi-synthetic dye produced in 1740 AD, was identified. Considering that the object was made in 1862, the use of indigo carmine by the dyers of the *Orarion* is reasonable. Finally, several components of weld (*Reseda luteola* L.) including glycosides (luteolin-3′, 7-di-O-glucoside, luteolin-7-O-glucoside, apigenin-7-O-glucoside) and aglycones (luteolin, apigenin, chrysoeriol) were identified in the sample, extracted from the *Orarion*. The results show therefore that glycosides are preserved with the TFA extraction method. Glycosides, however, decompose to their parent aglycones when HCl is used for dyestuff extraction, as shown in several previously published reports [16, 19, 23, 25].

It is interesting to note that the green colour of the sample removed from the *Orarion* was obtained by dyeing the threads with the blue indigo carmine and the yellow natural materials weld and old fustic.

Conclusion

Kaempferol is contained in natural colourants, which have been used to dye textiles since antiquity, in various civilizations of the European Continent, the New World and Egypt, as it has been proven through HPLC chemical analyses [6, 7, 11, 15, 16, 17, 18, 19, 20].

In the present article, the identification of kaempferol in three samples removed from Post Byzantine religious/ecclesiastical textiles (16th–19th c.), preserved in monasteries of Mount Athos (UNESCO World Heritage Site), is described. The study is carried out using HPLC-DAD. In one sample, kaempferol was identified as a component of buckthorn/Persian berries (*Rhamnus* trees) which was mixed with young fustic (*Cotinus coggygria* Scop.). In another sample, kaempferol was an ingredient of old fustic (*Chlorophora tinctoria* L.), which was mixed with weld (*Reseda luteola* L.) and indigo carmine. Finally, kaempferol and quercetin were reported in the HPLC results collected for the third sample.

Acknowledgments

We thank Professor Chris Cooksey for his critical review and very useful comments. Support by the Getty Foundation (USA), through the project Byz-tex-Athos (www.byztexathos.org), is gratefully acknowledged.

References

[1] Cardon D. (2007) *Natural Dyes*. Sources, Tradition, Technology and Science. Archetype Publications, London.
[2] Halpine S. (1996) *Studies in Conservation* 41:76-94.
[3] Ferreira E., Hulme A., McNab H., Quye A. (2003) *Dyes in History and Archaeology* 19:19-23.
[4] Ferreira E., Hulme A., McNab H., Quye A. (2004) *Chemical Society Reviews* 33:329-336.
[5] Deveoglu O., Karadag R., Yurdun T. (2009) *Jordan Journal of Chemistry* 4:377-385.
[6] Surowiec I., Orska-Gawryś J., Biesaga M., Trojanowicz M., Hutta M., Halko R., Urbaniak-Walczak K. (2003) *Analytical Letters* 36(6):1211-1229.
[7] Trojanowicz M., Orska-Gawryś J., Surowiec I., Szostek B., Urbaniak-Walczak K., Kehl J., Wróbel M. (2004) *Studies in Conservation* 49:115-130.
[8] Colombini M. P., Andreotti A., Baraldi C., Degano I., Łucejko J. J. (2007) *Microchemical Journal* 85:174-182.
[9] Peggie D. A. (2006) The development and application of analytical methods for the identification of dyes on historical textiles. *Ph.D. Thesis,* The University of Edinburgh.
[10] Surowiec I., Quye A., Trojanowicz M. (2006) *Journal of Chromatography* A 1112:209-217.
[11] Wouters J., Rosario-Chirinos N. (1992) *Journal of the American Institute for Conservation* 31(2):237-255.
[12] Bechtold T., Mahmud-Ali A., Mussak R. (2007) *Dyes and Pigments* 75:287-293.
[13] Leitner P., Fitz-Binder C., Mahmud-Ali A., Bechtold T. (2012) *Dyes and Pigments* 93:1416-1421.
[14] Shanker R., Vankar P.S. (2007) *Dyes and Pigments* 74:464-469.
[15] Petroviciu I., Vanden Berghe I., Cretu I., Albu F., Medvedovici A (2012) *Journal of Cultural Heritage* 13:89-97.
[16] Zhang X., Boytner R., Cabrera J.L., Laursen R. (2007) *Analytical Chemistry* 79: 1575-1582.
[17] Szostek B., Orska-Gawryś J., Surowiec I., Trojanowicz M. (2003) *Journal of Chromatography* A 1012:179-192.
[18] Quye A., Cheape H., Burnett J., Ferreira E., Hulme A., McNab H. (2003) *Dyes in History and Archaeology* 19:1-12.

[19] Mantzouris D., Karapanagiotis I., Valianou L., Panayiotou C. (2011) *Analytical and Bioanalytical Chemistry* 399:3065-3079.
[20] Karapanagiotis I., Mantzouris D., Kamaterou P., Lampakis D., Panayiotou C. (2011) *Journal of Archaeological Science* 38:3217-3223.
[21] Karapanagiotis I., Lakka A., Valianou L., Chryssoulakis Y. (2008) *Microchimica Acta* 160(4):477-483.
[22] Valianou L., Stathopoulou K., Karapanagiotis I., Magiatis P., Pavlidou E., Skaltsounis A-L., Chryssoulakis Y. (2009) *Analytical and Bioanalytical Chemistry* 394:871-882.
[23] Hofenk-de Graaff J. H. (2004) *The colourful past: origins, chemistry and identification of natural dyestuffs.* Archetype Publications, London.
[24] Wouters J. (1985) *Studies in Conservation* 30:119-128.
[25] Valianou L., Karapanagiotis I., Chryssoulakis Y. (2009) *Analytical and Bioanalytical Chemistry* 395:2175-2189.

Index

D

E

F

G

H

I

J

K

L

M

N

O

P

T

U

V

W

Y

Z